KT-104-005

Global Issues
An Introduction

THIRD EDITION

John L. Seitz

LIS - LIBRARY
Date
28/04/10
Fund
9
Order No.
2105160
University of Chester

© 1995, 2002, 2008 by John L. Seitz

BLACKWELL PUBLISHING
350 Main Street, Malden, MA 02148-5020, USA
9600 Garsington Road, Oxford OX4 2DQ, UK
550 Swanston Street, Carlton, Victoria 3053, Australia

The right of John L. Seitz to be identified as the Author of this Work has been asserted in accordance with the UK Copyright, Designs, and Patents Act 1988.

The publisher and the author make no representations or warranties with respect to the accuracy or completeness of the contents of this work and specifically disclaim all warranties, including without limitation warranties of fitness for a particular purpose. No warranty may be created or extended by sales or promotional materials. The advice and strategies contained herein may not be suitable for every situation. This work is sold with the understanding that the publisher is not engaged in rendering legal, accounting, or other professional services. If professional assistance is required, the services of a competent professional person should be sought. Neither the publisher nor the author shall be liable for damages arising herefrom. The fact that an organization or website is referred to in this work as a citation and/or a potential source of further information does not mean that the author or the publisher endorses the information the organization or website may provide or recommendations it may make. Further, readers should be aware that Internet websites listed in this work may have changed or disappeared between when this work was written and when it is read.

First edition published 1995
Second edition published 2002
Third edition published 2008 by Blackwell Publishing Ltd

6 2009

Library of Congress Cataloging-in-Publication Data

Seitz, John L., 1931–
Global issues : an introduction / John L. Seitz. — 3rd ed.
p. cm.
Includes bibliographical references and index.
ISBN 978-1-4051-5497-0 (pbk. : alk. paper) 1. Economic development. 2. Developing countries—Economic policy. 3. Developing countries—Economic conditions. 4. Economic history—1945– I. Title.

HD82.S416 2008
338.9—dc22

2007014212

A catalogue record for this title is available from the British Library.

Set in 10/12pt Palatino
by Graphicraft Limited, Hong Kong
Printed and bound in Singapore
by Markono Print Media Pte Ltd

The publisher's policy is to use permanent paper from mills that operate a sustainable forestry policy, and which has been manufactured from pulp processed using acid-free and elementary chlorine-free practices. Furthermore, the publisher ensures that the text paper and cover board used have met acceptable environmental accreditation standards.

For further information on
Blackwell Publishing, visit our website:
www.blackwellpublishing.com

Contents

Plates

Figures and Tables

Figures

Tables

Foreword

In the 1950s and 1960s I went as an employee of the US government to Iran, Brazil, Liberia, and Pakistan to help them develop. A common belief in those decades was that poverty causes people to turn to communism. As an idealist young person, I was pleased to work in a program that had the objective of helping poor nations raise their living standards. After World War II the United States was the richest and most powerful country in the world. Many countries welcomed US assistance since it was widely believed that the United States could show others how to escape from poverty.

Disillusionment came as I realized that we did not really know how to help these countries relieve their widespread poverty. The problem was much more complex and difficult than we had imagined. Also, one of the main political objectives of our foreign aid program – to help friendly, noncommunist governments stay in power – often dominated our concerns.

And more disillusionment came when I looked at my own country and realized that it had many problems of its own that had not been solved. It was called "developed" but faced major problems that had accompanied its industrialization – urban sprawl and squalor, pollution, crime, materialism, and ugliness, among others. So, I asked myself, what is development? Is it good or bad? If there are good features in it, as many people in the world believe, how do you achieve them, and how do you control or prevent the harmful features? It was questions such as these that led me to a deeper study of development and to the writing of this book.

I came to recognize that development is a concept that allows us to examine and make some sense out of the complex issues the world faces today. Many of these issues are increasingly seen as being global issues. Because the capacity human beings have to change the world – for better or for worse – is constantly growing, an understanding of global issues has become essential. The front pages of our newspapers and the evening TV news programs remind us nearly daily that we live in an age of increasing interdependence. (The Introduction explains the creation of global issues.)

In this book the term "development" will be defined as economic growth plus the social changes caused by or accompanying that economic growth. In the 1950s and 1960s it was common to think of development only in economic terms. For many economists, political scientists, and government officials, development meant an increase in the per capita national income of a country or an increase in its gross national product (GNP), the total amount of goods and services produced. Development and economic development were considered to be synonymous. In the 1970s an awareness grew – in both the less developed[1] nations and the developed industrialized nations – that some of the social changes which were coming with economic growth were undesirable. More people were coming to understand that for economic development to result in happier human beings, attention would have to be paid to the effects that economic growth was having on social factors. Were an adequate number of satisfying and challenging jobs being created? Were adequate housing, health care, and education available? Were people living and working in a healthy and pleasant environment? Did people have enough nutritious food to eat? Every country is deficient in some of these factors and, thus, is in the process of developing.

The definition of development I have given above is a "neutral" one – it does not convey a sense of good or bad, of what is desirable or undesirable. I have chosen this definition because there is no widespread agreement on what these desirable and undesirable features are. The United Nations now defines human development as the enlarging of human capabilities and choices, and in a yearly publication ranks nations on a human development index, which tries to measure national differences of income, educational attainment, and life expectancy.[2] The United Nations sees the purpose of development to be the creation of an environment in which people can lead long, healthy, and creative lives. Economists have traditionally used GNP or national income as the measures of economic development. My definition tries to combine both the economic and the social components into the concept of development. I find a neutral definition useful because development can be beneficial or harmful to people.[3]

In this book we will look at some of the most important current issues related to development. The well-being of people depends on how governments and individuals deal with these issues. We will first look at the issue of poverty in the world and then move on to issues related to population, food, energy, the environment, and technology, and will conclude with a consideration of the future.

This book is an introduction to a number of complicated issues. It is only a beginning; there is much more to learn. Readers who are intrigued by a subject or point made and want to learn more about it should consult the relevant note. The note will either give some additional information or will give the source of the fact I present. Consulting this source is a good place for the reader to start his or her investigation. After each chapter a list of readings gives inquisitive readers further suggestions for articles and books that will allow them to probe more deeply. The bibliography contains some additional books and articles. Appendix 1 gives the student some help in organizing the material the book covers and the teacher some suggestions for teaching this material. Appendix 2 offers suggestions of relevant video tapes, an important and interesting resource for those who want to understand these issues more deeply. Appendix 3 gives study and discussion questions for each chapter for use by students and by teachers. Appendix 4 gives internet sources. Many organizations on the internet now have a large amount of information related to many of the issues covered in this book. The glossary contains a definition of many of the uncommon terms used in the book.

The world is changing rapidly and significant developments have taken place in many of the topics covered in this book since the second edition was prepared. The third edition has been thoroughly updated and the section on climate change (global warming) has been expanded to give it the importance it deserves. An effort has been made to make clearer what are global issues. A new section explaining how geography affects the wealth and poverty of nations has been added, as has a new section on the UN's Millennium Goals and an assessment of the rate of achievement. A new glossary is included in this edition. Fresh attention to development assistance and foreign aid has been given. The new global issue of overweight people and obesity is examined and a new section on nuclear terrorism has been added.

I would like to thank the following teachers who made useful suggestions for improving this edition: Alan Gilbert, University College London, UK; Scott Anderson, State University of New York-Cortland, US; Jonathan Barton, Catholic University, Chile; Hans Holman, Linkoping University, Sweden; David Graham, Nottingham Trent University, UK; William Moseley, Macalester College, US; Rae McKay, University of Birmingham, UK; Edwin Clausen, Daemen College, US.

I would also like to thank Wofford College for an office. Offices are usually scarce on college campuses and I deeply appreciate Wofford allowing this retired professor to "hang around." Martin Aigner, of Wofford's Information Technology staff, performed great service in keeping my old computer running, and Chris Strauber and Ellen Tillett of Wofford's Library Reference Department provided valuable assistance.

Notes

1 The term "less developed" refers to a relatively poor nation in which agriculture or mineral resources have a large role in the economy while manufacturing and services have a lesser role. The infrastructure (transportation, education, health, and other social services) of these countries is usually inadequate for their needs. About 80 percent of the world's people live in nations such as this, which are also called "developing." (Some of these countries are highly developed in culture and many such regions of the world had ancient civilizations with architecture, religion, and philosophy which we still admire.) Since many of the less (economically) developed nations are in the Southern Hemisphere, they are at times referred to as "the South." During the Cold War these nations were often called the "Third World." Industrialized nations are called "developed" nations. Most of them are located in the Northern Hemisphere so they are called "the North." Some organizations such as the World Bank also divide countries according to their level of income. The Bank considers low and middle income countries to be "developing" and high income countries to be "developed."

2 United Nations Development Programme, *Human Development Report 2004* (New York: United Nations Development Programme, 2004), p. 127.
3 For a criticism of the Western concept of development see Ivan Illich, "Outwitting the 'Developed' Countries," in Charles K. Wilber (ed.), *The Political Economy of Development and Underdevelopment*, 2nd edn (New York: Random House, 1979), pp. 436–44. See also Lloyd Timberlake, "The Dangers of 'Development,'" in *Only One Earth: Living for the Future* (New York: Sterling, 1987), pp. 13–22.

Introduction: The Creation of Global Issues

What causes an issue to become a "global issue"? Are "global issues" the same as international affairs – the interactions that governments, private organizations, and peoples from different countries have with each other? Or is something new happening in the world? Are there now concerns and issues that are increasingly being recognized as global in nature? It is the thesis of this book that something new indeed is happening in the world as nations become more interdependent. While their well-being is still largely dependent upon how they run their internal affairs, increasingly nations are facing issues that they alone cannot solve, issues that are so important that the failure to solve them will adversely affect the lives of many people on this planet. In fact, some of these issues are so important that they can affect how suitable this planet will be in the future for supporting life.

The issues dramatize our increasing interdependence. The communications and transportation revolutions that we are experiencing are giving people knowledge of many new parts of the globe. We see that what is happening in far-off places can affect, or is affecting, our lives. For example, instability in the oil-rich Middle East affects the price of oil around the world and since many countries are dependent on oil as their main source of energy, the politics of oil becomes a global concern.

Many nations in the world are now dependent on other nations to buy their products and supply natural resources and goods they need to purchase in order for them to maintain their standard of living. An economic downturn in any part of the world which affects the supply and demand for products will affect the economic status of many other nations. This is an important part of globalization which will be discussed in chapter 1.

Even a global issue such as world hunger illustrates our increasing interdependence. A person might say that starving or malnourished people in Africa don't affect people in the rich countries, but even here there is a dependency. Our very nature and character depend on how we respond to human suffering. Some rich nations such as the Scandinavian nations in northern Europe give a much higher portion of their national wealth to poor nations for development purposes than do other rich nations such as the United States and Japan.

Global issues are often seen as being interrelated. One issue affects other issues. For example, climate change (an environmental issue) is related to an energy issue (our reliance on fossil fuels), the population issue (more people produce more greenhouse gases), the wealth and poverty issue (wealthy developed countries produce the most gases that cause climate change), the technology issue (technology can help us create alternative energy sources that produce less or no greenhouse gases), and the future issue (will the changes we are making in the earth's climate seriously harm life on this planet?). As we recognize these interrelationships, we realize that usually there are no simple solutions.

Interdisciplinary knowledge is required to successfully deal with the issues. The student or adult learner reading this book will be receiving information from multiple disciplines such as biology, economics, political science, environmental science, chemistry, and others. Neither the social sciences nor the physical sciences have the answers alone. Feel good about yourself reader because you are engaged in the noble task of trying to understand how the world really works. Complicated? Yes, of course. Impossible to discover? Certainly not. Just read seriously and carefully. It takes effort and you can keep learning throughout your life.

Perhaps, global issues were born on the day, several decades ago, when the earth, for the first time, had its picture taken. The first photograph of earth, which was transmitted by a spacecraft, showed our planet surrounded by a sea of blackness. Many people seeing that photograph realized that the blackness was a hostile environment, devoid of life, and that life on earth was vulnerable and precious. No national boundaries could be seen from space. That photograph showed us our home – one world – and called for us to have a global perspective in addition to our natural, and desirable, more local and national perspectives.

This book discusses *some* of the main current global issues of our time. The reader can probably identify others. During the reader's lifetime, humanity will have to face new global issues that will continue to surface. It is a characteristic of the world in which we live. Maybe our growing ability to identify such issues, and our increasing knowledge of how to deal with them, will enable us to handle the new issues better than we are doing with the present ones.

CHAPTER 1

Wealth and Poverty

The mere fact that opposing visions of economic development have grown to shape the international agenda is in one sense merely an indication that development concerns are receiving attention on a global scale for the first time in history.

Lynn Miller, *Global Order* (1985)

For most of history, human beings have been poor. A few individuals in many societies had a higher standard of living than their fellow humans, but the vast majority of people on earth have shared a common condition of poverty. The industrial revolution brought a fundamental change. New wealth was created in the industrializing nations in Europe and eventually shared by larger numbers of people. And the differences between the rich and the poor in the world began to increase. A few nations began to achieve higher living standards, and they began to pull away from the rest of the world, which had not yet begun to industrialize. It is estimated that the difference between the per capita incomes of the richest and poorest countries was 3 to 1 in 1820, 11 to 1 in 1913, 35 to 1 in 1950, 44 to 1 in 1973, and 72 to 1 in 1992.[1] Another way to show this trend is to note that the real (that is, controlling for inflation) per capita incomes for the richest one-third of countries increased nearly 2 percent annually from 1970 to 1995, while the middle third of countries increased only about 0.5 percent annually and the poorest one-third had no increase in incomes.[2] And one more way to demonstrate that the gap between the richest and poorest is increasing: in 1960 the richest 20 percent of the world's population had 30 times more income than the lowest 20 percent of the world's people. By 1997 the richest 20 percent had now nearly 75 percent more income than the poorest.[3] But this trend masks the fact that many countries have caught up with the richest country. In 2003 the United States had a gross national income[4] per capita of $38,000, but close behind were Norway with $37,000, Switzerland with $32,000, Denmark with $31,000 and Austria, Canada, and Ireland with $30,000.

Not only is the gap between the richer and poorer nations growing, but the gap between the rich and the poor in some rich countries is also growing. According to a United Nations International Labor Organization report in 2004 the gap between the rich and the poor was growing in Great Britain, Canada, and the United States. The gap was largest in the United States where the richest 1 percent of the population earned nearly 20 percent of the gross income, a situation, the report said, that had last occurred in the 1920s.[5]

By the way, don't let all these numbers make your head ache. You don't have to remember them all to understand the subject. But read them carefully as they illustrate the points being made. For example, do the numbers show that the rich are getting richer and the poor poorer, a statement most people, no doubt, would say is true? Well it's not. *Some people are getting poorer, but not the majority in the developing countries or the rich countries. The figures I have used don't support this false statement. If you think they do, you should read more carefully.*

The growing gap between the rich and the poor is only one part of the picture of worldwide economic conditions. Another important part of that picture is the vast number of people still living in poverty. What are you – an optimist, or a pessimist? When one thinks about the living standards of the world's people there are figures which support both positions. On the positive side the average level of per capita GNP (gross national product) for those developing countries for which data are available rose about 2 percent per year from 1960 to 1997.[6] The growth rate of per capita income in developing countries was relatively high in the 1960s and 1970s but stagnated in the 1980s. In the early 1990s rapid growth began again, especially in East Asia (from Indonesia to South Korea), but a financial crisis in the late 1990s stopped that growth. Overall the decade of the 1990s was one of impressive economic growth for some developing countries, such as China and India,[7] while other less developed nations became poorer.

Figure 1.1 shows the overall reduction in extreme poverty in developing countries during the final decades of the twentieth century. At the beginning of the 1980s China was one of the poorest countries in the world with about 60 percent of its people living in extreme poverty. In the decade of the 1980s China cut its poverty rate by about one-half and in the 1990s cut it by one-half again. We can see in figure 1.1 that China's impressive performance in cutting extreme poverty, along with its huge population, is largely responsible for the reduction in poverty by developing countries.

Continuing to focus on positive developments, one can find many reasons to feel optimistic when one reads the following assessment by the United Nations at the beginning of the twenty-first century:

> Progress in human development during the twentieth century was dramatic and unprecedented. Between 1960 and 2000 life expectancy in developing countries increased from 46 to 63 years. Mortality rates for children under five were more than halved. Between 1975, when one of every two adults could not read, and 2000 the share of illiterate people was almost halved. Real per capita incomes more than doubled from $2,000 to $4,200.[8]

Poverty in Indonesia (*World Bank*)

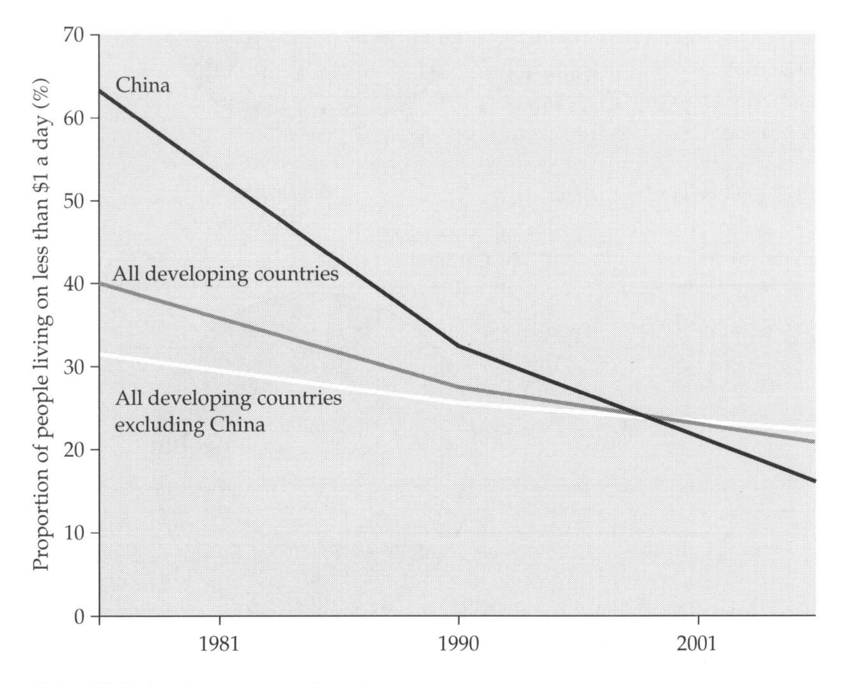

Figure 1.1 Global extreme poverty rate
Source: World Bank, *World Development Indicators*, 2005

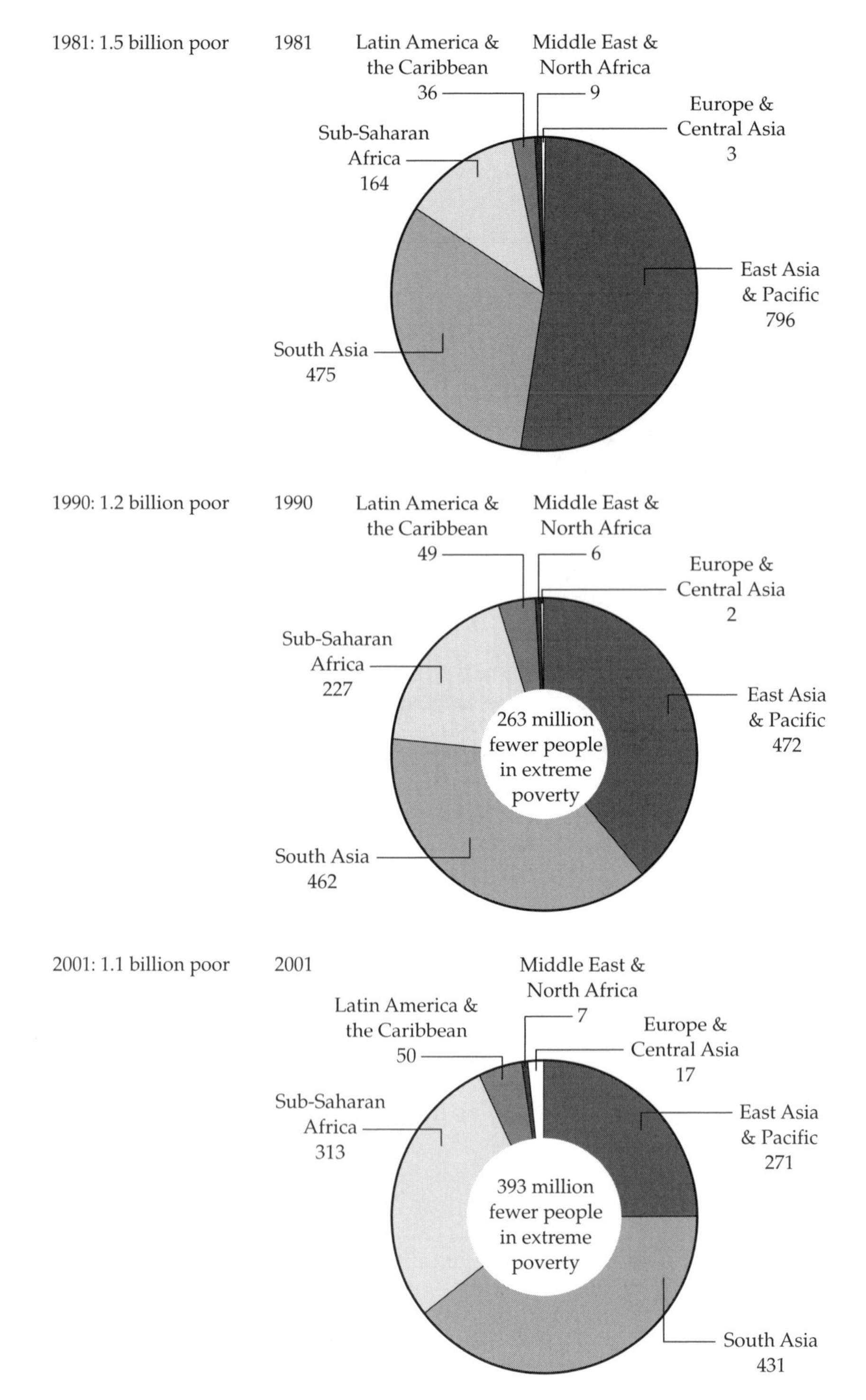

Figure 1.2 Fewer people in extreme poverty: people living on less than $1 a day, 1981, 1990, 2001 (millions)
Source: World Bank, *World Development Indicators*, 2005

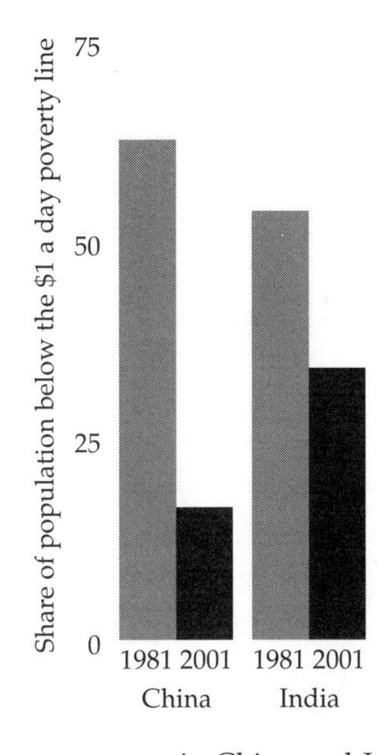

Figure 1.3 Reduction in extreme poverty in China and India, 1981–2001
Source: World Bank, *World Development Report 2005* (New York: Oxford University Press, 2004), p. 27

A closer look at the impressive progress made in the decades of the 1980s and 1990s in reducing the number of people in poor countries living in extreme poverty, that is, living daily on $1 or less, can be seen in figure 1.2. Note that much of that reduction took place in East Asia, where China is located, and in South Asia, where India resides. The impressive economic growth which both nations experienced in the late twentieth century, especially China, came after they introduced free market measures which made foreign investments welcome. The unprecedented reduction of extreme poverty in China and India is shown in figure 1.3.

Now, some information for the pessimist. The United Nations, on the same page it had the above quotation, also gives the following information: "But despite this impressive progress, massive human deprivation remains. More than 800 million people suffer from undernourishment. Some 100 million children who should be in school are not, 60 million of them girls. More than a billion people survive on less than $1 a day."

In the year 2000 about 1.2 billion people still had no access to improved water sources and about 2.7 billion had no adequate sanitation.[9] In 2004 about half the world's people lived on less than $2 a day[10] (see figure 1.4). A depressing number of countries (46) actually became poorer in the 1990s and in 2004 25 countries had more people going hungry than there were there a decade before.[11] As the reader can see by examining figure 1.4, many of these were in Africa and a few were in Latin America and in Europe and Central Asia. In Africa many of the countries growing poorer are in sub-Saharan Africa and are being hit by an HIV/AIDS epidemic, among other problems, while in Europe and Central Asia parts of the former Soviet Union found the transition from being part of the Soviet

The weight of poverty falls heavily on children in less developed nations (*United Nations*)

Union to an independent country difficult. For many the path from a planned, state-managed economy to a freer economy was filled with obstacles.

Who are the poor? According to the World Bank nearly half of them live in South Asia (e.g., India, Pakistan, Bangladesh), while a smaller but highly disproportionate number live in Africa, south of the Sahara desert. Within regions and countries the poor tend to be concentrated in rural areas with a high density of population, such as on the Ganges plain in India and on the island of Java in Indonesia. Many poor also live in areas with scarce resources such as in the Andean highlands in Latin America and in the Sahel region in Africa.[12] At least 200 million of the very poor live in China. The weight of poverty in the less developed nations falls

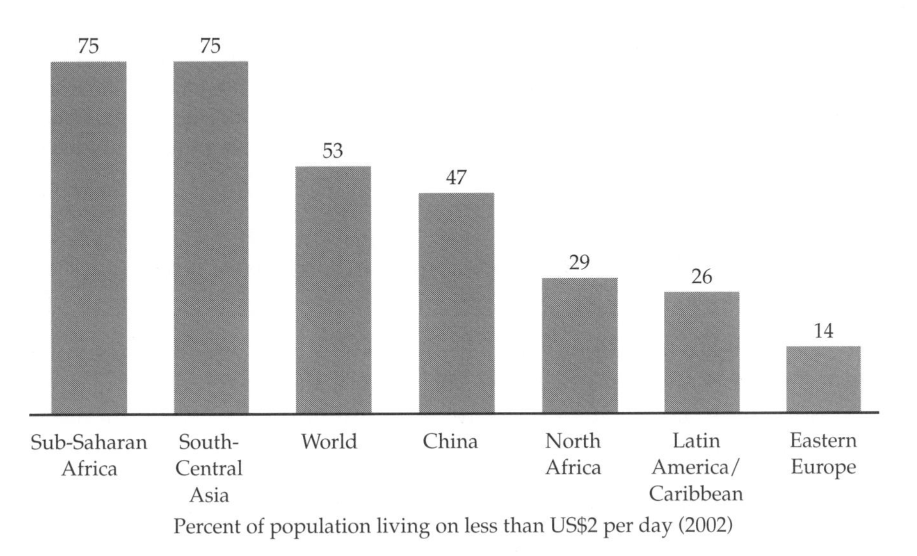

Figure 1.4 About half of the world's population lives on less than $2 per day
Source: Population Reference Bureau, *2005 World Population Data Sheet*

heaviest on women, children and minority ethnic groups. Figure 1.2 shows that the world's poor are concentrated in Africa, East Asia, and South Asia.

The Millennium Development Goals

In 2000, representatives of 189 nations met in a conference sponsored by the United Nations and adopted eight goals they would work to achieve in the new century. Each goal, which was stated in general terms, had specific targets to help measure progress in reaching the goal. The first goal was to eliminate extreme poverty and hunger in the world.[13] The target under this goal was to reduce, during the period 1990 to 2015, the number of people living on the equivalent of less than $1 a day by one-half.

How is the world doing in achieving the first goal, a goal, by the way, unprecedented in the world's history? As we have seen from the information presented above, China and India are doing quite well, but the same cannot be said for many other countries. Take the time to read the following two short paragraphs from a UN 2004 report for they present a good summary of the situation in the early part of the twenty-first century:

> More than 1.2 billion people – one in every five on Earth – survive on less than $1 a day. During the 1990s the share of people suffering from extreme income poverty fell from 30% to 23%. But with a growing world population, the number fell by just 123 million – a small fraction of the progress needed to eliminate poverty. And excluding China, the number of extremely poor people actually increased by 28 million.

South and East Asia contain the largest numbers of people in income poverty, though both regions have recently made impressive gains. As noted, in the 1990s China lifted 150 million people – 12% of the population – out of poverty, halving its incidence. But in Latin America and the Caribbean, the Arab States, Central and Eastern Europe and Sub-Saharan Africa the number of people surviving on less than $1 a day increased.[14]

Development Assistance and Foreign Aid

Do the rich have a responsibility to help the poor? It's an age-old question, isn't it, and individuals and countries have throughout time given different answers to it. Most major religions answer it with a resounding "yes," as do many moral philosophies. Now for the first time in history most of the nations of the world committed themselves in 2000 to work toward helping the neediest when they endorsed the Millennium Declaration with its objective of eradicating poverty in the world. Before this many nations had given aid to foreign nations both for political and humanitarian reasons. Foreign aid is still used regularly to help the donor nation achieve its political objectives and can include military aid as well as economic assistance. Development assistance is usually given with the objective of helping a nation improve its economy. The eighth Millennium Declaration goal focuses on the need for developed countries to increase their development assistance to developing nations. The United Nations has set an aid target for the rich countries of 0.7 percent of their wealth (as determined by their gross national product (GNP).

In addition, the Millennium Development Goals present the need for the rich countries to help the least developed by reducing tariffs and quotas on the poor country's exports, and granting debt relief. In return the nations receiving this aid are to show they are seriously working to reduce their poverty and improve their governmental administration, which, among other things, means reducing their corruption. Corruption, which is widespread in many poor countries, siphons off the aid for personal use by powerful governmental and private individuals.

How are the richer nations doing at present to help the least developed nations? Recently there have been some agreements to reduce the debt of developing nations, but the need for tariff and quota reductions remains. The average tariff by the high and middle income countries of the Organization for Economic Cooperation and Development (OECD) on manufactured goods from developing nations is four times higher than on manufactured goods from other members of the OECD.[15] Another big problem is the huge subsidies that developed nations give to their farmers, which make it very difficult for farmers in the developing world to compete with them.

The amount of money the developed nations have given to less developed nations in relation to the donors' wealth fell rather dramatically in the second half of the twentieth century, as can be seen in figure 1.5. There was a slight upturn at the end of the century, but the aid is still far below the United Nations target.

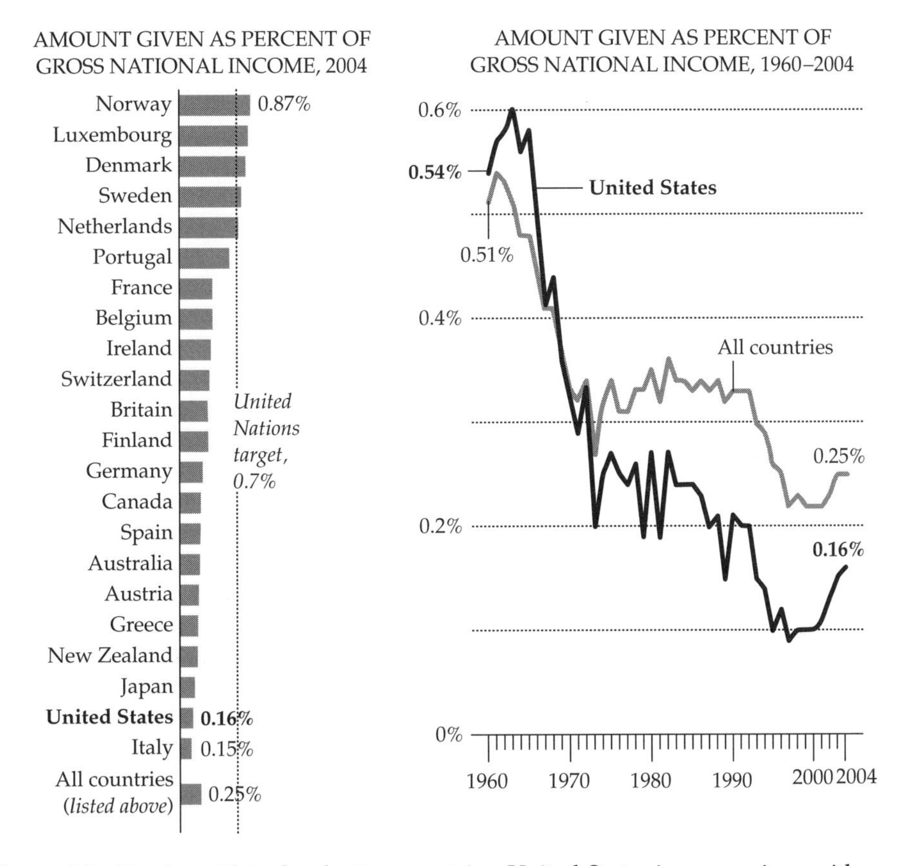

Figure 1.5 Foreign aid to developing countries: United States in comparison with other countries
Source: Organization for Economic Cooperation and Development, from *New York Times*, April 18, 2005, p. A10

While the United States gives the largest amount of aid, in relation to its wealth as measured by gross national income it is near the bottom of aid donors.[16] It shares this position with Japan and Italy, two other high income countries.

Let's end this section on a positive note. European Union nations have pledged to increase their foreign aid up to the United Nations target by 2015. As you can see in figure 1.5, several of the northern European countries had already reached, and even exceeded, that goal in 2004.

Now that we have made a brief examination of poverty in the world, let's focus on another question: Why are some countries rich and some poor? There is no agreement on the answer to that question, but various views have been presented over the years. Although vast differences among the nations of the world make generalizations hazardous, it can be useful to consider three of the most widely accepted approaches or views of economic development: (1) the market or capitalist approach, which is adhered to by many in the Western industrial countries;[17]

Consequences of poverty: the mothers who don't cry

In trying to understand what being poor means in a less developed country, a puzzling question comes to mind: Why don't many mothers cry when their children die in northeastern Brazil?

To answer this question, one needs to know first why the death of a child is fairly common in the Brazilian northeast, the poorest region in Brazil, and one of the poorest in the world. A North American anthropologist offers this explanation:

> The children of the Northeast, especially those born in shantytowns on the periphery of urban life, are at a very high risk of death. In these areas, children are born without the traditional protection of breast-feeding, subsistence gardens, stable marriages, and multiple adult caretakers that exists in the interior [of Brazil]. In the hillside shantytowns that spring up around cities . . . marriages are brittle, single parenting is the norm, and women are frequently forced into the shadow economy of domestic work in the homes of the rich or into unprotected and oftentimes "scab" wage labor on the surrounding sugar plantations, where they clear land for planting and weed for a pittance, sometimes less than a dollar a day. The women . . . may not bring their babies with them into the homes of the wealthy, where the often sick infants are considered sources of contamination, and they cannot carry the little ones to the riverbanks where they wash clothes because the river is heavily infested with schistosomes and other deadly parasites. . . . At wages of a dollar a day, the women . . . cannot hire baby sitters. Older children who are not in school will sometimes serve as somewhat indifferent caretakers. But any child not in school is also expected to find wage work. In most cases, babies are simply left at home alone, the door securely fastened. And so many also die alone and unattended.

The death of a child is thus a commonplace tragedy in northeast Brazil. But why is there a lack of mourning for the dead children? The anthropologist found that no tears were shed when an infant died and that mothers never visited the graves after the burials. The anthropologist concluded that in the face of the frequent deaths of children, mothers have learned to delay their attachment to any child until that child has proven to be a survivor, hardier than his or her weaker siblings. This reaction to the realities of their lives has allowed these women to continue living in a harsh situation and to not let grief make their lives unbearable. Actions similar to those of the Brazilian women have been observed in parts of Africa, India, and Central America.

Source: Nancy Scheper-Hughes, "Death without Weeping," *Natural History* (October 1989), pp. 8, 14

Consequences of poverty: the children who are killed

Another question also illustrates one of the many consequences of poverty. Why were children killed in some of the cities and rural areas of Brazil, some even by the police? The answer is that the children were often involved in begging or crime. They became the targets of drug dealers who were enforcing control of their gangs, or they were killed by assassins, some of them off-duty policemen who were hired by merchants or residents trying to control crime in their areas. A yearly average of about 1,200 children were killed in Brazil from 1993 to 1997. Some of these victims had been abandoned by their parents, while others were living with single parents or friends, many in the large shanty towns surrounding many Brazilian cities. And there are other depressing statistics. An estimated 10 million children and youths in Brazil were homeless or making a living off the streets in the mid-1990s. Poverty inflicts a harsh life for many children in other countries also. In the early twenty-first century it was estimated that between 100 million to 250 million children lived on the streets around the world, with half of them in Latin America, and their numbers were rapidly increasing. For example, it was estimated that in the Philippines in 1991 there were about 200,000 street children, while in 1999 the number had increased to 1.5 million. Forty-four million children reportedly work on the streets in India, some of the children begging having been mutilated by criminal gangs to make them more pitiable. Because of family poverty, greed, and the fear of AIDS there has been an increase in the use of children as prostitutes around the world, some as young as eight years old. In the early twenty-first century it was estimated that between 700,000 and 4 million children and women were trafficked across borders into the sex trade, often through coercion and abduction. Many faced high risk of HIV infection.

The Secretary-General of the United Nations presented a summary of the state of children in the world in a report to the organization at the beginning of the twenty-first century: "to enable families living in poverty to survive, a quarter of a billion children aged 14 and under, both in and out of school, now work, often in hazardous conditions or unhealthy conditions. They toil in urban sweatshops; on farms or as domestic servants; selling gum or cleaning shoes in urban streets; clambering down dangerous mine shafts; and – in distressing numbers – bonded or sold into sexual services."

Sources: Nancy Scheper-Hughes and Daniel Hoffman, "Brazil: Moving Targets," *Current History*, 106 (July/August 1997), pp. 34–43; Marion Carter, "Spotlight: Brazil," *Population Today* (Washington, DC: Population Reference Bureau, August 1996), p. 7; John Burns, "West Bengal Finds It Creates Child Beggars for Export," *New York Times*, national edn (March 13, 1997), p. A3; "Street Children Increasing," and "Sex Trade Is Enslaving Millions of Women, Youth," *Popline* (Population Institute, Washington, DC) (November/December 2003), pp. 1, 6; Kofi Annan, *We the Peoples: The Role of the United Nations in the 21st Century* (New York: United Nations, 2001) no. 99

Street children in Nepal (*Ab Abercrombie*)

(2) the state approach, varieties of which were commonly believed in communist countries and, in the past, were believed by many in the developing world; and (3) the civil society approach.[18]

The Market Approach

The market approach holds that nations can acquire wealth by following four basic rules: (1) the means of production – those things required to produce goods and services such as labor, natural resources, technology, and capital (buildings, machinery, and money that can be used to purchase these) – must be owned and controlled by private individuals or firms; (2) markets in which the means of production and the goods and services produced are freely bought and sold must exist; (3) trade at the local, national, and international levels must be unrestricted; and (4) a state-enforced system of law must exist to guarantee business contracts so as to ensure safe commercial relations between unrelated individuals.

Adam Smith, the eighteenth-century Scottish political economist and founder of the market approach, believed that the operations of labor are the key to increasing production. He argued that it is much more efficient for workers to specialize in their work, focusing on one product rather than making many different products. If workers do this, and if they are brought together in one location so their labor can be supervised, increased production will result. Smith also presented the idea that, if the owners of the means of production are allowed to freely sell their services or goods at the most advantageous price they can obtain, the largest amount of products and services will be produced and everyone will benefit. It is the prices in the markets that suggest to the businessman or businesswoman new profitable investment opportunities and more efficient production processes. (For example, when oil prices rose dramatically in the 1970s, new investments occurred in alternative energy sources and some industries came up with ways to reduce the amount of oil they needed to buy. Some business people saw the alternative energy investments as a way for them to make money in the energy field, and some industries cut their costs, thus increasing their profits, by becoming more efficient in their use of energy.)

Smith did not focus on the role of the entrepreneur, but later market theorists did, making the entrepreneur – the one who brought the means of production together in a way to produce goods and services – a key component in this approach. Finally, Smith and other market theorists emphasized the importance of free trade. If a nation concentrates on producing those products in which it has a comparative advantage over other nations, advantages that climate, natural resources, cheap labor, or technology give it, and if it trades with other nations that are also concentrating on those products that *they* have the greatest advantage in producing, then all will benefit.

The market approach holds that government has a crucial but limited role in maintaining an environment in which economic activities can flourish. Government should confine its activities to providing for domestic tranquility that would ensure that private property is protected; providing certain services, such as national defense, for which everyone should pay; enforcing private contracts; and helping to maintain a stable supply of money and credit. The reason some nations are poor, according to the market approach, is that they do not follow the basic rules listed above.

Advocates of the market approach point to the wealth of the United States and Western Europe as evidence of the correctness of their view. Even Karl Marx said that the hundred years of rule by capitalists were the most productive in the history of the world. And although an uneven distribution of income occurred in Western Europe during its early period of industrialization, the distribution of income later became much less uneven. This indicated that the new wealth was being shared by more and more people.

Nations such as Japan and West Germany, which came back from the devastation of World War II to create extremely strong economies by following the basic principles of the market approach, are also cited as evidence of the validity of the approach. Examples can also be found among developing nations that have achieved such impressive economic growth by following the principles of this

The market approach is followed on the streets of many developing nations (*Mark Olencki*)

approach that they have moved into a separate category of the less developed world: the newly industrializing states. Many of these states, such as China, South Korea, Taiwan, Singapore, and Hong Kong, achieved their high economic growth mainly by exporting light manufactured products to the developed nations.

Finally, advocates of the market approach point to the decisions of many ex-communist countries and developing countries, during the 1980s, to adopt at least some market mechanisms in their efforts to reform their economies. Even China – the largest remaining communist government – has adopted many important aspects of the market approach; it is this adoption that is widely believed to be responsible for China's impressive economic growth.

Critics of the market approach point to the high rates of unemployment that have existed at times in Western Europe and the US. At the present time, high unemployment exists throughout the developing world, even in a number of nations that follow the market approach and have had impressive increases in their GNP. Much of the industry that has come to the South has been capital intensive; that is, it uses large amounts of financial and physical capital but employs relatively few workers.

There is evidence from Brazil, which has basically followed the market approach for the past several decades, that the distribution of income within developing countries became more unequal during the period the countries were experiencing high rates of growth. The same thing happened in China in the 1990s. The rich got a larger proportion of the total income produced in these countries than they had before the growth began. And even worse than this is the evidence that the poor in these countries, such as Brazil, probably became absolutely poorer during the period of high growth, in part because of the high inflation which often accompanied the growth.[19] (High inflation usually hurts the poor more than the rich because the poor are least able to increase their income to cope with the rising prices of goods.) The economic growth that came to some developing nations following the market approach failed to trickle down to the poor and, in fact, may have made their lives worse. High inflation was halted in Brazil in the 1990s, as was the trend for income inequality to worsen. At the end of the century the distribution of incomes in Brazil continued to be highly unequal. The poorest 20 percent of the population received about 3 percent of the income in the country, and the richest 20 percent received about 62 percent.

Critics of the market approach have also pointed out that prices for goods and services set by a free market often do not reflect the true costs of producing those goods and services. Damage to the environment or to people's health that occurs in the production and disposal of a product is often a hidden cost, which is not covered by the price of the product. The market treats the atmosphere, oceans, rivers, and lakes as "free goods," and, unless prohibited from doing so by the state, it transfers the costs that arise because of their pollution to the broader community. Take for example a factory that pollutes the air while producing cars; the costs of treating illnesses caused by damaged lungs are not borne by the factory owner or the purchaser of the car, but rather by the community as a whole.

The State Approach

The state approach, which was founded mainly by Karl Marx, German philosopher and political economist, and Vladimir Ilyich Lenin, Russian revolutionary leader and first premier of the Soviet Union, has more to say about the causes of underdevelopment than it does about how development takes place. In a socialist country most of the means of production – land, resources, and capital – are publicly controlled to ensure that the profit obtained from the production of goods and services is plowed back to benefit the community as a whole. The prohibition on the private control or ownership of these so-called factors of production leads, according to the state approach, to a relatively equal distribution of income as everyone, not just a few individuals, benefits from the economic activity. The basic needs of all are provided for. The free market of the capitalist system is abolished and replaced with central planning. Prices are set by the central planners, and capital is invested in areas that are needed to benefit the society.

The explanation the state approach gives to the causes of poverty in the world focuses on international trade. According to the state approach, the root of the present international economic system, where a few nations are rich and the majority of nations remain poor, lies in the trade patterns developed in the sixteenth century by Western Europe. ("Dependency theory"[20] is the name given to this part of the state approach.) First Spain and Portugal and then Great Britain, Holland and France gained colonies – many of them in the Southern Hemisphere – to trade with. The imperialistic European nations in the Northern Hemisphere developed a trade pattern that one can still see clear signs of today. The mother countries in "the core" became the manufacturing and commercial centers, and their colonies in "the periphery" became the suppliers of food and minerals. Railroads were built in the colonies to connect the plantations and mines to the ports. This transportation system, along with the discouragement of local manufacturing competing with manufacturing in the mother countries, prevented the economic development of the colonies. The terms of trade – what one can obtain from one's export – favored the European nations, since the prices of the primary products from the colonies remained low while the prices of the manufactured products sent back to the colonies continually increased.

When most of the colonies gained their independence after World War II, this trade pattern continued. Many of the less developed countries still produce food and minerals for the world market and primarily trade with their former colonial masters. The world demand for the products from the poorer nations fluctuates greatly, and the prices of these products remain depressed. The political and social systems that developed in the former colonies also serve to keep the majority within these developing nations poor. A local elite, which grew up when these countries were under colonial domination, learned to benefit from the domination by the Western countries. In a sense, two societies were created in these countries: one, relatively modern and prosperous, revolved around the export sector, while the other consisted of the rest of the people, who remained in the traditional system and were poor. The local elite, which became the governing elite upon independence, acquired a taste for Western products, which the industrial nations were happy to sell them at a good price.

The present vehicle of this economic domination by the North of the South is the multinational corporation. Tens of thousands of these exist today, many of the largest with headquarters in the US but a growing number of others headquartered in Europe and Japan. These corporations squeeze out small competing firms in the developing nations, evade local taxes through numerous devices, send large profits back to their headquarters, and create relatively few jobs since the manufacturing firms they set up utilize the same capital-intensive technology that is common in the industrialized countries. Also, they advertise their products extensively, thus creating demands for things such as Coca-Cola and color television sets while many people in the countries in which they operate still do not have enough to eat.

The advocates of the state approach point to the adverse terms of trade that many developing nations face today. There is general agreement that there has been a long-term decline in the terms of trade for many of the agricultural and mineral products that the less developed nations export. The prices of primary products have fallen to their lowest level in a century and a half.[21] There has also been great volatility in the prices of some of these products, with a change of 25 percent or more from one year to the next not uncommon for some products. Such fluctuations make economic planning by the developing nations very difficult. There is also clear evidence that the industrialized countries, while primarily trading among themselves, are highly dependent on the less developed countries for many crucial raw materials, including chromium, manganese, cobalt, bauxite, tin, and, of course, oil.

Although international trade is still far from being the most important component of the US economy, it is a very important factor for many of the wealthiest corporations. In the early 1980s about one-half of the 500 wealthiest corporations listed in *Fortune* magazine obtained over 40 percent of their profits from their foreign operations.[22] Some multinational corporations have financial resources larger than those of many developing nations.

Finally, the defenders of the state approach argue that there is little chance for many poor nations to achieve as fair a distribution of income as that achieved by Europe after it industrialized. This situation has evolved because the controlling elites in underdeveloped nations today have repressive tools at their disposal (such as sophisticated police surveillance devices and powerful weapons) that the European elites did not have. This allows them to deal with pressures from the "have-nots" in a way the Europeans never resorted to.

Critics of the state approach point to the breakup of the Soviet empire in Eastern Europe in the late 1980s and to the collapse of communism in the former Soviet Union and the breakup of that country in the early 1990s as support for their view that the state approach cannot efficiently produce wealth. In fact, it was the dissatisfaction of Eastern Europeans with their economic conditions that played a large role in their massive opposition to the existing communist governments and their eventual overthrow. Dissatisfaction with economic conditions also played a large role in the overthrow of the Soviet government, a startling rejection of the state approach by a people who had lived under it for 70 years.

The state approach to development struggles to survive the collapse of communist regimes in Europe, as can be seen in the posters of a Communist Party conference in Nepal (*Ab Abercrombie*)

Critics of the state approach also point to the suppression of individual liberties in the former Soviet Union, China, and other communist states as evidence that the socialist model for development has costs that many people are not willing to pay. In fact, most revolutions have huge costs, leading to much suffering and economic deterioration before any improvement in conditions is seen; even after improvements occur, oppressive political and social controls are used by leaders to maintain power.

Central planning has proved to be an inefficient allocator of resources wherever it has been followed. Without prices from the free market to indicate the real costs of goods and services, the central planners cannot make good decisions. And if efficient central planning has proved to be impossible in a developed country such as the former Soviet Union, it has proved to be even worse in underdeveloped nations where governmental administrative capability is weak. A final negative feature of central planning is that it always leads to a large governmental bureaucracy.

Multinational corporations have created jobs in the developing world that would not have existed otherwise; they have brought new technologies to the less developed nations; and they have helped the balance of payments problems of those nations by bringing in scarce capital and by helping develop export industries that earn much needed foreign exchange. These advantages are well known in the South, and explain why multinational corporations are welcomed by many less developed countries.

Finally, the critics of the state approach argue that political elites in developing nations have used dependency theory, especially in Latin America where the theory is popular, to gain local political support among the bureaucracy, military, and the masses. To blame the industrial nations for their poverty frees them from taking responsibility for their own development and excuses their lack of progress. It also frees them from having to clean their own houses of governmental corruption and incompetence, and to stop following misguided economic development approaches. The newly industrializing countries have shown that when market principles are followed, economic progress can be made even by developing nations that have a dense population and few, if any, natural resources.

The Civil Society Approach

In the words of a Harvard anthropologist, civil society is "the space between the state and the individual where those habits of the heart flourish that socialize the individual and humanize the state."[23] In simpler terms, it is the activity that people engage in as they interact with other people and can be seen in neighborhoods, voluntary organizations, and in spontaneous grassroots movements. Although this activity can be directed toward economic gain, often it is not. It is the activity which makes a community, a connection between people, a realization that each one is dependent upon others and that they share life together. Without civil society isolation results and since human beings are social animals, that isolation can lead to illness, antisocial behavior, and even suicide.

The civil society approach to development emphasizes social development, how people act toward other people. But the approach can also have important political and economic aspects. The best way to demonstrate this is through examples. In 1973 a group of poor people in India rushed to the forests above their impoverished village and hugged the trees to prevent a timber company from cutting them down. This community action received worldwide publicity and helped to force some governments to reconsider their development policy regarding their nation's forests. The Chipko movement, which grew out of this action, is an example of self-help community action directed against threats and harm to the environment, harm that the local people realize will make their lives more difficult or even impossible.

Civil society can also be seen functioning in the efforts by some people in poorer countries to raise their low living standards. It is generated by the realization in many poor countries that neither their governments nor the market can be relied on to help their citizens obtain basic needs. Here are two examples. In Latin America after the bishops of the Catholic Church met in 1968 in Colombia and decided that the church should become active in helping the poor, many priests, nuns, and lay Christians helped form Christian Based Communities, self-help groups mainly made up of the poor themselves.

In Bangladesh a professor – Muhammad Yunus – concluded that the landless poor could never improve their conditions without some extra funds to help them start up an income-producing activity. Since no banks would lend them money, he set up the Grameen Bank. The bank's loans, some starting as small as $35, have been repaid at a much higher rate than loans at regular banks: the repayment rate is now over 95 percent! By 2004 the bank had lent about $4 billion and now serves about 3 million borrowers in Bangladesh, most of whom are poor women. This experience demonstrated that the poor can be good financial risks and has been imitated in 40 other countries, including the United States, where this idea is known as "microcredit" and "microfinance." Worldwide about 50 million people were receiving microcredit loans in 2004.[24]

Civil society can also be directed toward political goals. In Eastern Europe in the 1980s millions of citizens took to the streets to call for the end of their communist governments. This grassroots movement, which spread throughout Eastern Europe, and which was primarily peaceful, led to the end of the Soviet empire and to the collapse of communism in Europe. Western political scientists were amazed that such an occurrence could take place. Few, if any, had imagined that the end of a powerful totalitarian state could come from the nonviolent actions of average citizens.

A spontaneous grassroots movement also occurred in Argentina in the early 1980s when a group of mothers met daily in one of the main squares in the nation's capital to protest at the disappearance of their children (thousands of individuals who were abducted by the military government in its war against subversion and suspected subversion). The silent, nonviolent protest by the mothers helped undermine the internal and external support for the government.

In 2004 the Nobel Peace Prize was given to Wangari Maathai from Kenya who, despite being beaten and jailed by the government, had organized the Green Belt Movement. The movement got mainly very poor rural women to plant 30 million trees to help restore the overexploited land of the country.

The internet can be seen as technology making possible global civil society networks. With the vast amount of information now available to internet users, connections can be fostered among people around the world working for such common goals as monitoring the environment, and holding corporations responsible for their actions, and for economic and political purposes. By the early twenty-first century some 650 million people were using the internet, which represented a growth of nearly 600 percent over the previous five years.[25]

Advocates of the civil society approach to development say it is easy to show examples of failures by the market and by the state to make people's lives better. It is even easy to show examples where they have made people's lives worse. People have responded to the failures of the market and the state by undertaking self-help activities. Such individuals want to participate in controlling their lives and do not want to let the market or the state be the main determinants of how they should live. They believe that strong reliance on the market or the state can leave the individual stunted.

The advocates of civil society also point to flourishing voluntary efforts in many countries as evidence of the importance of their approach. Although it is impossible to know exactly how many such groups exist today, the Worldwatch Institute in the United States found much evidence of civil society in the world in the late 1980s. Among other things they found:

- In India tens of thousands of groups, many following the self-help tradition established by Mahatma Gandhi, were involved in promoting social welfare, developing appropriate technology, and planting trees.
- In Indonesia, 600 independent groups worked on environmental protection.
- In Sri Lanka citizens formed a village awakening movement in which one-third of the nation's villages participated: 3 million people were involved in work parties, education and health projects, and in cooperative crafts.
- In Kenya 16,000 women's groups with 600,000 members were registered in the mid-1980s, many starting as savings clubs.
- In Brazil 100,000 Christian Based Communities existed, their membership totaling 3 million.
- The women's self-help movement in the shantytowns surrounding the capital of Lima, Peru, operated 1,500 community kitchens.
- In the United States in the late 1980s an estimated 25 million people were involved in local actions to protect the environment.[26]

Finally, the advocates of the civil society approach point to the spread of democracy around the world. In the 1980s many developing nations adopted a democratic form of government and, with the collapse of the Soviet empire, many former communist countries became democratic. An estimated 2.5 billion people lived in fully or partially democratic countries in 1981 whereas in 2001 this number had grown to 3.9 billion people.[27] And it is in democracies that voluntary organizations flourish.

Critics of the civil society approach point out that while small may be beautiful, it can also be insignificant. Even the admirable Grameen Bank of Bangladesh provided only about 0.1 percent of the credit in the country in the mid-1990s.[28] The conclusion of a UN organization sympathetic to the efforts of self-help groups is that while nongovernmental organizations have helped transform the lives of millions of people throughout the world, "What seems clear is that even people helped by successful projects still remain poor."[29]

Efforts at the grassroots level directed toward community-managed economic development often fail. The worker cooperative is often the instrument used, but a majority of these survive only a few years.[30] The members of the cooperatives, where workers come together to purchase and operate a business, are usually inexperienced in management. They are plagued by outside economic forces, such as high inflation and uncertain markets, which they are unable to deal with.

Critics also point out that oppressive political and economic powers can block the efforts of community groups. One well-publicized example was the assassination of Chico Mendes, the leader of a group of rubber tree harvesters in the Brazilian Amazon region. The large landowners in this region and in other Latin

American countries have, with the support of local governments, traditionally used force against peasants' and workers' organizations.

Finally, critics of the civil society approach point to the spread of antidemocratic forces in the world at the same time as democracy is spreading. With the spreading of democracy, the end of the Cold War, and the collapse of communism in Europe, ethnic and regional hatreds surfaced in many countries, hatreds that had been suppressed by the former authoritarian and totalitarian governments. Yugoslavia entered into a cruel civil war and the world saw "ethnic cleansing" reemerge, an idea which it had incorrectly believed had been discredited in Europe with the defeat of Nazi Germany. Bitter ethnic hostilities also arose in Africa in the mid-1990s, with thousands slaughtered in horrifying civil wars. Sometimes incited by a few people for political reasons, group hatred toward "others," toward those outside one's group, unfortunately has become fairly common in the post-Cold War world. True civil society, where people have respect and tolerance for those outside their immediate group, does not exist in a number of countries today.[31]

Geography and Wealth, Geography and Poverty

Adam Smith had a second theory of why some nations are rich and some poor. Modern economists usually ignore this part of Smith's writings. Not only did Smith believe that a free market economy would lead to wealth, but he also believed that nations bordering a sea would usually be richer than inland, landlocked countries. Recent research shows that geography does matter. As table 1.1 shows, nations with access to the sea by coastal ports or by navigable rivers and those in the temperate climate zone are usually the wealthiest nations. Those nations landlocked and in the tropical zone or mainly desert or mountainous usually are the poorest.

Why does geography matter? The reasons are not hard to discover. First, shipping and receiving goods by sea is cheaper than shipping by land or air. For example, shipping a container to the Ivory Coast, which has a major coastal city, costs about $3,000, while shipping a similar container to the landlocked Central African Republic costs about $13,000. Also people and new ideas often arrive in coastal areas first. Second, tropical climates are plagued by infectious diseases, such as malaria, which debilitate the workforce. An estimated 300 million to 500 million new cases of malaria occur each year, nearly all of them in the tropics. Winter is the great natural controller of many diseases. In tropical countries many diseases flourish all year long, making them difficult to control. And because most of the tropical countries are relatively poor, pharmaceutical firms prefer to spend their research money on a condition such as erectile dysfunction rather than river blindness or malaria.

Agricultural production is also usually higher in temperate and subtropical climates than in tropical climates. For example, a hectare of land in the temperate zone produces about 6 tons of corn or maize, while the same amount of land in the tropics produces about 2 tons. Developed countries spend much more on

Table 1.1 The wealth of tropical, desert, highland, and temperate regions

Climate zone	*Percent of world total*	*Near**	*Far**
Tropical			
Population	40	22	18
GNP	17	10	7
Desert			
Population	18	4	14
GNP	10	3	7
Highland			
Population	7	1	6
GNP	5	1	4
Temperate			
Population	35	23	12
GNP	67	53	14

*"Near" means within 100 kilometers of sea coast or sea-navigable waterway; "Far" means otherwise.

Source: Jeffrey Sachs, Andrew Mellinger, and John Gallup, "The Geography of Poverty and Wealth," *Scientific American*, 284 (March 2001), p. 74

research to help their farmers in the temperate zone increase production than poorer developing countries spend on research that would benefit their tropical or semitropical farmers.

Geography alone does not explain why some countries are wealthier than others. While nearly all the wealthiest countries are in the temperate zone, such as North America, Western Europe, Northeast Asia, the economic system they follow is important also as shown by the fact that the former Soviet Union and Eastern Europe are still struggling to overcome their socialist pasts. This fact is shown even more dramatically by looking at present and past countries with the same geographical characteristics but with different economic systems and vastly different wealth: South Korea and North Korea, West and East Germany, Austria and the Czech Republic, and Finland and Estonia. In each case the first-mentioned state in the comparison followed the free market system and greatly outperformed the second socialist state.[32]

In addition to the difficulties caused by climate and lack of access to the sea, many landlocked countries face economic difficulties caused by borders, borders with their neighbors which restrict the easy flow of goods, capital, and people.

According to Ricardo Hausmann, professor of the practice of economic development at the John F. Kennedy School of Government at Harvard University, "If current trends persist, countries that face high transportation costs and a high dependence on tropical agriculture will be left far behind, mired in poverty and income inequality. Will the rest of the world find this outcome morally acceptable?" Hausmann believes that the world has tried to help these countries, but its efforts have been insufficient as shown by the widening gap between the rich and poor. He calls for more "globalized governance." By this he means more international agreements to make borders less of a barrier to people, goods, and capital. He also calls for international support for development projects that improve the transportation systems within and between countries, and lastly he calls for international aid in health and agricultural technology that benefits the tropical world.[33]

Globalization

After the killing in World War II ended in 1945 a number of world leaders asked "What should be done to prevent a person like Adolph Hitler coming to power again?" One of the answers given was to prevent an international economic collapse, such as the Great Depression, which created the conditions that led to the rise of Hitler. With that idea in mind, it was agreed that trade among nations should be encouraged so that, hopefully, prosperity would spread. In 1947, under the sponsorship of the United Nations, the General Agreement on Tariffs and Trade (popularly known as GATT) was signed by about 20 countries. These countries, later joined by about a hundred others, conducted a series of negotiations to promote free trade by reducing tariffs and other barriers to trade such as import quotas. The success of these efforts is clearly shown in figure 1.6. From 1950 to the end of the twentieth century world trade rose from about $500 billion to nearly $6 trillion. During a somewhat similar period, the amount of goods and services produced in the world (gross world product) rose from less than $10 trillion to nearly $60 trillion, as shown in figure 1.7.

In 1995 GATT evolved into the World Trade Organization (WTO). The WTO was given the task of implementing the many agreements reached under the GATT negotiations and of setting up an arbitrating mechanism to resolve trade disputes among its members.

The great expansion of international trade has created a world economy. That integration of the economies of many nations has been the main force in creating

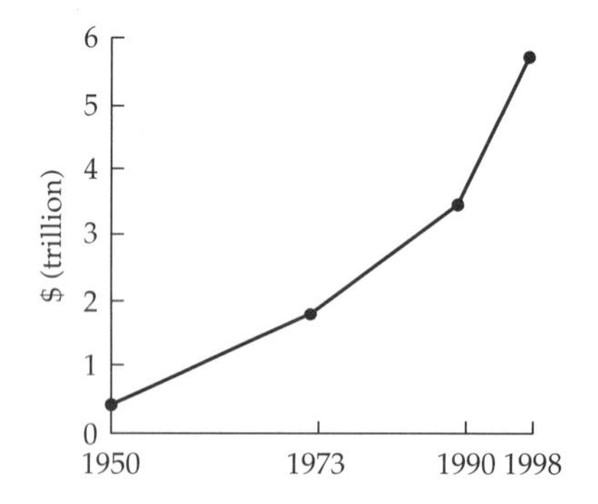

Figure 1.6 World trade, merchandise exports (in 1990 dollars)
Source: World Trade Organization

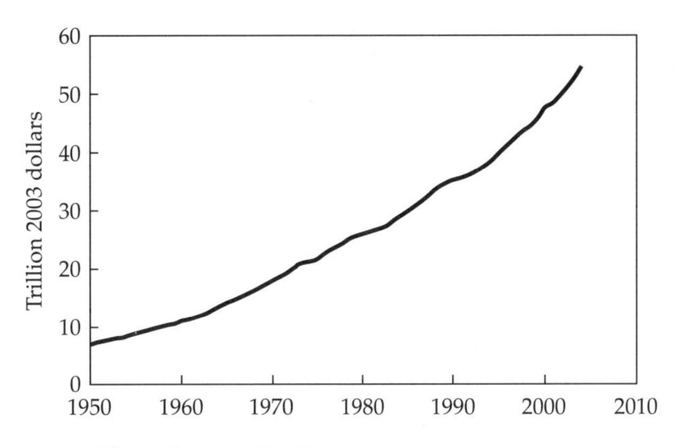

Figure 1.7 Gross world product, 1950–2004
Source: Worldwatch Institute, *Vital Signs 2005* (New York: W. W. Norton, 2005), p. 45

a new situation in the world called globalization. While globalization is mainly fueled by economic forces, it is fueled also by new political, social, and technical integrative forces in the world today. Politically, international governmental organizations such as the United Nations, the International Monetary Fund, and the World Bank, along with regional organizations and agreements such as the European Union and the North American Free Trade Agreement, are playing an increasingly important role in global governance.

Social integration is proceeding at such a rapid pace that one can say that there is the beginning of a world culture. Much of this culture is coming from the United States, but it is truly international as foods, music, dances, and fashions come from various countries. Technical integration comes from the information and transportation revolutions that are occurring in the world. Computers, fax machines, television, and airplanes now link the world.

Globalization is a process that is leading to the growing interdependence of the world's people. Like many things, if not most things in life, it has a positive side and a negative side. No one knows yet which side will dominate.[34] I feel it is best to inform the reader of some of its aspects – both positive and negative – and let the reader decide how he or she feels about it.

Positive aspects

The global economy has brought more wealth to both rich and poor nations. Although all nations have not benefited from it, "since 1950 there has been a close correlation between a country's domestic economic performance and its participation in the world's economy."[35] The United States, a country that has officially embraced globalization, had an unprecedented long period of economic growth and, as the information at the beginning of this chapter indicates, non-Western countries and areas such as China, the Republic of Korea, Taiwan, Singapore, and Hong Kong that also embraced globalization have obtained high levels of wealth.

The formation of a global community has started. Nations around the world now face common problems, both economic and environmental, that they are working together to solve. More and more individuals are taking advantage of the new communication and transportation technologies to learn about and enjoy the whole planet.

For nations to become wealthier in this new world, they must become freer, less corrupt, and more economically efficient. New ideas and more international contacts may even undermine the authoritarian rule in countries such as China as they did in other former communist states.

Hunger and crime rates are lowered as poverty is reduced. Many types of crime dropped in the United States during its recent long period of economic growth and "the number of hungry people [in the world] fell by nearly 20 million in the 1990s."[36]

New products are available and often at a lower price than if they had been produced locally.

New jobs are created, not just in the developed countries but also in many less developed nations. Jobs lead to the reduction of poverty. The World Bank reports that "There are almost no examples of countries experiencing significant growth without reducing poverty."[37]

Although it is true that increased production can cause more pollution, once nations become richer and solve their poverty problem, they tend to clean up their environments.

Negative aspects

Many people are losing their jobs in the rich countries as multinational corporations move some of their production facilities to less developed nations where labor costs are low. It is true that many new jobs are being created in the United States, fewer in Europe and Japan, but it is not easy for older workers who have been laid off to qualify for them.

Some corporations are moving facilities to developing nations to escape from complying with strict environmental laws and the demand for good working conditions in their home countries.

Rapid economic growth in countries such as China and India has led to major pollution of air, water, and land.

The authoritarian regime in China, with its dismal human rights record, is strengthened by economic growth.

Cultural imperialism by the United States, with its corresponding undermining of local cultures, is increasing. A world traveler can now literally dine only on Big Macs, fries, and shakes. The largest single export industry in the United States is not aircraft or automobiles but entertainment, especially Hollywood films.

The gap between rich and poor nations is growing. Some poor nations are going nowhere; they are being left behind. The shift to knowledge-based industries is accelerating and creating an even greater gap. The United States has more computers than the whole of the rest of the world combined. About 25 percent of people in the United States use the internet, whereas only 3 percent of Russians do and 1 percent of the population in South Asia. A citizen of the United States needs to save one month's salary to buy a computer, whereas a citizen of Bangladesh would have to save all his or her wages for eight years to do so.

Because nations' economies are so tied together today, an economic downturn in one can spread to others extremely quickly. We saw that happening in East Asia in the late 1990s when a financial crisis hit Thailand, Indonesia, Malaysia, the Republic of Korea and other countries. Economic recessions and depressions also come with the dominance of the market, not just growth. Capitalism has always had its cycles, and a "down" cycle can mean high unemployment and human suffering. Many of the fastest growing developing nations are tied to the US economy. If the United States goes into a period of slow or no growth it will affect many other countries whose wealth comes mainly from exports to the United States.

Nations are losing some of their national autonomy to institutions such as the International Monetary Fund, World Trade Organization, World Bank, and regional trade organizations.

The ease of transportation, of both people and goods, makes the transmission of diseases throughout the world easier than before. In the same way, rapid electronic communications and the huge number of people and goods moving through the world make criminal and terror activities more difficult to control.

An evaluation

Kofi Annan, as Secretary-General of the United Nations, had the following to say about globalization at the beginning of the twenty-first century:

> the central challenge we face today is to ensure that globalization becomes a positive force for all the world's people, instead of leaving billions of them behind in squalor. Inclusive globalization must be built on the great enabling force of the market, but market forces alone will not achieve it. It requires a broader effort to create a shared future, based upon our common humanity in all its diversity.[38]

In the box titled "The global village" he shows us what our world at the beginning of the twenty-first century looks like by asking us to imagine we are living in a small village with all the characteristics of our world.

Conclusions

The market approach to development places emphasis on the seemingly strong motivation individuals have to acquire more material goods and services. When people are freed from external restraints, the market allows them to use their initiatives to better their lives. The release of creative energy that comes with the market approach is impressive. At the beginning of the twenty-first century most countries throughout the world were following it, at least to some degree, as the Western capitalist countries became the models to imitate.

The global village

"Let us imagine, for a moment, that the world really is a 'global village' – taking seriously the metaphor that is often invoked to depict global interdependence. Say this village has 1,000 individuals, with all the characteristics of today's human race distributed in exactly the same proportions. What would it look like? . . .

Some 150 of the inhabitants live in an affluent area of the village, about 780 in poorer districts. Another 70 or so live in a neighbourhood that is in transition. The average income per person is $6,000 a year, and there are more middle income families than in the past. But just 200 people dispose of 86 per cent of all the wealth, while nearly half of the villagers are eking out an existence on less than $2 per day.

Men outnumber women by a small margin, but women make up a majority of those who live in poverty. Adult literacy has been increasing. Still, some 220 villagers – two thirds of them women – are illiterate. Of the 390 inhabitants under 20 years of age, three fourths live in the poorer districts, and many are looking desperately for jobs that do not exist. Fewer than 60 people own a computer, and only 24 have access to the Internet. More than half have never made or received a telephone call.

Life expectancy in the affluent district is nearly 78 years, in the poorer areas 64 years – and in the very poorest neighbourhood a mere 52 years. Each marks an improvement over previous generations, but why do the poorest lag so far behind? Because in their neighbourhoods there is a far higher incidence of infectious diseases and malnutrition, combined with an acute lack of access to safe water, sanitation, health care, adequate housing, education and work.

There is no predictable way to keep the peace in this village. Some districts are relatively safe while others are wracked by organized violence. The village has suffered a growing number of weather-related natural disasters in recent years including unexpected and severe storms, as well as sudden swings from floods to droughts, while the average temperature is perceptibly warmer. . . . The village's water table is falling precipitously, and the livelihood of one sixth of the inhabitants is threatened by soil degradation in the surrounding countryside."

Source: Kofi Annan, *We the Peoples: The Role of the United Nations in the 21st Century* (New York: United Nations, 2001), pp. 52–6

With the collapse of communism and the breakup of the former Soviet Union, the state approach to development received a serious blow. The reliance on the state to create wealth was discredited. Yet in no country of the world is a state without some significant functions relating to the economy. Within the capitalist world there is a debate among nations regarding how much involvement government should have in directing and guiding the economy. Traditionally Japanese and European capitalism relied on more government involvement than did capitalism in the United States. Even after the seemingly total victory of the market approach over the state approach in the 1990s, the state approach is not dead; what is dead is the total or near total reliance on it as the best way to create wealth.

The market approach relies on the materialistic self-interest of people to create wealth, while the state approach presents the government as the main actor in the creation of wealth and the reduction of poverty. The civil society approach presents a new participant in development and new motivations. By focusing on the benefits that occur when people exercise local initiative and function as a community, a new force is recognized as having a role in development. Civil society recognizes that social development – how people interact – is as important as economic and political development for the health of a community. The millions of voluntary efforts that can take place every day in a civil society, most of them not motivated by economic gain or the result of government intervention, can make the difference between a highly civilized community and one that is not.

The challenge societies face is to achieve the right balance among the three approaches. It appears that the sole reliance on only one, or even two, will not produce a society where the people are living up to their full potential. The United States can be considered a highly developed society from the perspective of the market and from the state – in other words, from an economic and political perspective. Although civil society is alive and well in the United States, some believe the country has yet to achieve the right balance among these three approaches. Social critics such as Alan Wolfe, former dean of the New School for Social Research, believe that the market has become such a dominant force in the United States that it is actually weakening the civil society.[39]

Today's globalization is mainly driven by market forces focusing on economic growth. That growth can do much to reduce world poverty. But large and vocal protests at international meetings dealing with aspects of globalization have drawn attention to some of the negative aspects of globalization. More emphasis on human governance appears to be needed. The United Nations Development Programme stated the need as follows:

> When the market goes too far in dominating social and political outcomes, the opportunities and rewards of globalization spread unequally and inequitably – concentrating power and wealth in a select group of people, nations, and corporations, marginalizing the others.

The UN Development Programme believes that markets should continue to expand but that more governance is needed:

> The challenge is to find the rules and institutions for stronger governance – local, national, regional and global – to preserve the advantages of global markets and competition, but also to provide enough space for human, community, and environmental resources to ensure that globalization works for people – not just for profits.[40]

Let's return to the central question addressed in this chapter: What makes some nations rich and some poor? The United States and Western Europe do not appear to offer many relevant lessons for the poorest nations. Industrialization in the West took place under conditions vastly different from those now experienced by many poor nations. The Western nations were generally rich in natural resources in relation to their needs, or if deficient, were able to get them from their colonies. Most developing nations today are not rich in the natural resources needed for industrialization (such as sources of energy), nor are they permitted to seize colonies. In addition, the West, while experiencing an expansion of its population during its early development, did not come close to experiencing the vast population growth that is common today in many of the poorest nations. Citizens of the United States, especially, need to remember that the unique features their nation possesses – fertile land, rich natural resources, an abundance of fresh water in its numerous lakes and rivers, and a temperate climate – make their country atypical in the world. North Americans have used these gifts to create an unprecedented amount of material wealth – but with real costs, many of which will be examined in this book. The policies advocated by the market approach generally worked well in the United States, where individual initiative was encouraged by a supportive government with limited powers.

Since much of the rest of the world does not share the assets Western Europe and the United States possess, what should they do to raise the living standards of the poorest? Some might advocate revolution to break the repressive bonds common in a number of societies, but the suffering that revolutions cause and the uncertainty of their accomplishments make it difficult to advocate this course. But to counsel moderation and slow reforms also is difficult since, as historian Barrington Moore Jr reminds us, it ignores the suffering of those who have not revolted. Moore believes that "the costs of moderation have been at least as atrocious as those of revolution, perhaps a great deal more." (Moore also makes some harsh criticisms of revolutions: "one of the most revolting features of revolutionary dictatorships has been their use of terror against little people who were as much victims of the old order as were the revolutionaries themselves, often more so.")[41]

Where does this leave us? It should leave us, I believe, with a sense of humility, if we are the rich, as we recognize how important factors outside of our personal control probably were in ensuring our richness. If we are the poor, it can leave us with an understanding of some of the causes of poverty and some suggestions of how nations might improve the lot of their poorest. And for both the rich and the poor, there should be an awareness that this is the first time in human history that there is a global concern with issues of development – why some are rich and many are poor. That we have not yet learned how to reduce the vast inequalities of wealth in the world should not be surprising. We may never learn how the South can catch up with the North. It seems likely now that within our lifetimes the gap between the rich and poor in the world will increase instead of decrease. But it is clear we are learning, through trial and error, how to improve the lot of the poorest. Whether the poor and rich nations will have the political will – and ability – to do what is necessary to help the world's poor is not known.

Notes

1 UN Development Programme, *Human Development Report 1999* (New York: Oxford University Press, 1999), p. 38.

2 Bruce Scott, "The Great Divide in the Global Village," *Foreign Affairs* (January/February 2001), pp. 162–3.

3 Ricardo Hausmann, "Prisoners of Geography," *Foreign Affairs* (January/February 2001), p. 46.

4 Gross national income calculated so the currency of each country has equal purchasing power. World Bank, *World Development Report 2005* (New York: World Bank and Oxford University Press, 2004), p. 266.

5 Elizabeth Becker, "UN Study Finds Global Trade Benefits Are Uneven," *New York Times*, national edn. (February 24, 2004), p. C5.

6 World Bank, *World Development Report 1999–2000* (New York: Oxford University Press, 2000), p. 14.

7 China increased its per capita gross domestic product (GDP) about tenfold from $440 in 1980 to $4,475 in 2002 (in international prices), while India's per capita GDP rose from

$670 in 1980 to $2,570 in 2002. World Bank, *World Development Report 2005*, p. 27.

8 UN Development Programme, *Human Development Report 2004* (New York: Oxford University Press, 2004), p. 129.

9 Ibid.

10 World Bank, *World Development Report 2005*, p. 31.

11 UN Development Programme, *Human Development Report 2004*, p. 132.

12 World Bank, *World Development Report 1990: Poverty* (New York: Oxford University Press, 1990), p. 2.

13 The other seven goals are: Goal 2, Achieve universal primary education; Goal 3, Promote gender equality and empower women; Goal 4, Reduce child mortality; Goal 5, Improve maternal health; Goal 6, Combat HIV/AIDS, malaria and other diseases; Goal 7, Ensure environmental sustainability; Goal 8, Develop a global partnership for development.

14 UN Development Programme, *Human Development Report 2003* (New York: Oxford University Press, 2003), p. 5.

15 World Bank, *World Development Report 2003*, p. 12.

16 According to the OECD, US private foundations and other private US organizations provide about $6 billion a year in foreign assistance, much more than private groups give in most other counties. But even if this amount is added to the US governmental aid, the total US aid given in relation to US GNP is still among the lowest ratios of all rich donor countries. Jeffrey Sachs, "Can Extreme Poverty Be Eliminated?" *Scientific American*, 293 (September 2005), p. 60.

17 Current thinking among Western economists about economic development is considerably more complex than the simplified view of the market approach presented below.

18 I am indebted to Alan Wolfe for this classification of the three main views of development. He presented his ideas in a paper titled "Three Paths to Development: Market, State, and Civil Society," which was prepared for the International Meeting of Nongovernmental Organizations (NGOs) and UN System Agencies held in 1991 in Rio de Janeiro. Some of his views on this subject are contained in his book *Whose Keeper? Social Science and Moral Obligation* (Berkeley: University of California Press, 1989).

19 Irma Adelman and Cynthia Taft Morris, *Economic Growth and Social Equity in Developing Countries* (Stanford, CA: Stanford University Press, 1973), p. 189. Censuses in Brazil have revealed that the percentage of national income going to the top 10 percent of the population was 40 percent in 1960, 47 percent in 1970, and 51 percent in 1980. During the same period the poorest 50 percent of the population received 17 percent of the national income in 1960, 15 percent in 1970, and 13 percent in 1980. Thomas E. Skidmore and Peter H. Smith, *Modern Latin America*, 2nd edn (New York: Oxford University Press, 1989), p. 180. In Latin America as a whole in the 1980s the poorest 10 percent suffered a 15 percent drop in their share of income. See UN Development Programme, *Human Development Report 1999*, p. 39.

20 For a fuller discussion of dependency theory see Bruce Russett and Harvey Starr, *World Politics: The Menu for Choice*, 2nd edn (New York: W. H. Freeman, 1985), ch. 16, and John T. Rourke, *International Politics on the World Stage*, 7th edn (New York: Dushkin/McGraw-Hill, 1999), p. 400.

21 UN Development Programme, *Human Development Report 1999*, p. 2.

22 Frederic S. Pearson and J. Martin Rochester, *International Relations: The Global Condition in the Twenty-First Century*, 4th edn (New York: Random House, 1998), p. 499.

23 David Maybury-Lewis, *Millennium: Tribal Wisdom and the Modern World* (New York: Viking, 1992), p. 265.

24 Muhammad Yunus, "The Grameen Bank," *Scientific American*, 281 (November 1999), pp. 114–19. Yunus believes that loans to the poor are best made by private organizations, not by governments. When the new president of Mexico, Vicente Fox, announced in 2000 that his government was going to begin a small loan program, Yunus talked him out of the idea, saying "Politicians are interested in the votes of the poor . . . not in getting the money back." Tim Weiner, "With Little Loans, Mexican Women Overcome," *New York Times*, national edn (March 19, 2003), p. A8. See also Celia Dugger, "Debate Stirs over Tiny Loans for World's Poorest," *New York Times*, national edn (April 29, 2004), p. A1, and Saritha Rai, "Tiny Loans Have Big Impact on Poor," *New York Times*, national edn (April 12, 2004), p. C3.

25 UN Development Programme, UN Environment Programme, World Bank, and World Resources Institute, *A Guide to World Resources 2002–2004* (Washington, DC: World Resources Institute, 2002), p. 21.
26 Alan Durning, "Mobilizing at the Grassroots," in *State of the World, 1989* (New York: W. W. Norton, 1989), pp. 157–8. The growth of civil society continued in the 1990s. See David Bornstein, "A Force Now in the World, Citizens Flex Social Muscle," *New York Times*, national edn (July 10, 1999), pp. A15, A17.
27 UN Development Programme et al., *A Guide to World Resources 2002–2004*, p. 21.
28 UN Development Programme, *Human Development Report 1993*, p. 95.
29 Ibid., p. 94. See also Barbara Crossette, "UN Report Raises Questions about Small Loans to the Poor," *New York Times*, national edn (September 3, 1998), p. A8, and Dugger, "Debate Stirs over Tiny Loans for World's Poorest."
30 Durning, "Mobilizing at the Grassroots," p. 163.
31 An interesting analysis of the causes of the many ethnic conflicts taking place in the world in the mid-1990s is contained in Daniel Coleman, "Amid Ethnic Wars, Psychiatrists Seek Roots of Conflict," *New York Times*, late edn (August 2, 1994), pp. C1, C13.
32 Most of the analysis on the relationship between geography and wealth and poverty is taken from Jeffrey Sachs, Andrew Mellinger, and John Gallup, "The Geography of Poverty and Wealth," *Scientific American* 284 (March 2001), pp. 70–5, and Ricardo Hausmann, "Prisoners of Geography," *Foreign Affairs* (January/February 2001), pp. 45–53.
33 Hausmann, "Prisoners of Geography," p. 53.
34 For a good discussion of the potential for globalization doing good or harm see "Overview: Globalization with a Human Face," in UNDP, *Human Development Report 1999*, pp. 1–13.
35 Peter F. Drucker, "Trade Lessons from the World Economy," *Foreign Affairs*, 73 (January/February 1994), p. 104.
36 UN Development Programme, *Human Development Report 2003*, p. 6.
37 World Bank, *World Development Report 2005*, p. 31.
38 Kofi Annan, *We the Peoples: The Role of the United Nations in the 21st Century* (New York: United Nations, 2001).
39 Wolfe, *Whose Keeper?*
40 UN Development Programme, *Human Development Report 1999*, p. 2.
41 Barrington Moore Jr, *Social Origins of Dictatorship and Democracy: Lord and Peasant in the Making of the Modern World* (Boston: Beacon Press, 1966), pp. 505–7.

Further Reading

Bhagwati, Jagdish, *In Defense of Globalization* (Oxford and New York: Oxford University Press, 2004). The argument of this economics professor at Columbia University is that globalization has been overwhelmingly a good thing and its few downsides can be mitigated. His thesis that globalization leads to economic growth and economic growth leads to the reduction of poverty is the foundation for his belief that poor nations are not hurt by globalization but actually need more of it.

Chua, Amy, *World on Fire: How Exporting Free Market Democracy Breeds Ethnic Hatred and Global Instability* (New York: Anchor Books, 2003). A professor of law at Yale University, the author, who is a friend of globalization, argues that as the market and democracy have spread into the less developed nations, ethnic hatred and violence have increased along with anti-Americanism. Chua explains why and identifies the urgent need for a greater sharing of the economic wealth that globalization has brought to various minorities.

De Soto, Hernando, *The Other Path: The Invisible Revolution in the Third World* (New York: Harper & Row, 1989). De Soto focuses on the informal part of the economy in Peru, those illegal street vendors, private bus owners, and slum dwellers who have built houses on illegally occupied land. His provocative thesis is that these people are the real and only capitalists in the country and are the most productive part of the economy.

Falk, Richard, *Predatory Globalization: A Critique* (Cambridge: Polity, 1999). Falk believes that the present international system heavily favors the richest nations. It has created a permanent elite and a permanent underclass. He advocates reform from below through nongovernmental organizations and individual citizens.

Farmer, Paul, *Pathologies of Power: Health, Human Rights, and the New War on the Poor* (Berkeley, CA: University of California Press, 2003). Farmer presents case studies to support his three main points: the poor are not responsible for their situation, but have been hurt by their circumstances; the poor can be successfully treated and cured of disease, even those in the most dire conditions; good health is a human right for without it all other human rights are meaningless.

Friedman, Thomas L., *The World is Flat: A Brief History of the Twenty-First Century* (New York: Farrar, Straus & Giroux, 2005). Friedman believes we are entering the next phase of globalization, a so-called flat world where individuals and companies anywhere in the world can participate in the global economy. The laying of fiber-optic cables across the ocean floors and the development of new computer software made "outsourcing" possible. For example, engineers and technicians in India and manufacturers in China have now become important participants in the US economy.

Menzel, Peter, *Material World: A Global Family Portrait* (San Francisco: Sierra Club Books, 1994). A remarkable book which seeks to show, mainly through photographs, the differences in material goods owned by average families around the world. The author traveled to 30 nations.

Persaud, Avinash, "The Knowledge Gap," *Foreign Affairs*, 80 (March/April 2001, pp. 107–117. Persaud says, "It is doubtful that the knowledge revolution will let developing countries leapfrog to higher levels of development, as many technologists and Internet evangelists assert. In fact, the knowledge gap will likely widen the disparities between rich and poor, imprisoning many developing countries in relative poverty."

Roberts, J. Timmons, and Nikki D. Thanos, *Trouble in Paradise: Globalization and Environmental Crises in Latin America* (London and New York: Routledge, 2003). Focuses on "the other side of development," where short-term foreign investments have led to a crushing foreign debt, deforestation, destruction of habitats, disease, pollution, and dislocation of peoples.

Sachs, Jeffrey D., *The End of Poverty: Economic Possibilities for Our Time* (New York: Penguin Press, 2005). Sachs presents a plan to rid the world of extreme poverty by 2025. He does not dismiss the effectiveness of the market approach but believes that it is incomplete by itself. Poor countries that are weighed down by harmful geography, an inadequate healthcare system, and weak infrastructure (e.g. roads, ports, power and communication facilities) cannot improve without significant, wisely given, foreign aid.

Singer, Peter, *One World: The Ethics of Globalization* (New Haven, CT and London: Yale University Press, 2002). Called one of the most provocative philosophers of our time, Singer writes, "How well we come through the era of globalization (perhaps whether we come through it at all) will depend on how we respond ethically to the idea that we live in one world."

Thurow, Lester C., *Building Wealth: The New Rules for Individuals, Companies, and Nations* (New York: Harper Business, 1999). Thurow believes the world is in the Third Industrial Revolution, one based on wealth from the control of knowledge. Industries such as microelectronics, computers, robotics, and biotechnology are at the forefront. A pessimist at present, Thurow sees the revolution as transferring wealth from the many to the few. Thurow presents recommendations that he believes can help make twenty-first century capitalism benefit the many and not just the few.

United Nations Development Programme, *Human Development Report 2005* (New York: Oxford University Press, 2005). This uncharacteristically interesting annual report rates nations according to their level of human development.

United Nations Development Programme, United Nations Environment Programme, World Bank, and World Resources Institute, *World Resources 2005: The Wealth of the Poor: Managing Ecosystems to Fight Poverty* (Washington, DC: World Resources Institute, 2005). An attractive, easy-to-read reference source giving environmental, social, and economic trends of about 150 nations. In this volume the focus is on how the natural world can be utilized in a sustainable manner to benefit the rural poor.

World Bank, *World Development Report 2000–2001: Attacking Poverty* (New York: Oxford University Press, 2001). This particular report focuses on poverty. It presents the policies, politics and programs affecting poverty and presents a strategy to reduce it. The book's many statistics are presented in interesting charts.

Yergin, Daniel, and Joseph Stanislaw, *The Commanding Heights: The Battle Between Government and the Marketplace That Is Remaking the Modern World* (New York: Simon & Schuster, 1998). The authors argue in a book that reads like a novel that the shift of power from government to markets is reshaping the modern world.

CHAPTER 2
Population

Prudent men should judge of future events by what has taken place in the past, and what is taking place in the present.
Miguel de Cervantes (1547–1616), *Persiles and Sigismunda*

The Changing Population of the World

The population of the world is growing. No one will be startled by that sentence, but what is startling is the rate of growth, and the fact that the present growth of population is unprecedented in human history. The best historical evidence we have today indicates that there were about 5 million people in the world about 8000 BC. By AD 1 there were about 200 million, and by 1650 the population had grown to about 500 million. The world reached its first billion people about 1850; the second billion came about 1930. The third billion was reached about 1960, the fourth about 1974, and the fifth about 1987. The sixth came in 1999. These figures indicate how rapidly the population is increasing. Table 2.1 shows how long it took the world to add each billion of its total population. Projections are also given for the next three billions.

There is another way to look at population growth, one that helps us understand the uniqueness of our situation and its staggering possibilities for harm to life on this planet. Because most people born can have children of their own, the human population can – until certain limits are reached – grow exponentially: 1 to 2; 2 to 4; 4 to 8; 8 to 16; 16 to 32; 32 to 64; 64 to 128; etc. When something grows exponentially, there is hardly discernible growth in the early stages and then the numbers shoot up. The French have a riddle they use to help teach the nature of exponential growth to children. It goes like this: if you have a pond with one lily in it that doubles its size every day, and which will completely cover the pond

Table 2.1 Time taken to add each billion to the world population, 1800–2048

Date	*Estimated world population (billions)*	*Years to add 1 billion people*
1800	1	2,000,000
1930	2	130
1960	3	30
1974	4	14
1987	5	13
1999	6	12
2012 (projected)*	7	13
2027 (projected)*	8	15
2048 (projected)*	9	21

**Source*: *UN World Population Prospects: The 2004 Revision*

in 30 days, on what day will the lily cover half the pond? The answer is the twenty-ninth day. What this riddle tells you is that if you wait until the lily covers half the pond before cutting it back, you will have only one day to do this – the twenty-ninth day – because it will cover the whole pond the next day.

If you plot on a graph anything that has an exponential growth, you get a J-curve. For a long time there is not much growth but when the bend of the curve in the "J" is reached, the growth becomes dramatic. Figure 2.1 shows what the earth's population growth curve looks like.

The growth of the earth's population has been compared to a long fuse on a bomb: once the fuse is lit, it sputters along for a long while and then suddenly the bomb explodes. This is what is meant by the phrases "population explosion" and "population bomb." The analogy is not a bad one. The world's population

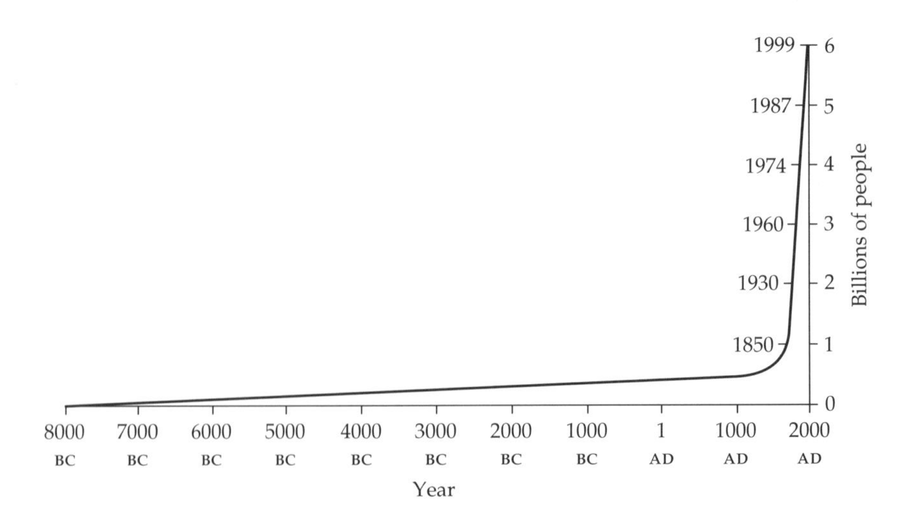

Figure 2.1 Population growth from 8000 BC to 2000 AD

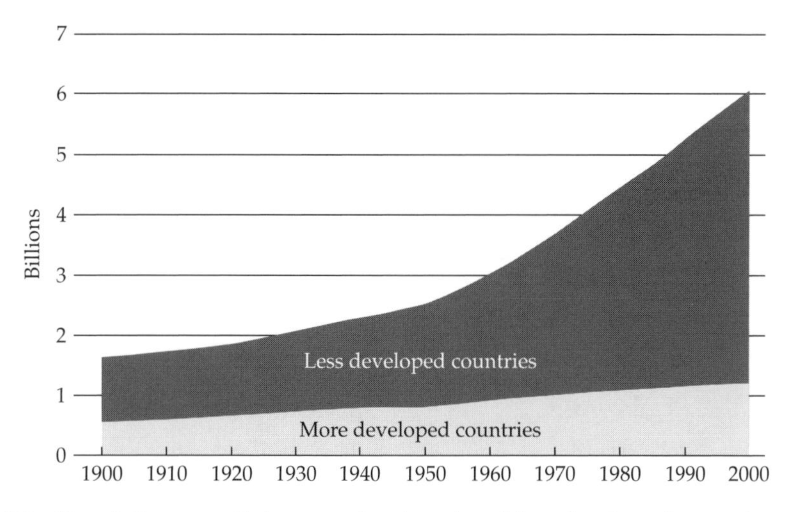

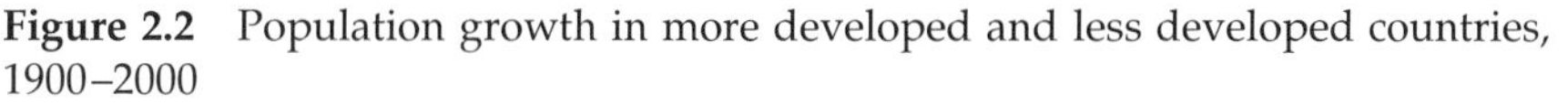

Figure 2.2 Population growth in more developed and less developed countries, 1900–2000

Note: Developed countries include Australia, Canada, Japan, New Zealand, United States, and Europe. All other countries are included in less developed.

Sources: UN Population Division, *World Population Prospects: The 2002 Revision* (2003), and Population Reference Bureau estimates; as presented in "Transitions in World Population," *Population Bulletin*, 59 (Population Reference Bureau, Washington, DC) (March 2004), p. 5

has passed the bend of the J-curve and is now rapidly expanding. The United Nations estimates the world's population reached 6 billion in 1999, shooting up from the estimated 2 billion in 1930, a tripling in just 70 years.

Figure 2.2 shows that the largest growth in the future will be in the poorer countries of the world. At the end of the twentieth century about 80 percent of the people in the world lived in the less developed countries. During the present century, nearly all of the growth in population will occur in the less developed countries. An ever larger percentage of the world's population will be nonwhite and relatively poor. In 1950 about two-thirds of the world's people lived in the less developed countries. By 2000 this percentage had increased to about 80 percent and the United Nations projects that by 2050 about 85 percent of the earth's population will be residing in the poorer nations.

Because no one knows for sure what the size of the earth's population will be in the future, the United Nations gives three projections: a high, medium, and low one, based on the possible number of children the average woman will have. Projections are educated guesses. The United Nations believes the middle projection is the most likely and most authors writing on the subject use that number. The population in the developed regions is expected to remain stable at 1.2 billion, while the growth takes place in the less developed countries. About one-half of the annual growth is expected to occur in nine countries – India, Pakistan, Nigeria, Democratic Republic of the Congo, Bangladesh, Uganda, the US, Ethiopia, and China. The biggest growth is expected in India, which is likely

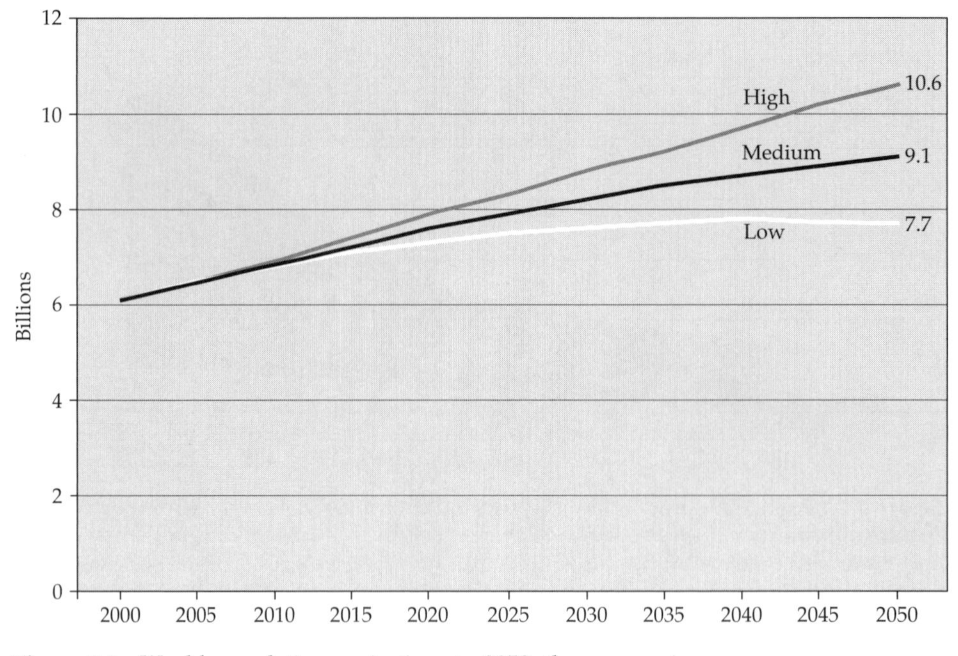

Figure 2.3 World population projections to 2050: three scenarios
Source: UN Population Division, *World Population Prospects: The 2004 Revision*

to pass China by 2035 to be the largest country in the world, with over 1.5 billion people. Figure 2.3 gives the three projections by the United Nations up to the year 2050.

High growth rates will take place in the less developed countries because a large percentage of their population consists of children under the age of 15 who will be growing older and having children themselves. If we plot the number of people in a country according to their ages, we can see clearly the difference between rapidly growing populations, which most less developed nations have, and relatively stable or slowly growing populations, which the more developed nations have. Figure 2.4 shows the difference between the populations of more developed Western Europe and less developed West Africa in 2000. The age structure of the more developed countries is generally column shaped, while the age structure of the less developed countries is usually pyramid shaped. The HIV/AIDS epidemic will have an effect on the population distribution of some countries, especially in sub-Saharan Africa where in 2003 about 10 percent of the adult population was infected with the virus.

Another major change occurring in the world's population is the movement of people from rural areas to urban areas. Although this is happening throughout the world, the trend is especially dramatic in the developing world, where people are fleeing rural areas to escape the extreme poverty that is common in those areas, and because the cities seem to offer a more stimulating life. Mostly it is the young people who go to the cities, hoping to find work and better living

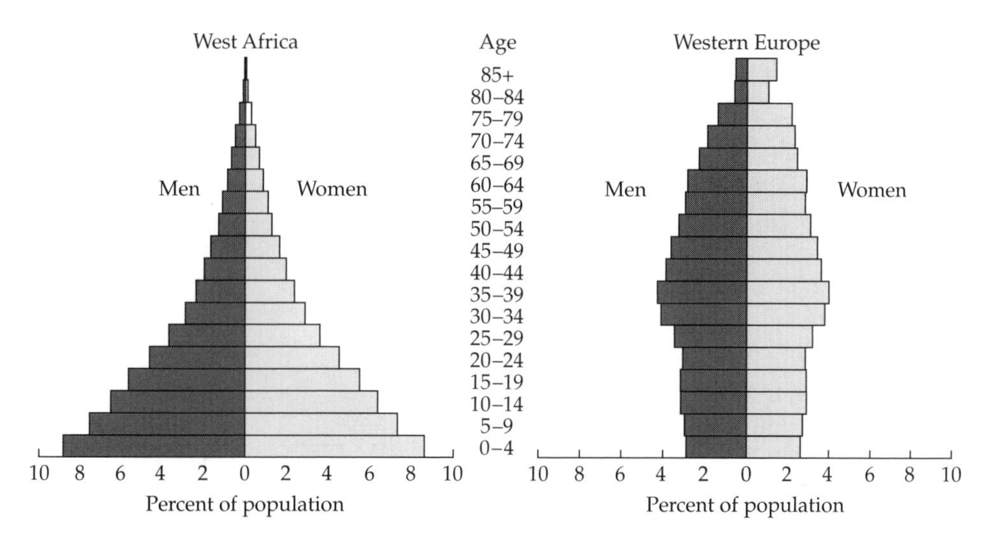

Figure 2.4 Population by age and sex in less developed West Africa and in more developed Western Europe, 2000
Source: UN Population Division, *World Population Prospects: The 2002 Revision* (2003)

conditions. But all too often jobs are not available in the cities either. These rural migrants usually settle in slums on the edges of the big cities. It is estimated that from 30 to 60 percent of the urban populations of less developed countries live in such shanty towns (called "uncontrolled settlements" in government reports).[1] At current rates the populations in these informal settlements will double every 10 to 15 years. It is hard to imagine a city like Calcutta getting any bigger. In 1950 it had a population of about 4 million, and thousands of these people lived permanently on the streets; in 1990 it had a population of about 13 million and an estimated 400,000 lived on the streets.[2] If the present rate of growth continues, it will have a population of about 17 million by 2015. Table 2.2 gives the world's ten largest cities in 1950 and 2000 and the projected ten largest for the year 2015.[3] Note that eight of the ten largest cities in the year 2015 are expected to be in the less developed countries – Tokyo, and New York are the exceptions – whereas in 1950 only three of the ten were in the poorer countries (Shanghai, Buenos Aires, and Calcutta).

Note also the increased size of the cities. Cities with over 5 million people are sometimes called "megacities." In 1950 Buenos Aires and Shanghai were the only cities in the developing world with at least 5 million people. By 2000 there were about 35 megacities in the developing world. Many of these cities had vast areas of substandard housing and serious urban pollution, and many of their residents lived without sanitation facilities, safe drinking water, or adequate healthcare facilities.

The world's population is becoming increasingly urban. Although countries differ on their definitions of "urban" (the United States defines urban as places of 2,500 or more, Japan uses 50,000, and Iceland 200), in 2000 about 75 percent of

Table 2.2 Ten largest cities* in the world, 1950, 2000, and 2015 (projection)

Rank	*Population in 1950 (millions)*		*Population in 2000 (millions)*		*Population in 2015 (projected) (millions)*	
1	New York, US	12	Tokyo, Japan	26	Tokyo, Japan	27
2	London, UK	9	Mexico City, Mexico	18	Dhaka, Bangladesh	23
3	Tokyo, Japan	7	São Paolo, Brazil	18	Mumbai (Bombay), India	23
4	Paris, France	5	New York, US	17	São Paolo, Brazil	21
5	Moscow, USSR	5	Mumbai (Bombay), India	16	Delhi, India	21
6	Shanghai, China	5	Los Angeles, US	13	Mexico City, Mexico	20
7	Essen, Germany	5	Calcutta, India	13	New York, US	18
8	Buenos Aires, Argentina	5	Shanghai, China	13	Jakarta, Indonesia	17
9	Chicago, US	5	Dhaka, Bangladesh	13	Calcutta, India	17
10	Calcutta, India	4	Delhi, India	12	Karachi, Pakistan	16

* Formally called "urban agglomerations" in UN publications.
Source: UN Population Division, *World Urbanization Prospects: The 2001 Revision*, table 56

Rural migrants often settle in urban slums in developing nations (*United Nations*)

Growing cities in less developed nations often have a mixture of modern and substandard housing (*United Nations*)

The most innovative city in the world

Curitiba, Brazil, has been called the most innovative city in the world. City officials from around the world visit Curitiba to learn how this city, with relatively limited funds, has been tackling urban problems. By using imaginative, low cost solutions and low technology, Curitiba has created a pleasant urban life that many cities in the more developed nations have yet to achieve. Here is how the city achieved this.

Transportation The city has made public transportation attractive, affordable, and efficient. Instead of building a subway, which the city could not afford, it established a system of extended, high-speed buses, some carrying as many as 275 passengers, on express routes, connecting the city center with outlying areas. Many people own cars in Curitiba but 85 percent of the commuters use public transportation. This has reduced traffic congestion and air pollution. There are 30 percent fewer cars on city streets than you would expect from the number of cars owned by its residents.

Trash collection Poor people are encouraged to collect recyclable trash in the areas where they live and turn it in for food. In 1995 the city gave out about 350,000 Easter eggs, 25,000 Christmas cakes, and 2 million pounds of food in exchange for trash. About 70 percent of its trash is recycled, one of the highest rates in the world.

Education Small libraries have been built throughout the city in the shape of a lighthouse. Called Lighthouses of Learning, they provide books (many schools in Brazil have no books), an attractive study room, and, in a tower, a strong light and guard to make the area safe.

Health Curitiba has more health clinics – that are open 24 hours a day – per person than any other city in Brazil.

Environmental education The Free University for the Environment was built out of recycled old utility poles next to a lake made from an old quarry. Short courses on how to make better use of the environment have been designed for contractors, merchants, and housewives. Taxi drivers are required to take a course there in order to get their licenses.

Governmental services Colorful, covered Citizenship Streets have been built throughout the city to bring government offices to where the people live and shop. Here people can pay their utility bills, file a police complaint, go to night court, and get a marriage license. Vocational courses are offered for $1 a course.

The main credit for this innovative city has been given to its former mayor Jaime Lerner. Lerner, an architect and planner, headed an honest and very capable government for many years. Lerner is now governor of the state.

Sources: Robin Wright, "The Most Innovative City in the World," *Los Angeles Times* (June 3, 1996). Curitiba's accomplishments are also described in Jonas Rabinovitch and Josef Leitman, "Urban Planning in Curitiba," *Scientific American*, 274 (March 1996), pp. 46–53; Eugene Linden, "The Exploding Cities of the Developing World," *Foreign Affairs* (January/February 1996), p. 62; Jim Motavalli, "All Aboard: Good Public Transportation Need Not Be High Tech or High Cost," *Sierra*, 87 (January/February 2002), pp. 46–8; Tran Reed McManus, "Imagine a City with 30 Percent Fewer Cars," *Sierra*, 91 (January/February 2006), pp. 48–9

the people in the more developed nations lived in urban areas whereas about 40 percent of the people in the less developed countries were urban. In 1950 only about 20 percent of the population in the poorer nations was urban. By 2000 this number had growth to 40 percent. By the early twenty-first century, for the first time in history, more people in the world will live in urban areas than in rural areas.[4]

Causes of the Population Explosion

Although it is easy to illustrate that the human population has grown exponentially, it is not so easy to explain why we are in a situation at present of rapidly expanding population. Exponential growth is only one of many factors that determine population size. Other factors influence how much time will pass before the doublings – which one finds in exponential growth – take place. Still other factors influence how long the exponential growth will continue and how it might be stopped. We will consider these last two matters later in the chapter, but we will first look at some of the factors that drastically reduced the amount of time it took for the world's population to double in size.

The agricultural revolution, which began about 8000 BC, was the first major event that gave population growth a boost. When humans learned how to domesticate plants and animals for food, they greatly increased their food supply. For the next 10,000 years until the industrial revolution, there was a gradually accelerating rate of population growth, but overall the rate of growth was still low because of high death rates, caused mainly by diseases and malnutrition. As the industrial revolution picked up momentum in the eighteenth and nineteenth centuries, population growth was given another boost as advances in industry, agriculture, and transportation improved the living conditions of the average person. Population was growing exponentially, but the periods between the doublings were still long because of continued high death rates. This situation changed drastically after 1945. Lester Brown of the Worldwatch Institute explains why that happened:

LIBRARY, UNIVERSITY OF CHESTER

> The burst of scientific innovation and economic activity that began during the forties substantially enhanced the earth's food-producing capacity and led to dramatic improvements in disease control. The resulting marked reduction in death rates created an unprecedented imbalance between births and deaths and an explosive rate of population growth. Thus, while world population increased at 2 to 5 percent *per century* during the first fifteen centuries of the Christian era, the rate in some countries today is between 3 and 4 percent *per year*, very close to the biological maximum.[5]

It was primarily a drastic reduction in the death rate around the world after World War II that gave the most recent boost to population growth. The spreading of public health measures, including the use of vaccines, to less developed countries enabled these countries to control diseases such as smallpox, tuberculosis, yellow fever, and cholera. Children and young adults are especially vulnerable to infectious diseases; thus, the conquering of these diseases allowed more children to live and bear children themselves.

While death rates around the world were dropping rapidly, birth rates remained generally high. Birth rates have been high throughout human history. If this had not been true, you and I might not be here today since high birth rates were needed to replace the many people who died at birth or at an early age. (If you walk through a very old cemetery in the United States or especially in Europe,

Sons preferred in India

Sons are preferred in many less developed countries. An explanation for this preference in India is given by a North American anthropologist:

> In India, a strong preference for sons has increased the ratio of men to women over the past century. Girls are more likely than boys to be neglected or mistreated, and India has a history of higher death rates and lower life expectancy for women than for men. Most recently, medical advances have enabled expectant parents to abort female fetuses, which has pushed the sex ratio at birth well above 105 boys to 100 girls, the normal ratio throughout the world.

Families in India, as in China, Korea, and a number of other East and South Asian countries, value sons because sons usually live with their parents after marriage and contribute to family income. Sons provide vital financial support to elderly or ill parents, who often have no other source of income. Daughters move away at marriage and transfer their allegiance to their husband's family. Parents can expect little financial or emotional support from daughters after they leave home.

In many parts of India, daughters mean an additional disadvantage to parents – the obligation of paying her prospective husband's family a large dowry. Dowries often require parents to go into debt, and the amount families must pay has been increasing over the years.

The financial and social disadvantages of having a daughter prompt some women to abort their pregnancies if they are carrying a daughter. Pregnant women can determine the sex of their fetus through ultrasound and other examinations. As this technology becomes more widely available, more parents are using it to choose the number and sex of their children. Nearly all aborted fetuses in Indian hospitals are female.

The national government has passed laws prohibiting sex-selective abortion, as have many Indian states, but abortion practices are difficult to regulate.

Source: Nancy E. Riley, "Gender, Power, and Population Change," *Population Bulletin*, 52 (Population Reference Bureau, Washington, DC) (May 1997), pp. 14–15

you can see evidence of this fact for yourself as you pass the family plots with markers for the many children who died in infancy and in adolescence.) Birth rates remained high right up until the late 1960s, when a lowering of the rate worldwide was seen, which was the beginning of a gradual lowering of the birth rate around the world.

The birth rate has dropped significantly in the developed nations but remains high in most developing countries. There are a number of reasons for this. First,

Children take care of children in many poorer countries, as this girl is doing in Mexico (*Mark Olencki*)

many poor people want to have many children. If many of them die in infancy, as they still do in countries such as India, Pakistan, Bangladesh, and in tropical Africa, many births are needed so that a few children survive. If the poor families are peasants, as many of them are, sons are needed to work in the fields and to do chores. And since there are rarely old age pension plans in less developed countries, at least one son (and preferably two) is needed to ensure that the parents have someone to take care of them when they are old and can no longer work. These needs are reflected in the commonly heard greeting to a woman in rural India: "May you have many sons."

Poor families also want children to provide extra income to the family. Before child labor laws severely restricted the use of children in factories in the United States and Europe, it was common for children to take paying jobs to help the family gain income. A study of an Indian village found that the poorest families had many children since they felt that increased numbers meant more security and a better chance for prosperity.[6]

Other reasons for continued high birth rates in poor countries are tradition and religion. The unusually high birth rates in large parts of Africa today are primarily caused, according to one study, by family patterns and religious beliefs that developed over thousands of years in response to conditions in the region.[7] Tradition is very important in most rural societies, and traditionally families have been large in rural settings and among poor people. One does not break with such a tradition easily. Also, religion is a powerful force in rural societies and some religions advocate large families. The influence of Islamic fundamentalism is strong in some Islamic states and it is a major force discouraging the use of contraceptives by women. The Catholic religion is a powerful force in rural Latin America, especially with the women, and some – although fewer than in earlier times – obey the Catholic Church's prohibition against the use of contraceptives. In some Latin American countries men have commonly regarded a large number of children as proof of their masculinity.

The unavailability of birth control devices is also a reason for the high birth rates in developing nations. It has been estimated that about 100 million married women of reproductive age worldwide are not using contraceptives even though they do not want more children. It is believed that these women have an unmet need, or demand, for family planning services.[8]

How Population Growth Affects Development

How does population growth affect development? While there is no easy answer to the question of what is "too large" or "too small" a population for a country – a question we will return to in the final section of this chapter – we can identify some obvious negative features of a rapidly growing population, a situation which would apply to many less developed countries today.

Too rapid

Let's look again at the age distribution of the population in less developed regions in figure 2.4. It is striking that a large percentage of people are below the age of 15. This means that a large proportion of the population in these countries is mainly nonproductive. Although children do produce some goods and services, as mentioned above, they consume more than they produce. Food, education, and health care must be provided for them until they are old enough to become productive themselves. Obviously, if a nation has a large portion of its population in the under-15 age group, its economy will be faced with a huge burden.

A rapidly growing population also puts a great strain on the resources of the country. If the population is too large or the growth too rapid, people's use of the country's resources to stay alive can actually prevent the biological natural resources from renewing themselves. This can lead to the land becoming less fertile, and the forests being destroyed. An example of this is the making of patties out of cow droppings and straw by women in India and Pakistan. These patties are allowed to dry in the sun and are then used for fuel. In fact, dung patties are the only fuel many peasants have for cooking their food. But the use of animal droppings for fuel prevents essential nutrients from returning to the soil, thus reducing the soil's ability to support vegetation.

A large population of young people also means that there will be a terrific demand for jobs when these children grow old enough to join the labor force – jobs that are unlikely to exist. The ranks of the unemployed and underemployed will grow in many poorer nations, and this can easily lead to political and social unrest. As we saw earlier in this chapter, people from the rapidly growing rural areas of the South are heading for the cities hoping to find work. What they find though is a scarcity of jobs, undoubtedly a contributing factor in the high rates of urban crime.

Rapid population growth also has a harmful effect on the health of children and women. Malnutrition in infancy can lead to brain damage, and childbearing frequently wears women down. This is what happens to many poor women:

> After two decades of uninterrupted pregnancies and lactation women in their mid-thirties are haggard and emaciated, and appear to be in their fifties. As researchers Erik Eckholm and Kathleen Newland point out, such women are "Undernourished, often anemic, and generally weakened by the biological burdens of excessive reproduction," they "become increasingly vulnerable to death during childbirth or to simple infectious diseases at any time," and "their babies swell the infant mortality statistics."[9]

A rapidly growing population also puts a tremendous strain on the ability of a nation to provide housing for its people. The poor condition of much of the

Urban crime in the developing world: a personal experience

An experience in Liberia helped me to understand that urban areas in less developed nations are often less safe than rural areas. I lived at different times in Monrovia, the capital city of that country, and in a small village in a rural area. Once while I was in Monrovia, a thief entered my bedroom and stole my wallet and watch from under my pillow, which was under my sleeping head at the time. Such an event was unheard of in the rural areas, but was not that uncommon in the city. After the theft happened, I was happy to return to my "primitive" village, where I felt much safer.

housing in the less developed nations is something that makes a lasting impression on foreign visitors to these countries – that is, if they venture beyond the Hilton hotels where they sometimes stay. Overcrowding also is produced by an excessive and rapidly growing population, and that leads to a scarcity of privacy and to limited individual rights.

Not surprisingly, a study sponsored by the US National Academy of Sciences to explore the relationship between population growth and economic development concluded in the mid-1980s that slower population growth would aid economic development in most of the less developed nations.[10]

Too slow

A country's population growth rate can also be too slow to support a high level of economic growth. Partly because of low birth rates, a number of European countries had to import unskilled workers during the 1950s and 1960s from Turkey, southern Italy, and other relatively poor areas of Europe and North Africa. Within the world of business, there is concern if the population stops expanding since a growing population is seen as representing more consumers of products. But a number of the industrial countries have shown in the post–World War II period that a high level of economic growth can be obtained even when population growth is low.

Japan is a good example to look at. Even though the country has experienced impressive economic growth, the Japanese show some ambivalence regarding its extremely slow population growth. A survey conducted in 1990 by the Japanese government revealed that many people believed that a recent decline in fertility rates was undesirable because it could lead to an aging population and fewer young people entering the labor market. At the same time the survey revealed that many people (nearly one-half) believed that the country was overpopulated.[11] This latter view is probably not surprising given the fact that much of the land is mountainous, with only about 15 percent of the land being suitable for cultivation. About one-quarter of the nation's population lives in the Tokyo metropolitan area, with the consequence that the cost of land and housing in that area is extremely high.

An aging population and low birth rates

We saw earlier the types of problems that are created when a country has a large portion of its population aged 15 or under. But special problems are also created when the proportion of a population that is over 65 starts to expand. As can be seen in table 2.3 this is happening in many areas of the world. As Table 2.3 shows it is especially happening in Europe and North America. The United States, for example, is facing such a problem with its social security system, which provides financial support to retired persons. As the percentage of the US population that is over 65 expands because of advances in health care and healthier lifestyles by

Table 2.3 Regional trends in aging: percentage of total population 65 years or older, 2000, 2015 (projection), 2030 (projection)

Region	*Year*	*65 years or older*
Asia	2000	6
	2015	8
	2030	12
Europe	2000	15
	2015	18
	2030	24
Latin America/Caribbean	2000	6
	2015	8
	2030	12
Middle East/North Africa	2000	4
	2015	6
	2030	8
North America	2000	12
	2015	15
	2030	20
Oceania	2000	10
	2015	12
	2030	16
Sub-Saharan Africa	2000	3
	2015	3
	2030	4

Source: US Census Bureau, International Data Base, 2004

some, and the number of new workers is reduced because of low birth rates, the ratio of working-age people to retired people declines and puts a strain on the social security system. (It is the payments from the current workers that provide money for the retirement benefits.)

There are also increased governmental healthcare costs as a population ages. More funds are needed to care for the medical and social needs of the aged since most developed countries believe it is the whole community's responsibility to help families pay for these services. This is a common concern in Europe, where by the year 2015 it is expected that about 20 percent of its people will be 65 or older, and in 2030 this age group will increase to about 25 percent. At the beginning of the twenty-first century only about 15 percent were of that age. Also of concern in Europe is the expected doubling of the number of people 80 or over, from 3 percent in 2000 to over 6 percent in 2030.[12]

Caring for the aged is a concern in nearly all developed countries. It certainly is for Japan. Except for Italy, Japan is graying faster than any other industrialized country. (Greece and Germany are close behind.) In 2004 about 20 percent of the population was 65 or older, and by 2020 it is expected that this group will increase to about 25 percent in a population that will probably be smaller than it was in 2000. By contrast, because of immigration and a relatively higher birth rate, in 2000 about 13 percent of the population in the United States was 65 or older and that group will grow to about 17 percent by 2020.

Some developing countries, such as China, will also face an aging problem in the twenty-first century. Mainly because of the dramatic reduction in its birth rate, the percentage of people aged 60 or over in China is expected to increase from 10 percent in 2000 to about 25 percent by mid-century.[13]

When a country has a low birth rate, and the number of young people entering the labor market is reduced – a situation now common throughout Europe and Japan – often the result is conflict over immigration policies. Hostility to foreign workers by extreme nationalists in Germany in the early 1990s led to fatal attacks on some foreigners in the country. Japan, a country that traditionally has been wary of outsiders, is also concerned about having to rely on foreign workers. (The Japanese are as worried as the Americans and Europeans that a shrinking workforce will be unable to support the increasing healthcare costs and welfare costs of an aging population.)

A number of European nations and Japan had such low birth rates in the mid-1990s that their populations had started to decline or would soon do so. Declining populations became common in Russia and the former Eastern European satellites, no doubt because of the harsh economic conditions these countries were facing as they tried to replace their planned economies with market economies. Long-term decline in population for most of Europe appears inevitable. The population of Europe is expected to decline between 2000 and 2050 by about 100 million, going from about 730 million to 630 million.[14]

Projections have been made that about 50 countries, most of them developed, will lose population between now and 2050. Germany is expected to go from 83 million to 79 million, Italy from 58 million to 51 million, Japan from 128 million to 112 million, and a great reduction will occur in the Russian Federation, which will go from 143 million to 112 million.[15] The United States would also face a declining population in the twenty-first century were it not for its high immigration levels. Declining populations raised fear about the loss of national power, economic growth, and even national identities by some people in these countries. But most population experts believe that if population decline is gradual, the negative social and economic consequences can be handled. Much more difficult to manage, they believe, are situations where the decline is rapid.[16] It is possible that some nations will find a smaller population easier to maintain in a sustainable manner, a concept which will be discussed in the final chapter.

Some governments have tried different measures to encourage families to have more children – such as direct financial payments for additional children, tax benefits, subsidized housing preferences, longer maternity and paternity leave, childcare, and efforts to promote gender equality in employment – but these policies have had only modest effects in authoritarian states and minimal effects in liberal democracies such as France and Sweden.[17]

International conferences on population

The first international conference on population was held in 1974 in Romania under the sponsorship of the United Nations. It was anticipated that this conference would dramatize the need for population control programs in the less developed countries, but instead a debate took place between rich and poor countries over what was causing poverty: population growth or underdevelopment. The United States and other developed nations argued for the need for birth control measures in the poorer countries, while a number of the poorer countries argued that what was needed was more economic development in the South. Some developing countries called for a new international economic order to help the South develop. They advocated more foreign aid from the richer countries, and more equitable trade and investment practices. The conference ended with what seemed to be an implicit compromise: that what was needed was both economic development and population control, that an emphasis on only one factor and a disregard of the other would not work to reduce poverty.

In 1984 the United Nations held its second world population conference in Mexico City. The question of the relationship between economic growth and population growth was raised again. The United States, represented by the Reagan administration, argued that economic growth produced by the private enterprise system was the best way to reduce population growth. The United States did not share the sense of urgency that others felt at the conference concerning the need to reduce the world's increasing population. It announced that it was cutting off its aid to organizations that promoted the use of abortion as a birth control technique. (Subsequently the United States stopped contributing funds to the United Nations Fund for Population Activities and the International Planned Parenthood Federation, two of the largest and most effective organizations concerned with population control.)[18] The United States stood nearly alone in its rejection of the idea that the world faced a global population crisis, as well as in its advocacy of economic growth as the main population control mechanism. The conference endorsed the conclusion reached at the first conference ten years earlier that *both* birth control measures *and* efforts to reduce poverty were needed to reduce the rapidly expanding population of the less developed nations.

In 1992 the United Nations Conference on the Environment and Development – the so-called Earth Summit held at Rio de Janeiro, Brazil, which will be discussed in detail in chapter 5 – did not directly address the need for population control measures. The Rio Declaration says only that "states should . . . promote appropriate demographic policies," and Agenda 21, the action plan to carry out the broad goals stated in the declaration, does not mention family planning. The weak treatment of the population issue by this conference was the result of North/South conflicts over whether the poor nations or the rich nations were mainly responsible for the destruction of the environment. (When the population issue was raised, attention was focused on the harm to the environment that large numbers of poor people in the South could inflict, whereas the South held that overconsumption by the North caused most of the pollution that was harming the environment.) The failure to directly address the connection between rapid population growth and environmental damage was also a result of opposition by the Vatican to any declarations which could be used to support the use of contraceptives and abortion to control population growth. Opposition was also voiced by some countries with conservative social traditions concerning issues that could raise the subject of the status of women in their countries.

In spite of the failure to strongly address the population issue in its formal statements, the Rio conference, and the multitude of meetings around the world held to prepare for it, did cause increased attention to be placed on population, especially bringing to the forefront the perspectives of women.

The United Nations held its third conference on population – formally called the International Conference on Population and Development – in Cairo, Egypt in 1994. Although the Vatican and conservative Islamic governments made abortion and sexual mores the topic of discussion in the early days of the conference, the conference broke new ground in agreeing that women must be given more control over their lives if population growth was to be controlled. The conference approved a 20-year plan of action whose aim was to stabilize the world's population at about 7.3 billion by 2015. The plan called for new emphasis to be placed on the education of girls, providing a large range of family planning methods for women, and providing health services and economic opportunities for females. The action plan called for both developing and industrial nations to increase the amount they spent on population-related activities to $17 billion by the year 2000, a significant increase over the $5 billion that was then being spent. Whether or not countries would actually follow through with the conference's recommendations and commit these funds for this purpose was unknown. As the secretary-general of the conference, Dr Nafis Sadik, stated at its conclusion, "Without resources the Program of Action will remain an empty promise."[19]

Five years after the Cairo conference the UN found that a number of new approaches to controlling population had been initiated around the world, but that scarce resources and many needs led to population programs not receiving top priority in all developing nations. It found also that pledges of aid from developed nations were seriously underfunded.[20]

In 2003 the UN General Assembly voted to end the automatic holding of international conferences. Because of their large expense of funds and human energy and the danger that they were becoming routine, the UN decided that the decision to hold an international conference should be made on a case-by-case basis when there was a special need for international cooperation.

How Development Affects Population Growth

How does development affect the growth rate of population? There is no easy answer to that question, but population experts strongly suspect that there *is* a relationship, since the West had a fairly rapid decline in its population growth rate after it industrialized. In the nineteenth century Europe began to go through what is called the "demographic transition."

Demographic transition

The demographic transition, which is shown in figure 2.5, has four basic stages. In the first stage, which is often characteristic of preindustrial societies, there are high birth rates and high death rates which lead to a stable or slowly growing population. Death rates are high because of harsh living conditions and poor health. In the second stage, there is a decline in the death rate as modern medicine and sanitation measures are adopted and living conditions improve. Birth rates continue to be high in this stage because social attitudes favoring large families take longer to change than do the technological advances and innovations that come with modern medicine. This situation ignites what is known as the population explosion. In the third stage birth rates finally drop in response to the lower death rate. Population growth remains high during the early part of the third stage but falls to near zero during the latter part. Most industrial nations passed through the second and third stages from about the mid-1800s to the mid-1900s. In the final and fourth stage, both the death and birth rates are low, and they fluctuate at a low level. As in the first stage, there is a stable or slowly growing population.

The more developed nations are already in the fourth stage of the demographic transition but most less developed nations are still in the second stage or the early parts of the third stage. There have been some significant differences between the developed and developing nations' second and third stages. For the developed nations, the reduction in the death rate was gradual as modern medicines were slowly developed and the knowledge of germs gradually spread. The birth rate dropped sharply, but only after a delay that caused the population to expand. For the developing countries, the drop in the death rate has been sharper than it was for the developed nations as antibiotics were quickly adopted, but because poverty lingered for many the reduction in the birth rate has lagged

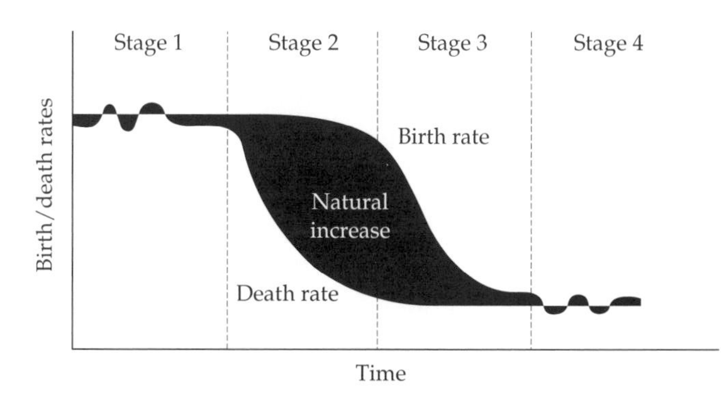

Figure 2.5 The classic stages of demographic transition
Note: Natural increase is produced from the excess of births over deaths.
Source: Joseph A. McFalls Jr, "Population: A Lively Introduction," *Population Bulletin*, 53 (Population Reference Bureau, Washington, DC) (September 1998), p. 39

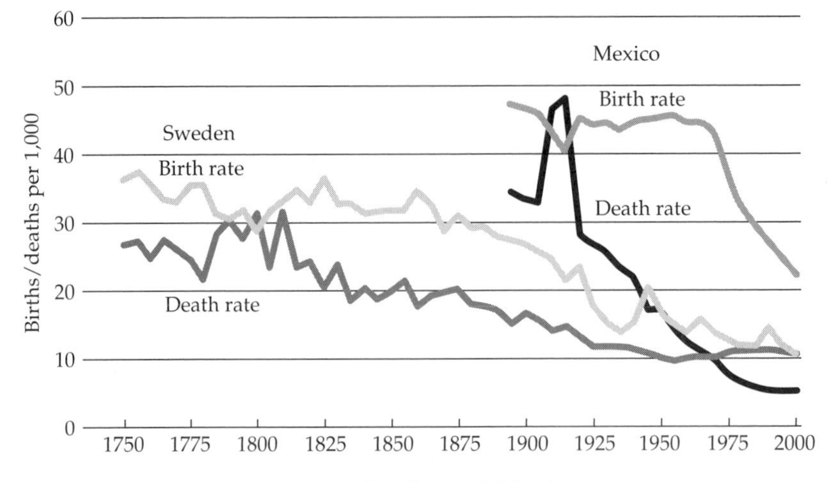

Figure 2.6 Demographic transition in Sweden and Mexico
Sources: B. R. Mitchell, *European Historical Statistics 1750–1970* (1976), table B6; Council of Europe, *Recent Demographic Developments in Europe 2001* (2001), tables T3.1 and T4.1; CELADE, *Boletin demográfico*, 69 (2002), tables 4 and 7; Francisco Alba-Hernandez, *La poblacion de México* (1976), p. 14; UN Population Division, *World Population Prospects: The 2002 Revision* (2003), p. 326. As presented in "Transitions in World Population," *Population Bulletin*, 59 (Population Reference Bureau, Washington, DC) (March 2004), p. 7

more than it did for the developed nations. Both of these facts have caused a much larger increase in the population of the less developed nations than had occurred in the more developed nations. These two facts can be seen in figure 2.6 which compares the demographic transitions of the more developed Sweden and the less developed Mexico. Sweden, which has completed the demographic transition and is now in the fourth stage with zero population growth, once sent its excess population to the US as immigrants and still growing Mexico is doing the same.

The differences in the experiences of many developed and developing nations have led many demographers to change the opinion they had in the 1950s that economic development would cause less developed nations to go through the same demographic transition as the North – and thus achieve lower population growth. There are obviously important differences between the Northern experience and that of the developing world. Probably as important as the fact that death rates have dropped much faster in the South than they did for the North is the fact that the industrialization that is taking place in much of the South is not providing many jobs and is not benefiting the vast majority of people in that region. A relatively small, modern sector *is* benefiting from this economic development, and the birth rate of this group is generally declining; but for the vast majority in the rural areas and in the urban slums, the lack of jobs and continued poverty are important factors in their continued high birth rates.

Factors lowering birth rates

If industrialization as it is occurring in the less developed world is not an automatic contributor to lower birth rates, what factors do cause birth rates to decline? Certainly, better health care and better nutrition, both of which lower infant mortality and thus raise a family's expectations of how many children will survive, are important factors. (The irony here, of course, is that these advances, at least in the short run, tend to worsen the population problem since more children live to reproduce.) Another factor tending to lower birth rates is the changing role of women. Better educated women are more likely to use some sort of contraception than are those women with little or no education.[21] Figure 2.7 shows the relationship between female education and childbearing in selected countries. Education for women enables them to delay marriage, to learn about contraceptives, and to acquire different views of their role in society. Their education also allows them to obtain jobs that can often be of more benefit to the family than a larger family is.

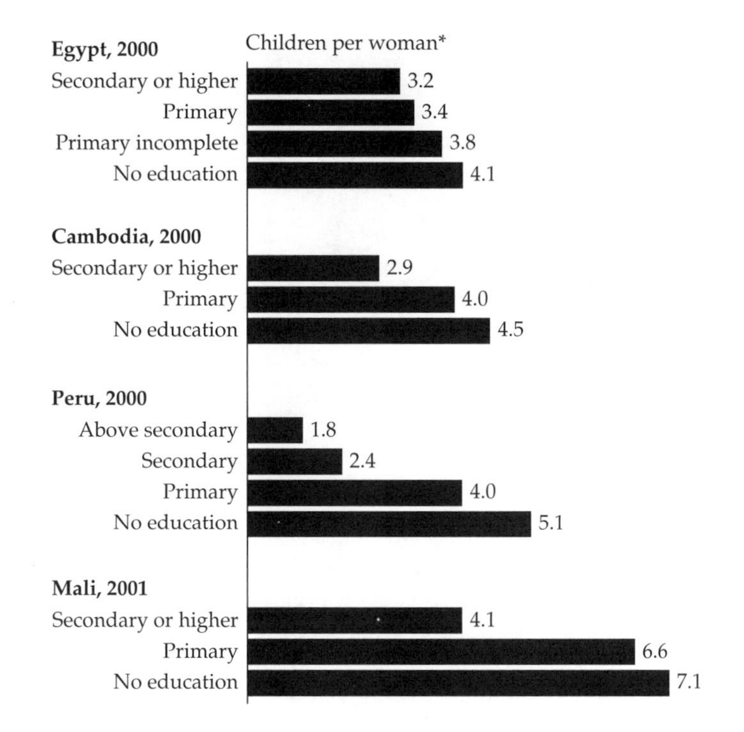

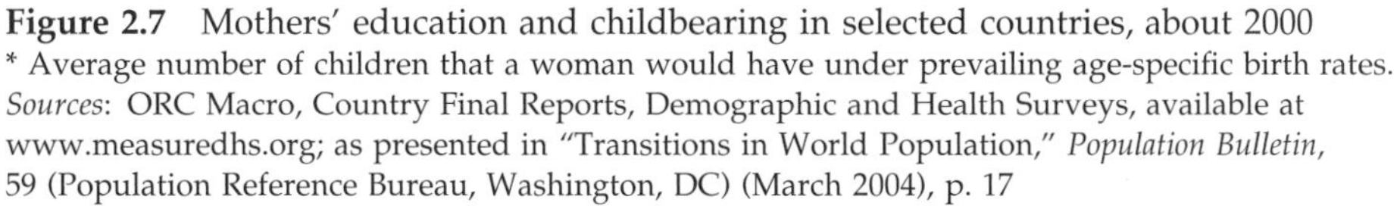

Figure 2.7 Mothers' education and childbearing in selected countries, about 2000
* Average number of children that a woman would have under prevailing age-specific birth rates.
Sources: ORC Macro, Country Final Reports, Demographic and Health Surveys, available at www.measuredhs.org; as presented in "Transitions in World Population," *Population Bulletin*, 59 (Population Reference Bureau, Washington, DC) (March 2004), p. 17

Breast-feeding delays a woman's ability to conceive and provides the most healthful food for a baby (*United Nations*)

Actually, however, a little bit of modernization can be a bad thing if birth rates are to be reduced. In large parts of the developing world women with a little exposure to Western ways are giving up breast-feeding their babies and switching to bottle-feeding. Their fertility goes up when they do this since prolonged breast-feeding naturally delays – sometimes for years – a woman's ability to conceive again.

As Western nations industrialized, child labor laws, compulsory education for children, and old age pension laws reduced the economic incentive for having many children. Children changed from being producers on the farms and in the early factories and became instead consumers and an economic burden on their families. Also, as the West industrialized, it became more urban, and living space in urban societies is scarcer, and more expensive, than it was in rural societies. The availability of goods and services increased, which caused families to increase their consumption of these rather than spend their income on raising more children. Traditional religious beliefs, which often support large families, also tended to decline.

There is little debate today that economic development, especially if it benefits the many and not just the few, can lead to lower birth rates. There is ample evidence that improving the social and economic status of women can lead to lower birth rates, even in areas which remain very poor – such as in the southern state of Kerala in India, where birth rates are significantly lower than in the rest of India. But there is now evidence that birth rates can decrease and are decreasing in poor countries – even in some where there has been little or no economic growth and where the education and social status of women remains very low, such as in Bangladesh – if an effective family planning program exists and modern contraceptives are available.[22] The conclusion of some researchers who have reviewed the results of fertility studies conducted in various less developed countries is that "although development and social change create conditions that encourage smaller family size, contraceptives are the best contraceptive."[23] These researchers found that three factors are mainly responsible for the impressive decline in birth rates that have occurred in many less developed countries since the mid-1960s: more influential and more effective family planning programs; new contraceptive technology; and the use of the mass media to educate women and men about birth control.[24]

In the past several decades fertility has declined significantly in the world, although as figure 2.8 shows, the decline has been much greater in some regions

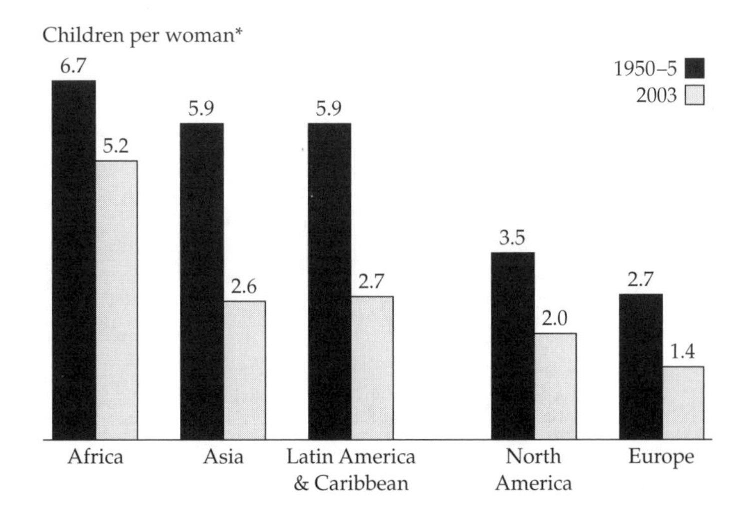

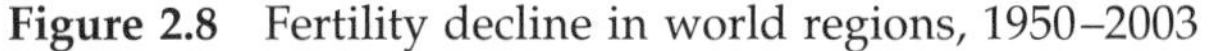

Figure 2.8 Fertility decline in world regions, 1950–2003
* Average number of children that a woman would have under prevailing age-specific birth rates.
Sources: UN Population Division, *World Population Prospects: The 2002 Revision* (2003); C. Haub, *2003 World Population Data Sheet*

than in others. Note that Europe is below replacement level, which is generally considered to be on the average 2.1 children per woman (the extra one-tenth compensates for the death of some girls and women before the end of their child-bearing years). Also note that Africa still has high fertility. Women throughout the world now have on average fewer than three children per woman whereas in the 1950s they had five.

Governmental Population Policies

Controlling growth

Many governments today have some policies that try to control the growth of their populations, but this is a very recent trend. Traditionally, governments have sought to increase their populations, either through encouraging immigration (as the United States did in its early years) or through tax and other economic assistance to those families with many children. As late as the mid-1970s, many governments had no population control programs. A survey of developing nations taken in conjunction with the 1974 UN population conference found that, out of 110 developing countries, about 30 had population control programs, another 30 had information and social welfare programs, and about 50 had no population limitation programs at all.[25] This UN conference ended with no explicit consensus among the participants that there was a world population problem at all. The delegates at the conference did pass a resolution stating that all families

Advertisement for contraceptives in Costa Rica (*George Shiflet*)

have the right to plan their families and that it is the responsibility of governments to make sure all families have the ability to do so.

The ability to control the number and timing of children a couple has is called family planning. Family planning services provide health care and information on contraceptives. The expansion of family planning services around the world in the past 40 years has been truly revolutionary. By 2000 about 60 percent of married couples worldwide were using contraceptives, a dramatic increase from the approximately 10 percent in the 1960s. The average number of children per woman dropped from six to three. But in the developing nations – excluding China – women averaged four children, twice the number needed for a population to stabilize. In sub-Saharan Africa the average number of children per woman was still six. Most developed countries, a few rapidly industrializing countries, and China had achieved a birth rate at or below two children per couple, the replacement level.

Around 2000 about one-half of married women in developing countries were using contraceptives. The rate among countries varied greatly. In about ten sub-Saharan countries the contraceptive rate among married women was about 10 percent while in Mexico and Thailand the use was closer to 70 percent.[26] The more developed world had a use rate of about 70 percent at this time. As mentioned before in this chapter, at the beginning of the twenty-first century about 100 million women of reproductive age in the poor nations wanted no further children but were not using contraceptives. They are considered to be potential family planning users if the services were made available to them. Figure 2.9 gives increases in contraceptive use in selected countries. A three-year study by a Princeton University professor of demographic studies of 12 Central Asian and Eastern European countries has shown that as contraceptive use increased, the number of abortions decreased.[27]

Requests by developing nations for foreign aid to help them control their population growth now, for the first time, exceed the international assistance available for this activity. Providing family planning services to the approximately 100 million women whose potential demand remained unmet at present would cost an estimated $2 billion annually. While this seems like a huge amount, relative to other expenditures being made at present it is not. (The cost of one modern submarine in the US is over $2 billion, and the US tobacco industry spends about that amount yearly on advertising.) The US government has been the largest single donor of aid for population and family planning activities in the developing world, providing about $4 billion during the 1970s and 1980s. In real dollars (controlling for inflation), US aid in 1990 was actually lower than it was in the late 1970s.

In 1999, the first five-year review of progress toward accomplishing the goal made at the 1994 UN conference on population in Cairo of stabilizing world population at about 7 billion by 2015 revealed that the developing countries had reached two-thirds of their goal of contributing $11.3 billion to population-related programs by 2000, while the donor countries had reached only one-third of their goal of $5.7 billion.

Mexico is a country that has had rather dramatic success with its family planning program. The government began this program only in 1972, when it had

Figure 2.9 Increases in modern contraceptive use in selected countries, 1960s to 2000
Note: US figures are for women ages 15 to 44. Modern contraceptives include sterilization, oral contraceptives, IUDs, condoms, diaphragms, Depoprovera, Norplant, and other barrier and chemical methods.
Sources: UN Population Division, *Levels and Trends in Contraceptive Use, 1998* (2000); C. Haub and B. Herstad, *Family Planning Worldwide* (2002); ORC Macro, Demographic and Health Survey data, available at www.measuredhs.org. As presented in "Transitions in World Population," *Population Bulletin*, 59 (Population Reference Bureau, Washington, DC) (March 2004), p. 8

one of the highest rates of population growth in the world. In the early 1970s the annual population growth rate was estimated to be above 3 percent, while in 2005 it was estimated to be down to about 2 percent.[28] The average number of children per woman in Mexico dropped from about 7 in 1965 to about 2.5 in 2000.

In 1972 Mexico's President Luis Echeverria Alvarez announced a reversal of governmental policy on the population issue. His decision to support a strong effort to control the rapid growth of the Mexican population led the government to use MEXFAM, the local affiliate of the International Planned Parenthood Federation, to set up family planning clinics throughout the country. (By the early 1990s MEXFAM had set up 200 of these clinics.) Besides making contraceptives readily available, the government and MEXFAM mounted a large propaganda campaign using television soap operas, popular songs, billboards, posters on buses and in subway stations, and spot announcements on radio and television. The leaders of the Catholic Church in Mexico did not oppose the government's efforts.

But if the present birth rate is not reduced further, Mexico's population will increase 30 per cent by 2050. To increase the use of contraceptives the National Population Council is now focusing its efforts on the rural population, adolescents, and men.[29] Men are an especially important target since the rate of contraceptive use by men in Mexico is low. The use of reliable contraceptives by males accounts for only about 15 percent of contraceptive use in less developed countries,[30] and reportedly one-half of the women in Mexico using family planning services do not tell their husbands because they fear physical abuse.[31] Partly because of difficult economic conditions in the country, men are receptive to the message that controlling family size makes sense.

A few countries have adopted more forceful measures than family planning to try to reduce their population growth. Japan drastically reduced its population growth by legalizing abortion after World War II, by some accounts in order to reduce the number of Japanese children fathered by American military men stationed in Japan.

India, which had disappointing results with its voluntary family planning programs, enacted more forceful measures in the mid-1970s, such as the compulsory sterilization of some government workers with more than two children. Several Indian states passed laws requiring sterilization and/or imprisonment for those couples who bore more than two or three children. A male vasectomy program was also vigorously pursued, with transistor radios and money being given as an incentive to those agreeing to have the sterilization operation. Public resentment against these policies mounted and helped lead to the defeat of Prime Minister Indira Gandhi's government in 1977. Birth control efforts slackened after that event. The Indian government has now returned to voluntary measures to try to limit the growth of its population. Fertility has declined in India from about five children per woman in 1970 to about three children in 2000.[32] Even with this decline, population is now increasing by about 18 million a year, an increase which leads the world. India now has 17 percent of the world's population on only

Family planning class (*United Nations*)

2.5 percent of the world's land. India's population reached 1 billion in 1999 and is projected to approach 2 billion before growth ends.[33]

China, which has about 20 percent of the world's population but only about 7 percent of its arable land, has launched a vigorous program to limit its population growth and has drastically reduced its birth rate. For many years the communist government, under the leadership of Mao Zedong, encouraged the growth of the population, believing that there was strength in numbers. The policy was eventually reversed and the average number of children per woman dropped from about six in 1970 to about two in the year 2000. In 2000 China's population was about 1.3 billion. The population continues to increase by about 10 million per year and it is projected to grow by another 100 million by 2050.[34]

A wide assortment of measures is being used to limit the growth. Contraceptives are widely promoted, sterilization is encouraged, and abortion is readily available. The government, through extensive publicity efforts, began promoting the one-child family as the ideal. Late marriages are strongly encouraged and couples who have only one child receive better jobs and housing, whereas couples with more than two children are taxed more and receive reduced pay. Sociologist Ronald Freedman describes why the Chinese efforts have been successful:

> The massive Chinese national birth-planning program . . . has been organized through the network of political and social organization which mobilizes the masses of the population in primary groups at their places of work and residence. . . . That system is used to promote priority objectives – such as birth planning – by persistent and repetitive messages, discussions, and both peer and authority pressure, which is so awesome in its extent that it is hard for us to comprehend.[35]

Partly because of a concern that there will not be enough adult children to care for their aging parents, the one-child policy is being moderated. It has been widely enforced in the urban areas, but in rural areas, where 70 percent of the people still live, couples were usually allowed to have a second child if the first child was a female. There is evidence that there was widespread disregard of the policy in some rural areas where it is not uncommon today to find families with three, four, five, or even more children.[36] A male child is still strongly desired in these areas to carry on the family name, to take care of his parents when they get old (an old age social security system still does not exist in the rural areas), and to help with agricultural work. The one-child policy was also not applied to ethnic groups in the country, partly because many of them live in strategic border areas and the government did not want to cause resentment among them.

China's birth control policies have been both admired and criticized in other countries. Admiration has been given for the spectacular accomplishment, for producing "one of the fastest, if not the fastest, demographic transitions in history."[37] The policy has been credited with having prevented the birth of an additional 300 million children which some demographers believe could have caused a demographic disaster. The one-child policy was criticized because of the means used to enforce it, which included the use of abortions as a backup to contraceptives – sometimes on women who strongly preferred not to have one. The use of the coercive techniques of the past has mainly ended and emphasis is now placed on education and "family planning fees" for women who have unapproved children.[38] Concern has also been expressed with the unnaturally low numbers of female births being reported. Because of stringent family planning policies, and to guarantee having a male child, many couples have either aborted a pregnancy if the fetus was female – ultrasound equipment is now widespread throughout China and ultrasound tests can be used to indicate the sex of the fetus – or practiced female infanticide, abandoned their child, or not reported the birth.[39] While there is still disagreement about how many girls are "missing" and why, one respected report contends that female infanticide is probably relatively rare and that the most likely cause is the use of abortion for the sex selection of the newborn.[40]

Population-related problems in China's future include an aging population because of low fertility; fewer children to take care of aging parents; single children with no siblings, aunts or uncles; and a shortage of females for males to marry. Dissatisfaction is also spreading in parts of the country because of the increasing social and financial inequality that have come with China's increasing economic prosperity as it follows the market approach. Also, as the population continues to grow and industrialization expands, there will be increasing stress on the environment.

Promoting growth

Although most countries now seem to realize the need to limit population growth, a few have openly favored increasing their populations, among them the military governments that ruled in Argentina and Brazil in the 1960s and 1970s. Both countries have large areas that are still sparsely populated and both are rivals for the role of being the dominant power in Latin America. A few Brazilian military officers even advocated encouraging population growth so that Brazil could pass the United States in size and become the dominant nation in the Western hemisphere. It is doubtful that a larger population could ever put Brazil in this position unless the economy makes great advances. A major region of the country with a relatively dense population already – the Northeast – is one of the poorest areas in the world, and vast tracts of Brazilian land, such as in the Amazon River basin, cannot support large populations.

Aside from some pro-growth statements, the Brazilian military governments did not effectively promote population growth. They became basically neutral on the issue of population and gradually made it possible for the main nongovernment family planning organization to operate in the country. After the military left power in Brazil in the mid-1980s, a new constitution acknowledged the right of women to family planning. This provision had the tacit approval of the Brazilian Catholic Church. About 75 percent of Brazilian women now use some form of contraception. From 1960 to 1994 the average number of children born to a Brazilian woman dropped from about six to about three. The right to an abortion is limited by law but it is widely used. One estimate is that 30 percent of all pregnancies in Brazil are terminated by abortion.[41]

Other countries, such as Mongolia and some in sub-Saharan Africa, have at times advocated larger populations both for strategic reasons and because of the belief that a large population is necessary for economic development. Even the US government, which generally recognizes the need for a check on population growth, has some policies that promote large families, such as income tax laws that allow deductions for children. Many developed nations have contradictory policies, some encouraging population growth while others discourage it. Some developing nations

Romania: a disastrous pro-birth policy

Romania is an example of a country that tried to promote the growth of its population. After World War II, the birth rate there fell so sharply that within a few years the population of the country would actually have started declining. In the mid-1960s the communist government, headed by Nicolae Ceausescu, decided to try to reverse this trend, not only to ward off a possible decline of population but to actually increase the number of people. Ceausescu believed that a large population would improve Romania's economic position and preserve its culture since Romania was surrounded by countries with different cultures. "A great nation needs a great population," said Ceausescu. He called on all women of childbearing age to have five children. Monthly – and in some places, even weekly – gynecological exams were given to all working women 20 to 30 years old. If a woman was found to be pregnant, a "demographic command body" was called in to monitor her pregnancy to make sure she did not interrupt it. A special tax was placed on those who were childless.

The main techniques the government used to promote its pro-growth policy were to outlaw abortion, which was one of the main methods couples had used in the postwar period to limit the size of their families, and to ban the importation and sale of contraceptives. The birth rate immediately shot up, but within a few years it was nearly back to its previous low as couples found other means to limit their families. One of the means was secret abortions, and many women either died or ended up in hospital after abortions were performed or attempted by incompetent personnel. Another tragic result of Ceausescu's pro-birth policy (as well as of his failed economic policies) was the abandoning of unwanted children. Tens of thousands of these children ended up in understaffed and ill-equipped orphanages.* Many babies were even sold for hard currency to infertile Western couples. The pro-growth policy ended in 1989 with the overthrow of the Ceausescu regime and with his execution.

* Two photo essays of this subject are contained in James Nachtwey, "Romania's Lost Children," *New York Times Magazine* (June 24, 1990), pp. 28–33, and Jane Perlez and Ettore Malanca, "Romania's Lost Boys," *New York Times Magazine* (May 10, 1998), pp. 26–9

also have such contradictions, although the greater agreement now in these countries about the need to limit growth often causes these contradictions to be exposed and eliminated.

A generalization one can make about governmental policies that are aimed at influencing population growth is that, aside from drastic measures, governmental policies have not been very successful in either promoting or limiting birth rates very much if these policies are out of line with what the population desires. One can also generalize that matters pertaining to reproduction are still considered to be basically the subject for private decisions and not matters for public policy to control.

As the twenty-first century begins a growing number of industrialized countries are faced with a rapidly expanding retirement-age population and a shrinking labor force that will have to support its elderly citizens. Some of these countries, such as Sweden, Hungary, South Korea, and Japan, have tried various policies to encourage women to have more children. The policies have included paid maternity and paternity leave, free childcare, tax breaks for large families, family housing allowances, and even cash paid to parents raising a child. The conclusion of a study of these efforts is as follows:

> As we enter the next century, a growing number of countries will have near-zero growth or will decline in size. Experience in Europe, Japan, and other countries suggests that governments can encourage people to have more children, but at a high price and not enough to affect long-term trends.[42]

The Future

The growth of the world's population

The United Nations projects that the world's population will continue to grow to about 7.5 to 10.5 billion by 2050 depending on the success of efforts to control population growth. The most likely total, according to the UN, is around 9 billion.[43] The UN bases its projection on the assumption that the world's population growth rates will continue the decline which started in the late 1960s. In 2005 the world's population was estimated to be about 6.5 billion.

Unusual for a developed country, the United States's population is expected to continue to grow significantly, increasing from about 300 million in 2005 to about 400 million around 2050. Rather typical of Northern Europe, the United Kingdom's population is projected to grow modestly from about 60 million in 2005 to about 67 million in 2050.[44]

The carrying capacity of the earth

Will the earth be able to support a population of 9 billion, about one-half more than the present size, or will catastrophe strike before that figure is reached?

Understanding the concept of "carrying capacity" will help answer that question. Carrying capacity is the number of individuals of a certain species that can be sustained indefinitely in a particular area. Carrying capacity can change over time, making a larger or smaller population possible. Human ingenuity has greatly increased the carrying capacity of earth to support human beings, for example, by increasing the production of food. (This was unforeseen by Thomas Malthus, who wrote about the dangers of overpopulation in the late 1700s.) But carrying capacity can also change so that fewer members of the species can live. A climate change might do this. Care must be exercised when using the concept of carrying capacity because, in the past, its definition implied a balance of nature. Many ecologists no longer use the concept of balance of nature because numerous studies have shown that nature is much more often in a state of change than in a balance.[45] Populations of different forms of life on earth are usually in a state of flux as fires, wind storms, disease, changing climate, new or decreasing predators, and other forces make for changing conditions and thus changing carrying capacity.

There are four basic relationships that can exist between a growing population and the carrying capacity of the environment in which it exists. A simplified depiction of these is given in figure 2.10. Graph (a) illustrates a continuously growing carrying capacity and population. Although human ingenuity as seen in the agricultural revolution (to be discussed in chapter 3) and in the industrial/scientific revolution has greatly increased the capacity of the earth to support a larger number of human beings, it is doubtful the human population can continue to expand indefinitely. A basic ecological law is that the size of a population is limited by the short supply of a resource needed for survival. The scarcity of only one of the essential resources for humans – which would include air, energy, food, space, nonrenewable resources, heat, and water – would be enough to put a limit on its population growth. It is unknown how much farther the carrying capacity can be expanded before one of the limits is reached.

Graph (b) of figure 2.10 illustrates a population that has stabilized somewhat below the carrying capacity. (In actuality the population may fluctuate slightly above and below the carrying capacity, but the carrying capacity remains basically unchanged.) Examples of this are seen in the undisturbed tropical rainforests where many species are relatively stable in an environment where average temperature and rainfall vary little.[46] Graph (c) portrays a situation where the population has overshot the carrying capacity of the environment and then oscillates above and below it. An example of this situation may be the relationship between the great gray owls and their prey, lemmings and voles, in northern forests. Lemmings and voles are an important food source for the owls. Their populations rapidly increase over a period of four to six years and then, as predators increase their consumption of them, their numbers crash catastrophically, causing the owls to flee the area to escape mass starvation.[47] Graph (d) illustrates a situation in which the overshooting of the carrying capacity leads to a precipitous decline in the population, or even to its extinction, and also to a decline in the carrying capacity. Such a situation has occurred with deer on the north rim of the USA's Grand Canyon,[48] and with elephants in Kenya's Tsavo National Park.[49] In both cases, the

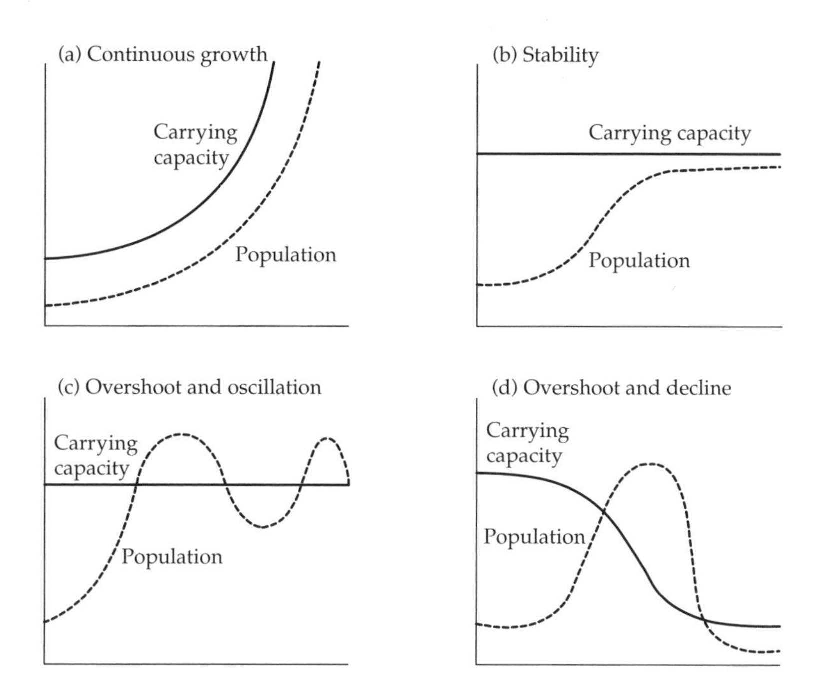

Figure 2.10 A growing population and carrying capacity

number of animals increased to a point where they destroyed the vegetation they fed upon.

It is my hope that the human species with its unique mental powers will create a situation which combines elements of graphs (a) and (b), using its abilities to increase the carrying capacity of earth, where possible, and where not, making sure its numbers do not exceed that capacity. But there are many indications that the species has not yet recognized its danger and is not yet taking effective efforts to prevent either situation (c) – which would mean the death of millions – or situation (d), which could lead to the decline of the human race. There are places in the world where population expansion has already passed the carrying capacity of the land and the land itself is now being destroyed; in sub-Saharan Africa, for example, fertile land is turning into desert and in the Himalayan mountain area, land is being destroyed by human-made erosion and floods. There are many other examples of the reduction of the carrying capacity of the earth that is taking place at unprecedented rates today around the world – the result of uncontrolled overgrazing, overfishing, overplanting, overcutting of forests, and the overproduction of waste which leads to pollution. (Some of this reduction of carrying capacity is being caused by population pressures and some by economic forces, for example the desire to increase short-term profits.) This deterioration has led many ecologists to believe that unless there is a rapid and dramatic change in many governmental policies, the human species may indeed be headed for the situations depicted in either the oscillation or decline graphs in figure 2.10.

There is one other aspect of the carrying capacity concept that makes it very difficult to use in an intelligent manner. Joel Cohen, a distinguished biologist, head of the Laboratory of Populations at Rockefeller University and Columbia University and author of the influential book *How Many People Can the Earth Support?*, persuasively argues that when asking the question "how many people can the earth support?" an attempt must be made to answer questions such as the following:

- *How many at what average level of well-being*? What type of diet, transportation, and health system will be provided?
- *How many with what distribution of well-being*? If we are content to have a few rich and a large number poor, the earth can probably support more than if the income distribution is fairly equal.
- *How many with what physical, chemical, and biological environments*? How much clean air, water, and wilderness do we want?
- *How many with what kinds of domestic and international political institutions*? How will conflicts be settled at home and internationally?
- *How many with what technology*? How food and goods and services are produced affects the earth's carrying capacity.
- *How many with what values, tastes, and fashions*? Are we vegetarians or meat eaters? Do we commute to work by car, mass transport or bicycle?
- *How many for how long*? How long can that number of people be supported?[50]

The concept "sustainable development," which will be discussed in chapter 7, is related to carrying capacity and is now being used more frequently than carrying capacity to convey some of the same concerns.

Optimum size of the earth's population

What is the optimum size of the earth's population? That question, like others we have asked in this chapter, is not going to be easy to answer, but it is worth asking. Paul Ehrlich, professor of population studies and of biology at Stanford University, defines the optimum size of the earth's population as that "below which well-being per person is increased by further growth and above which well-being per person is decreased by further growth." What does "well-being" mean? Ehrlich explains what he believes it means:

> The physical necessities – food, water, clothing, shelter, a healthful environment – are indispensable ingredients of well-being. A population too large and too poor to be supplied adequately with them has exceeded the optimum, regardless of whatever other aspects of well-being might, in theory, be enhanced by further growth. Similarly, a population so large that it can be supplied with physical necessities only by the rapid consumption of nonrenewable resources or by activities that irreversibly degrade the environment has also exceeded the optimum, for it is reducing Earth's carrying capacity for future generations.[51]

Ehrlich believes that, given the present patterns of human behavior – behavior that includes the grossly unequal distribution of essential commodities such as food and the misuse of the environment – and the present level of technology, we have already passed the optimum size of population for this planet.

The late Julian Simon, professor of business administration at the University of Maryland and author of *The Ultimate Resource*, believed that the ultimate resource on earth is the human mind. The more human minds there are, said Simon, the more solutions there will be to human problems. Simon admitted that population growth in the less developed countries could lead to short-term problems since more children will have to be fed. But in the long run these children will become producers, so the earth will benefit from their presence. Simon agreed that rapid population growth could harm development prospects in poor nations, but he was not disturbed by moderate growth in these countries. Larger populations make economies of scale possible; cheaper products can be made if there are many potential consumers. Also, services can improve, as seen by the development of efficient mass transportation in Japan and Europe in areas of dense population.[52]

Simon's views won favor in the Reagan and Bush administrations (father and son) in the United States and were used to give academic support to a new US policy on population – popularly called the Mexico City policy. This policy, which has been discussed earlier in this chapter, basically saw the effect of population growth as a "neutral phenomenon . . . not necessarily good or ill," a position which Marxist ideology also held.[53] While many economists in the United States do not share Simon's view that "more is better," many do share his view that human ingenuity, especially new technology and resource management practices, can increase the carrying capacity of the earth as it has in the past.[54]

Joel Cohen believes there is no way to estimate the optimum size of the human population on earth because no scientifically based answers have been given to the questions he presented above. In simpler terms, no one has answered the fundamental question: "How many people can the earth support with what quality of life?" Obviously the earth can support a large number of people if they are all living at a subsistence level – with barely enough to eat – or if a relatively few are rich and the rest poor, or if they accept frequent risks of violent storms and droughts. But Cohen believes that even without an agreed-upon "optimum number," the many things that governments and individuals can do to improve conditions for the present generation and future generations are worth doing.[55]

Population-related problems in our future

Throughout this book we are going to be looking at many current problems related to overpopulation and its effects. Here I will mention a few of the most important ones. Hunger is an obvious problem in which overpopulation plays a key role, and the number of hungry people is huge. The news media are used to dramatizing this problem only when there are many children with bloated bellies to be photographed, but much more common than the starving child today, and probably in the future, will be the child or adult who is permanently debilitated or who dies because of malnutrition-related diseases. Pollution and the depletion of nonrenewable resources will increase as the world's population grows. Migration of people to lands which do not want them will probably increase in the future and this would cause international tension. At least 300,000 immigrants, and probably more, enter the United States illegally annually, many of whom are Mexicans looking for work.

In Assam, India, several thousand unwanted immigrants were massacred in 1983. Wars have taken place in the past in which overpopulation played an important role and they will probably occur in the future. In the 1960s a border war broke out between El Salvador and Honduras over unwanted Salvadorians in Honduras. In the 1990s numerous brutal civil wars occurred in Africa. While we cannot identify overpopulation as the main cause of these conflicts, it is likely that increasing population pressures made the ethnic conflicts more likely.

Growing populations in countries situated in regions with serious water shortages are a direct cause of competition and conflict over the scarce water. The most dangerous areas are the Middle East and North Africa where population doubled between 1970 and 2001 thus reducing the amount of water available per person by one-half. More than 1 billion people today do not have access to safe drinking water. According to the United Nations and the World Bank, some 80 countries with about 40 percent of the world's population already face severe water shortages. In some of these regions droughts have been common throughout

A more frequent picture in the future? A crowded train in Bangladesh (*World Bank*)

history. What is not common in these regions is the population density which is present and projected. While water scarcity can obviously promote conflict, it also has the potential of promoting cooperation as nations are forced to devise ways to conserve and share scarce water. It is projected that that by 2030 about one-half of the world's population will live in water-stressed areas.[56]

Another problem that has the potential of making the problem worse is climate change, which will be discussed in chapter 4. An increase in the average global temperature could intensify the water cycle. There could be more rain in some locations but also more droughts.

One bright development is that the industrialized countries are learning to conserve water and to use it more efficiently. Water use in the United States has actually declined – about 20 percent since 1980. This decline came because of new water-saving technologies and practices such as less wasteful irrigation techniques and water-stingy toilets. (In the US toilets used about 6 gallons of water per flush. After a law passed by Congress in 1992 setting new standards, new toilets now use 1.5 gallons per flush, a 70 percent reduction.) Japan has made major reductions in water use in its industry. Some of these water-saving practices have spread to the poorer nations.[57]

Conclusions

How should I end this chapter? As I look over the statements by a number of population experts, they seem to share the conclusion that the earth faces an overwhelming problem with its current population growth of about 1 million people every five days. This is a problem that is second only to the threat of nuclear weapons for having the potential for causing untold human misery. But many of these experts also emphasize that human thinking and governmental policies are starting to change and impressive reductions in birth rates are taking place in various countries around the world. We know how to reduce birth rates. What is lacking at present is the political will to do what needs to be done to address the problem. We will do it if we take seriously the warning given in a joint statement by the US National Academy of Sciences and its British counterpart, the Royal Society of London:

> If current predictions of population growth prove accurate and patterns of human activity on the planet remain unchanged, science and technology may not be able to prevent either irreversible degradation of the environment or continued poverty for much of the world.[58]

Notes

1 Malin Falkenmark and Carl Widstrand, "Population and Water Resources: A Delicate Balance," *Population Bulletin*, 47 (Population Reference Bureau, Washington, DC) (November 1992), p. 23.

2 Edward Gargan, "On Meanest of Streets, Salvaging Useful Lives," *New York Times*, national edn (January 8, 1992), p. A2.

3 The United Nations uses the term "urban agglomeration" in place of "city." An urban agglomeration covers a much larger area than that of the administrative boundaries of the central city. It includes the central city and the continuous, densely populated area surrounding the city. It is the urban area as if seen from an airplane and is generally

recognized to be a more useful concept than the idea that a "city" is only that area within the administrative boundaries. Table 2.2 uses the UN's broader definition of a city.

4 "Transitions in World Population," *Population Bulletin*, 59 (2004), p. 12.

5 Lester R. Brown, *The Twenty-Ninth Day* (New York: W. W. Norton, 1978), p. 73.

6 Cited in Paul R. Ehrlich, Anne H. Ehrlich, and John P. Holdren, *Ecoscience: Population, Resources, Environment* (San Francisco: W. H. Freeman, 1977), pp. 777–8.

7 John C. Caldwell and Pat Caldwell, "High Fertility in Sub-Saharan Africa," *Scientific American*, 262 (May 1990), pp. 118–25.

8 "Transitions in World Population," p. 9.

9 Brown, *The Twenty-Ninth Day*, p. 77.

10 National Research Council, *Population Growth and Economic Development: Policy Questions* (Washington, DC: National Academy Press, 1986), p. 90.

11 Machiko Yanagishita, "Japan's Declining Fertility: '1.53 Shock,'" *Population Today*, 20 (Population Reference Bureau, Washington, DC) (April 1992), pp. 3–4.

12 Kevin Kinsella and David Phillips, "Global Aging: The Challenge of Success," *Population Bulletin*, 60 (March 2005), p. 6.

13 Nancy Riley, "China's Population: New Trends and Challenges," *Population Bulletin*, 59 (June 2004), p. 21.

14 Roger-Mark De Souza, John Williams, and Frederick Meyerson, "Critical Links: Population, Health, and the Environment," *Population Bulletin*, 58 (September 2003), p. 6.

15 Joel Cohen, "Human Population Grows Up," *Scientific American*, 293 (September 2005), p. 50.

16 Joseph A. McFalls Jr, "Population: A Lively Introduction," *Population Bulletin*, 46 (October 1991), p. 36. See also the third edition of this publication in *Population Bulletin*, 53 (September 1998), p. 42.

17 Kinsella and Phillips, "Global Aging: The Challenge of Success," p. 15.

18 A discussion of the political forces that were instrumental in the Reagan and Bush administrations in developing this policy and in keeping it in force for ten years is contained in Michael S. Teitelbaum, "The Population Threat," *Foreign Affairs* (Winter 1992/1993), pp. 63–78. The policy was reversed in 1994 when the Clinton administration took office but reinstated when Bush's son became President.

19 Alan Cowell, "UN Population Meeting Adopts Program of Action," *New York Times*, national edn (September 14, 1994), p. A2.

20 "Transitions in World Population," p. 25.

21 Bryan Robey, Shea Rutstein, and Leo Morris, "The Fertility Decline in Developing Countries," *Scientific American*, 269 (December 1993), p. 63.

22 Ibid., pp. 60–7. See also Susan Kalish, "Culturally Sensitive Family Planning: Bangladesh Story Suggests It Can Reduce Family Size," *Population Today*, 22 (February 1994), p. 5; Lina Parikh, "Spotlight: Bangladesh," *Population Today*, 26 (January 1998), p. 7; and Malcolm Potts, "The Unmet Need for Family Planning," *Scientific American*, 282 (January 2000), p. 91.

23 Robey et al., "The Fertility Decline in Developing Countries," p. 65.

24 Ibid., p. 60.

25 *New York Times*, late city edn (August 18, 1974), p. 2.

26 "Transitions in World Population," p. 9, and "Twenty Percent of Married Women Unserved by Contraception," *Popline*, 25 (Population Institute, Washington, DC) (November/December 2003), p. 7.

27 Elizabeth Leary, "As Contraceptive Use Rises, Abortions Decline," *Popline*, 25 (November/December 2003), p. 3.

28 Population Reference Bureau, *2005 World Population Data Sheet* (Washington, DC: Population Reference Bureau, 2005).

29 "Mexican Men Get the Message about Limiting Family Size," *Christian Science Monitor* (July 8, 1992), p. 11.

30 Ajoa Yeboah-Afari, "Male Responsibility: Still a Missing Link," *Popline*, 13 (May/June 1991), p. 6.

31 "Mexican Men Get the Message about Limiting Family Size," p. 11.

32 "Transitions in World Population," p. 27.

33 "Spotlight: India," *Population Today* (May 1996).

34 Riley, "China's Population," p. 3.

35 Ronald Freedman, "Theories of Fertility Decline: A Reappraisal," in Philip M. Hauser (ed.), *World Population and Development* (Syracuse, NY: Syracuse University Press, 1979), p. 75.

36 Elisabeth Rosenthal, "Rural Flouting of One-Child Policy Undercuts China's Census," *New York Times*, national edn (April 14, 2000), p. A10.

37 H. Yuan Tien et al., "China's Demographic Dilemmas," *Population Bulletin*, 47 (June 1992), p. 6.
38 Rosenthal, "Rural Flouting of One-Child Policy Undercuts China's Census."
39 Nancy E. Riley, "Gender, Power, and Population Change," *Population Bulletin*, 52 (May 1997), p. 39.
40 Riley, "China's Population," p. 18.
41 Marion Carter, "Spotlight: Brazil," *Population Today* (August 1996), p. 7.
42 Mary Mederios Kent, "Shrinking Societies Favor Procreation," *Population Today*, 27 (December 1999), pp. 4–5.
43 UN Population Division, Department of Economic and Social Affairs, "World Population Prospects: The 2004 Revision," in *Population Database 2005*, at http://esa.un.org/unpp.
44 Population Reference Bureau, "2005 World Population Data Sheet."
45 Daniel B. Botkin, "A New Balance of Nature," *Wilson Quarterly* (Spring 1991), pp. 61–72, and William K. Stevens, "New Eye on Nature: The Real Constant Is Eternal Turmoil," *New York Times*, national edn (July 31, 1990), pp. B5–6.
46 G. Tyler Miller Jr, *Living in the Environment: Principles, Connections, and Solutions*, 8th edn (Belmont, CA: Wadsworth, 1994), p. 153.
47 E. Vernon Laux, "In a Vast Hungry Wave, Owls Are Moving South," *New York Times*, national edn. (March 8, 2005), p. D2; Peter Hudson and Ottar Bjornstad, "Vole Stranglers and Lemming Cycles," *Science*, 302 (October 31, 2003), p. 797.
48 Edward J. Kormondy, *Concepts of Ecology*, 2nd edn (Englewood Cliffs, NJ: Prentice-Hall, 1976), pp. 111–12.
49 Botkin, "A New Balance of Nature," pp. 61–3.
50 Joel E. Cohen, *How Many People Can the Earth Support*? (New York: W. W. Norton, 1995).
51 Ehrlich et al., *Ecoscience*, p. 716.
52 Julian L. Simon, *The Ultimate Resource* (Princeton, NJ: Princeton University Press, 1981). See also Julian L. Simon, *Population Matters: People, Resources, Environment, and Immigration* (New Brunswick, NJ: Transaction, 1990).
53 Teitelbaum, "The Population Threat," pp. 71–2.
54 An interesting wager that Ehrlich and Simon made about whether the world's growing population was running out of natural resources (and which was won by Simon) is reported in John Tierney, "Betting the Planet," *New York Times Magazine* (December 2, 1990), pp. 52–81.
55 Cohen, "Human Population Grows Up," p. 55.
56 De Souza et al., "Critical Links: Population, Health, and the Environment," p. 26.
57 Peter Gleick, "Safeguarding Our Water," *Scientific American*, 284 (February 2001), pp. 38–45.
58 "A Warning on Population," *Christian Science Monitor* (March 24, 1992), p. 20.

Further Reading

Cohen, Joel E., *How Many People Can the Earth Support*? (New York: W. W. Norton, 1995). A scholarly, but also entertaining, nonideological discussion of population issues. Written by a population biologist at Rockefeller University and Columbia University, it covers a wide range of subjects related to population including resources, ecology, climate, social organization, and technology. He presents an extensive history of what various thinkers have said, over the years, about carrying capacity.

Cohen, Joel E., "Human Population Grows Up," *Scientific American*, 293 (September 2005), pp. 48–55. A relatively short summary of where we are at present and the challenges humanity faces as the earth moves to a new stage of life.

De Souza, Roger-Mark, John Williams, and Frederick Meyerson, "Critical Links: Population, Health, and the Environment," *Population Bulletin*, 58 (Population Reference Bureau, Washington, DC) (September 2003). Three critical questions about population, health, and the environment are examined. First, the relationships among the three; second, how these relationships affect human well-being; and three, what communities and policymakers can do about these situations.

Ehrlich, Paul R., and Anne H. Ehrlich, *The Population Explosion* (New York: Simon & Schuster, 1990). Taking the neo-Malthusian position that the world is heading for disaster because of overpopulation, the Ehrlichs argue that the earth's population is already too large for the earth's ecosystem. Continuing the argument they first made two decades earlier in *The Population Bomb*, they state that unless major efforts are made to rein in our numbers, natural forces (such as widespread death) will soon bring the human population back into balance.

Hardin, Garrett, *Living within Limits: Ecology, Economics, and Population Taboos* (New York: Oxford University Press, 1993). Hardin argues that we are heading for an ecological catastrophe unless Draconian steps are taken to stop the growth of human population. He believes that each nation must take care of its own population problem and that some form of governmental coercion is needed to address this problem.

Kinsella, Kevin, and David Phillips, "Global Aging: The Challenge of Success," *Population Bulletin*, 60 (Population Reference Bureau, Washington, DC) (March 2005). Kinsella and Phillips examine the causes of global aging and the changes it will bring to society. They also explore various public policies related to this development.

Linden, Eugene, "The Exploding Cities of the Developing World," *Foreign Affairs* (January/February 1996), pp. 52–65. Linden believes the fate of cities determines the fate of nations. He explores the deplorable conditions in many of the huge cities in the less developed countries. He contrasts the despair in cities such as Kinshasa, Zaire, with the hope in cities such as Curitiba, Brazil.

Livi-Bacci, Massimo, *A Concise History of World Population: An Introduction to Population Processes*, 3rd edn (Cambridge, MA and Oxford, UK: Blackwell, 2001). A readable history of the growth of the human population from prehistoric times to the present. The author takes a position in between that of the optimists and those predicting a catastrophe; he presents an alternative way to understand and deal with population growth.

Longman, Phillip, "The Global Baby Bust," *Foreign Affairs*, 83 (May/June 2004), pp. 64–79. Longman does not believe that overpopulation is one of the worst dangers facing the world. Rather he argues that an aging population and falling birthrates, which we see now in many rich countries and some developing countries, should be our major concern.

Moffett, George D., *Critical Masses: The Global Population Challenge* (New York: Viking, 1994). A journalist, the author takes us to Egypt, Kenya, Guatemala, and Bangladesh. In highly readable prose, he allows the reader to witness the impacts of overpopulation and listen to the people who must grapple with it.

O'Neill, Brian, and Deborah Balk, "World Population Futures," *Population Bulletin*, 56 (Population Reference Bureau, Washington, DC) (September 2001). This publication looks closely at population projections, how they are made, and their uncertainties. Based on these projections, the authors discuss what kind of world is likely in the future.

Potts, Malcolm, "The Unmet Need for Family Planning," *Scientific American*, 282 (January 2000), pp. 88–93. The author argues that falling birth rates have led some people to declare that overpopulation is no longer a threat, but it recently took just 12 years to add an additional billion people to the earth. Many people still lack access to contraceptives. Unless they are given access to these, severe environmental and health problems will occur in large parts of the world in the twenty-first century.

Ratzan, Scott C., Gary L. Filerman, and John W. LeSar, "Attaining Global Health: Challenges and Opportunities," *Population Bulletin*, 55 (Population Reference Bureau, Washington, DC) (March 2000). Social and economic factors play a major role in determining the health of people in rich and poor nations. Threats to global health, such as HIV/AIDS, are identified and ways to combat them are discussed.

Salgado, Sebastiao, *Migrations: Humanity in Transition* (New York: Aperture, 2000). The Brazilian photographer set out to record why people leave their communities, what happens to them en route, and where they end up. After six years and visits to 35 countries, Salgado presents his findings in 360 black-and-white photographs.

Simon, Julian L., *Population Matters: People, Resources, Environment, and Immigration* (New Brunswick, NJ: Transaction, 1990). This challenge to the doomsayers' vision of food shortages, environmental damage, and economic disaster looks at population growth from a positive viewpoint. Disagreeing with many social scientists and ecologists about the population issue, Simon believes that population growth has long-term benefits and that the human condition is much better than the neo-Malthusians admit.

CHAPTER 3

Food

The day that hunger is eradicated from the earth, there will be the greatest spiritual explosion the world has ever known.

Federico Garcia Lorca (1899–1936), Spanish poet and dramatist

One way a civilization can be judged is by its success in reducing suffering. Development can also be judged in this way. Is it reducing the misery that exists in the world? Throughout human history, hunger has caused untold suffering. Because food is a basic necessity, when it is absent or scarce humans need to spend most of their efforts trying to obtain it; if they are not successful in finding adequate food, they suffer, and, of course, can eventually die. In this chapter we will look at hunger and also at a problem the more developed countries face: how their own level of development affects the food they eat.

World Food Production

How much food is produced in the world at present? Is there enough for everyone? The answer, which may surprise you, is that, yes, there is enough. Food production has more than kept up with population growth. At the beginning of the twenty-first century, food supplies were about 25 percent higher per person than they were in the early 1960s and the real price of the food (taking inflation into account) was about 40 percent lower.[1] Impressive gains were made in the less developed nations where, from about 1960 to 2002, the average daily food calories available per person rose from about 1,900 to 2,700. (What was available for consumption does not indicate what individuals actually consumed.) In the developed countries the daily calorie supply increased from about 3,000 to 3,300 during the same period.[2] Enough food was available at the beginning of the twenty-first century to provide every person with 2,350 calories, the amount needed daily for a healthy and active life.

In the past 35 years the world's output of major food crops increased significantly – the most dramatic increase came in the production of cereals – as improved seeds, irrigation, fertilizers, and pesticides were used to increase production and new land was cultivated. (Most of this growth in production came from an increase in yield per acre rather than from an increase in the amount of cropland. In 2004 record average grain yields of about 3 tons per hectare were about three times what they were in 1960, with nearly perfect weather in major growing areas helping to make this possible even with a stable use of fertilizers and irrigation.)[3] This impressive performance was counterbalanced, however, by the rapid growth of population that was also taking place in the world at this time. But food production increased rapidly enough in the 35 years so that, except for sub-Saharan Africa and the former Soviet Union, the output of food in the world kept up with population growth. There was a decline in per capita food output in the former Soviet Union after the collapse of that country in 1991 and in sub-Saharan Africa from the mid-1970s because of poor performance in agriculture (which was caused in part by droughts, civil wars, and nonsupportive government policies) and because of very rapid population growth.

The exception to this generally rosy picture is rice, which is the most important food for about one-half of the world's people. The UN warned of a "pending crisis" as the production of rice in the early twenty-first century failed to keep up with demand. Although rice production has increased recently, 2004 was the fourth consecutive year that demand exceeded production. At that time the storage of rice by governments was down to a two-months supply.[4]

Figure 3.1 gives the long-term trend in per capita food production for the world, as well as for developed and developing countries. Figure 3.2 shows that during

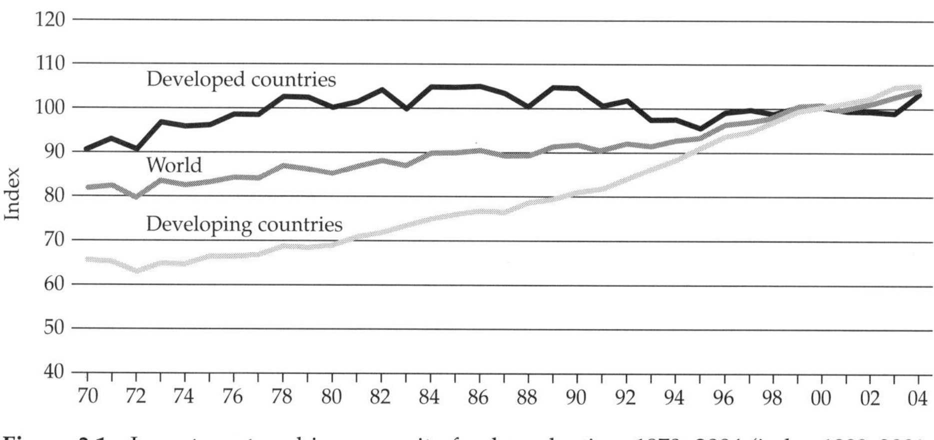

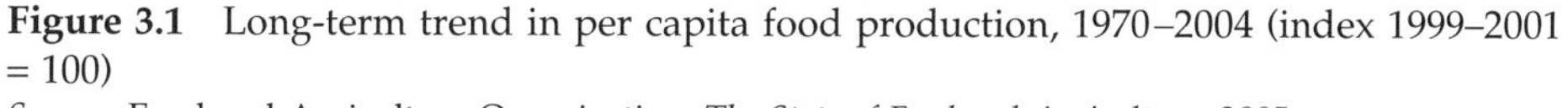

Figure 3.1 Long-term trend in per capita food production, 1970–2004 (index 1999–2001 = 100)
Source: Food and Agriculture Organization, *The State of Food and Agriculture, 2005*

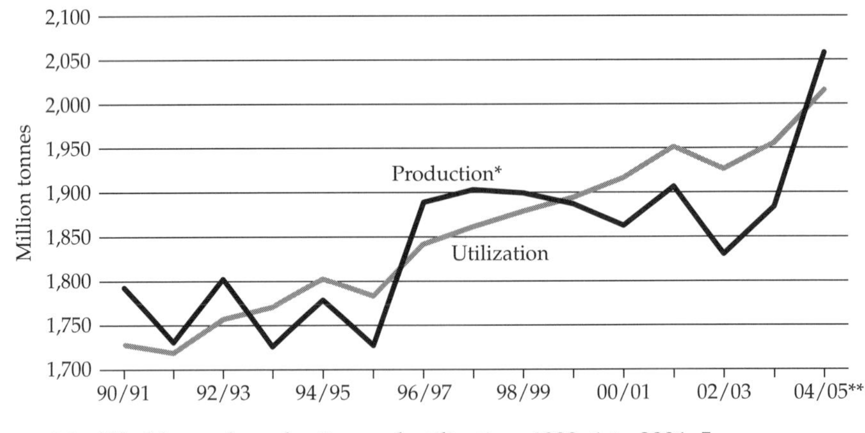

Figure 3.2 World cereal production and utilization, 1990–1 to 2004–5
* Data refer to the calendar year of the first year shown.
** Forecast.
Source: Food and Agriculture Organization, *The State of Food and Agriculture, 2005*

the 1990s and the early years of the twenty-first century cereal production swung above and below what was actually utilized. Cereals make up the bulk of the diet of most of the world's people. When production dips below demand, stored surpluses must be utilized, if available. China was responsible for most of the depletion of global inventories in the early years of the twenty-first century.

How Many are Hungry?

Unprecedented amounts of food in the world do not mean, unfortunately, that everyone is getting enough food. The UN Food and Agriculture Organization (FAO) estimates that about 850 million people are hungry[5] today, 800 million of them in developing countries. Two hundred million of these are children. The World Health Organization estimates that about one-third of the children in developing countries are malnourished.[6] About one-third of the population in sub-Saharan Africa is malnourished, as are about 15 percent of those in Asia and the Pacific, and about 10 percent in both Latin America and the Caribbean and the Near East and North Africa.[7] Nearly one-half of the children in large parts of South Asia (mainly India and Bangladesh) have stunted growth.[8] These depressing statistics can be countered somewhat by the fact that there has been real progress in reducing the number of hungry people even with the rapidly growing population in the less developed world. As can be seen in figure 3.3, the number of hungry people is decreasing in most of the major regions in the developing world except in sub-Saharan Africa. Sub-Saharan Africa is the only area in the world where the prevalence of hunger is over 30 percent and the absolute number of hungry

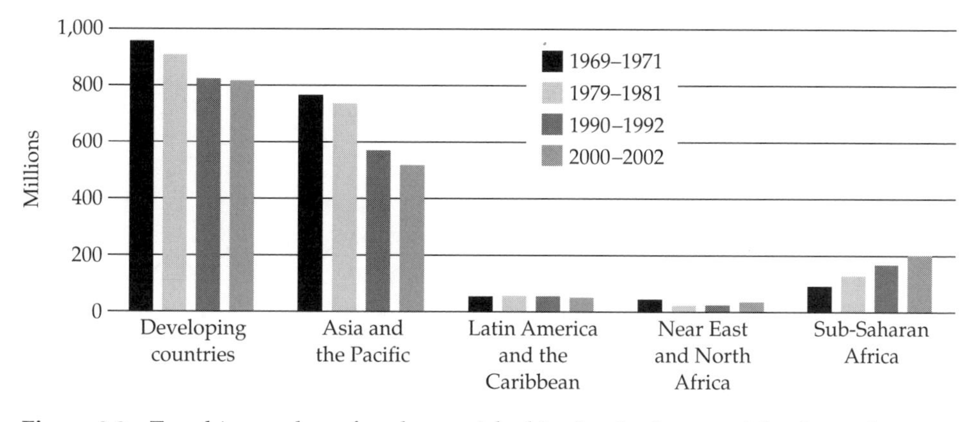

Figure 3.3 Trend in number of undernourished in developing countries, by region
Source: Food and Agriculture Organization, *The State of Food and Agriculture, 2005*

people is increasing.[9] In the 1990s the number of hungry people worldwide decreased by 20 million.[10]

Except for Africa, actual starvation is uncommon in the present world. A much larger number of people die today because of malnutrition, a malnutrition that weakens them and makes them susceptible to many diseases. Children die from diarrhea in poor countries – a situation nearly unheard of in rich countries – partly because of their weakened condition.

Who are the hungry and where do they live? The answer to the first question is that, according to World Bank estimates, 80 percent are women and children. Because the men are needed to acquire food for the family, they are fed the best. A UN sponsored Hunger Task Force estimates that about 50 percent of the hungry are small farmers; 20 percent are landless rural people; 10 percent are herders, fishers and forest dwellers; and the remaining 20 percent are the urban poor.[11] Although urban poverty is a growing problem in the less developed world because of the rapid urbanization taking place there, 80 percent of extreme poverty occurs in the rural areas of the poorer countries.

In the year 2000 about 40 percent of the hungry lived in South Asia (mainly India, Pakistan, and Bangladesh), 25 percent in East Asia and the Pacific, and 20 percent in sub-Saharan Africa; about 5 percent each lived in Latin America and the Caribbean, in Central and Eastern Europe and the former Soviet Union, and in the Arab states.[12]

There are indications that the number of hunger-related deaths in the world has decreased during the past 30 years in spite of the world's growing population. It was estimated in the mid-1980s that about 15 million people were dying each year from hunger-related causes.[13] One estimate is that in the early 2000s about 200,000 people died yearly during famines, while about 11 million people died early from hunger-related causes.[14] Five million of these were reported to be children.[15]

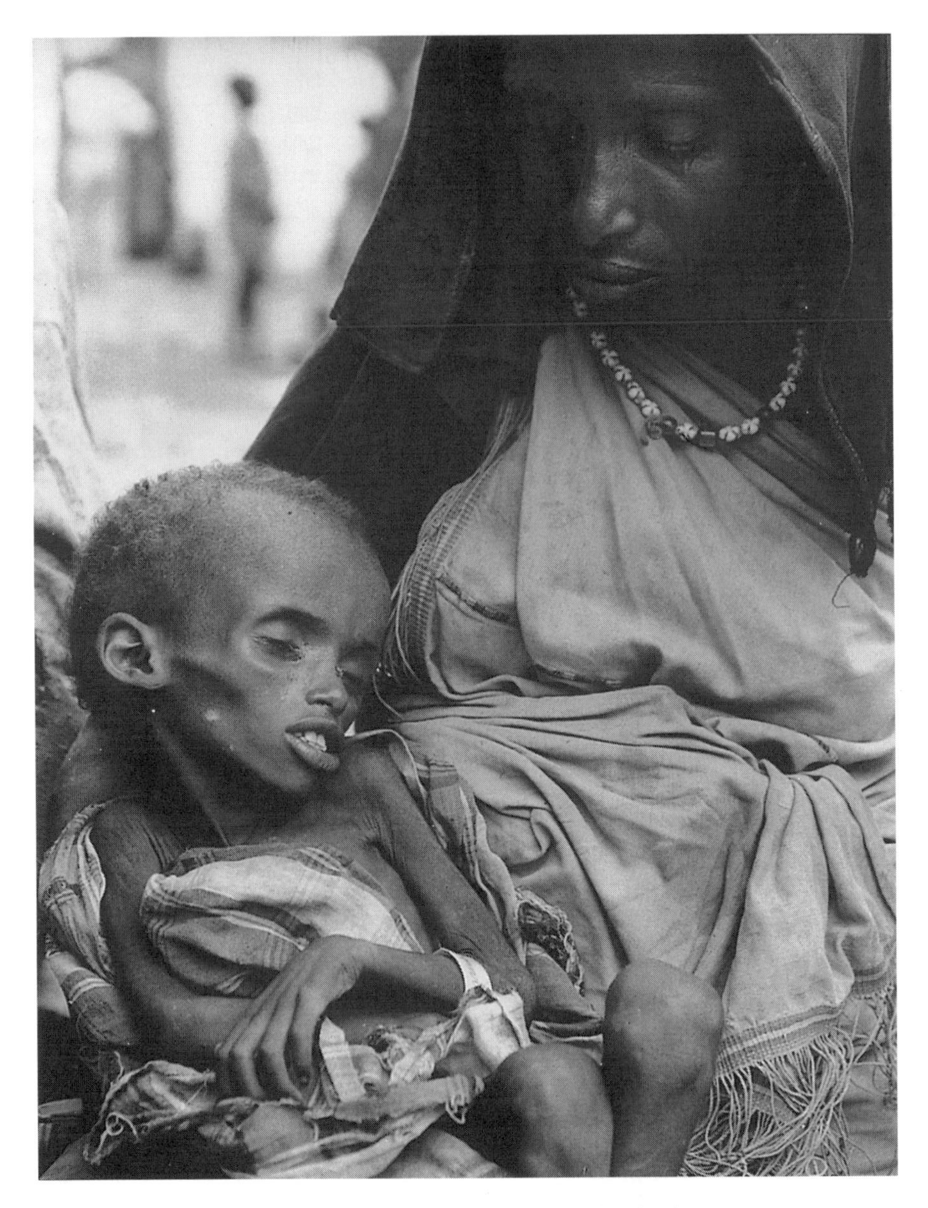

Starvation in Somalia (*CARE: Zed Nelson*)

Causes of World Hunger

If there is more than enough food being grown at present for the world's population but about 15 percent of the earth's people are malnourished, what is causing hunger in the world? Food authorities generally agree that poverty is the main

What the poor eat in Calcutta

Here is what some of the poor eat in Calcutta, as described by a US journalist who was invited to share a meal:

> Ahmed . . . invited me for tea at the kiosk, where he shined shoes. A monsoon storm broke, and about a dozen of us ended up spending an hour or two talking together. All of them were either shoeshine men, beggars, or pickpockets. Chowringhee Road was their universe. A meal was served in a big tiffin (luncheon) can. It was rice and curry mixed together – left-overs scraped off the plates at a nearby government canteen by some enterprising Bengali. One portion cost about five cents. It was dumped in a big pile on a newspaper and everyone squatted around in a circle, avidly eating with his fingers.

Source: Richard Critchfield, *Villages* (Garden City, NY: Anchor Press/Doubleday, 1981), pp. 285–6

cause of world hunger. Millions of people do not have enough money to buy as much food as they need, or better kinds of food. This is the reason one food expert has written "Malnutrition and starvation continue more or less unchanged through periods of world food glut and food shortage."[16] The world's poorest cannot afford to purchase the food they need, whatever its price. In tropical Africa and remote parts of Latin America and Asia, low agricultural productivity (not enough food produced) tends to be the main reason for hunger.[17]

Other low income people in the South suffer during food shortages when the price of food increases dramatically, as it did during the early 1970s when world prices of rice, wheat, and corn doubled in just two years. The poor traditionally spend 60 to 80 percent of their income on food. If world demand is high for certain foods, such as beef for the US fast-food market, then the large landowners in developing countries grow food or raise cattle for export rather than for domestic consumption. This tends to cause domestic food prices to increase since the supply of local foods is reduced.

According to a recent report by the World Bank, malnutrition in children is more often caused by a lack of nutritional information among mothers in less developed countries than by the actual lack of food. The failure to understand that exclusively breast-feeding their babies for six months is the best food for them, and their unawareness that giving infants food mixed with unsafe water can cause diarrhea, lead to more malnutrition than the lack of food. This malnutrition during the infant's first two years of life can lead to permanent damage to their mental capabilities in addition to stunted growth.[18]

With spreading economic development, famines are becoming rarer than they were in the past. But a number of major famines occurred in the twentieth century. In the Soviet Union in the early 1930s Stalin forcibly collectivized agriculture and deliberately caused a famine in the area where most of the grain was grown – the Ukraine and Northern Caucasus – in order to break the resistance of the peasants. An estimated 7 million people – 3 million of them children – died in that famine.[19] Another country with a communist government experienced the worst famine in the twentieth century. Although it was kept secret from the outside world while it was occurring, China had a famine in the late 1950s and early 1960s that led to an estimated 30 million deaths. The famine was caused mainly by misguided governmental policies during the period known as The Great Leap Forward.[20]

Much land was being used to grow export crops such as cotton and peanuts in the Sahel, a huge area in Africa just south of the Sahara desert, when a famine hit that area in the early 1970s. Six years of drought, rapid population growth, and misuse of the land led to widespread crop failures and livestock deaths. It is estimated that between 200,000 and 300,000 people starved to death in the Sahel and in Ethiopia before international aid reached them.[21]

Famine also hit Cambodia in the 1970s. Years of international and civil war, coupled with the genocidal policies of the communist government under the leadership of Pol Pot, led to an estimated 10,000 to 15,000 people dying every day during the worst of the famine in 1979. A highly successful international aid effort, first organized by private organizations and then joined by governmental agencies, saved the Cambodian people from being destroyed.

Famine hit again in Africa in the mid-1980s, and early and late 1990s. Television pictures of starving people in Ethiopia led to a large international effort by private organizations and by governments to provide food aid. The famine in Ethiopia, Somalia, and the Sudan, and in other sub-Saharan African countries, was not caused only by the return of a serious drought to the region. Many of the causes of these famines were the same as those that brought on the famine in Africa in the early 1970s. In addition to the reasons stated above, the extensive poverty in the region, a world-wide recession which seriously hurt the export-oriented economies of the African countries, civil wars, and governmental development policies that placed a low priority on agriculture have been identified as likely causes.[22]

North Korea experienced a famine for about four years in the mid and late 1990s. An estimated 2 to 3 million people died, about 10 percent of its population. The famine led to stunted growth in about two-thirds of the children under five. This made it, relatively, one of the worse famines in the twentieth century, comparable to the famines in the other two communist totalitarian regimes. Like the Soviet Union and China, North Korea was a closed society at the time of the famine and evidence of the famine was kept secret. Although a flood and drought were partial causes, the main causes appear to be the inflexible political and economic systems and the downfall of the country's long-term patron – the Soviet Union. Serious food shortages continue to the present. Food donors are now reluctant to continue to help the country when the food shortage continues past ten years, and North Korea has admitted making costly efforts to develop nuclear weapons at the same time as its people were starving.[23]

The Secretary-General of the United Nations appointed a Hunger Task Force in 2002 to recommend ways the Millennium Development Goal to reduce the number of hungry people in the world by one-half by 2015 could be reached. The Task Force concluded that this goal can be achieved and hunger eventually eliminated by using our present financial resources and technological capabilities. To do so will require action by both rich and poor countries to honor the commitments they all made when they endorsed the Millennium Goals in 2000.

Among various recommendations, the Task Force specified the need for about $10 billion in development assistance yearly from the developed world aimed at the reduction of hunger. From developing countries, the Task Force said hunger reduction policies at all levels of government were needed, as well as the reduction of corruption, the establishment of the rule of law, and respect for human rights. The Task Force's message to political leaders of both rich and poor nations was that "halving hunger is within our means; what has been lacking is action to implement and scale up known solutions."[24]

How Food Affects Development

The availability of food has a direct effect on a country's development. Possibly the most destructive and long-lasting is the effect the absence of food – or, more often, of the right kinds of food – has on the children of the less developed nations. As mentioned in chapter 2, the death of many children in poor nations at birth or in their first few years is one of the causes of high birth rates. As mentioned previously, the World Health Organization estimates that about one-third of the children in less developed nations are malnourished. Also, malnutrition contributes to over one-half of the deaths of children under the age of five in poorer countries.[25]

A deficiency of vitamin A leads to the blindness of about 100,000 children a year in developing countries.[26] More common than blindness are the harmful effects malnutrition has on the mental development of the children. Eighty percent of the development of the human brain occurs before birth and during the first two years after birth. Malnutrition of the pregnant mother or of the child after birth can adversely affect the child's brain development and, along with limited mental stimulation, which is common in poor homes, can lead to a reduced capacity for learning.

Malnutrition also reduces a person's ability to ward off diseases since it reduces the body's natural resistance to infection. Measles and diarrhea, which are not generally serious illnesses in the developed nations, often lead to the death of children in the developing nations; in fact, diarrhea is the single greatest cause of death of children in the less developed world. When a child has been weakened by malnutrition, sickness is likely to come more frequently and to be more serious than that experienced by the well-nourished child.

Some North American citizens, when they first visit less developed nations, come away with a feeling that the people are lazy since they are likely to see a number of people sitting around, not doing much of anything. Aside from the absence of jobs or of land they can farm, malnutrition also may be playing a role here

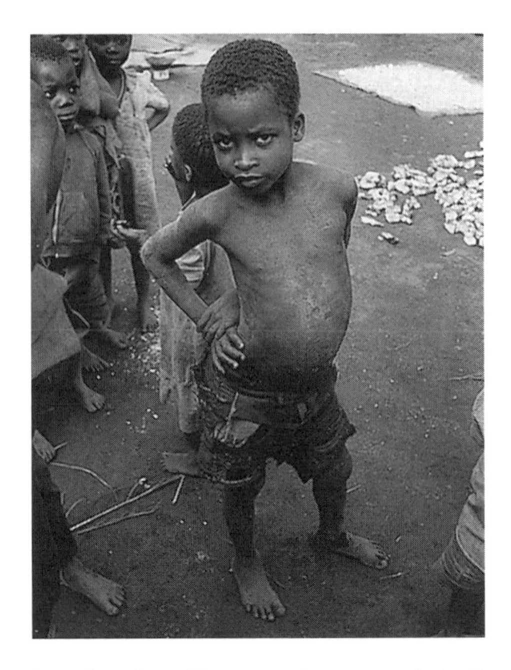

The bloated belly is a sign of malnutrition, a major cause of stunting and death in the children of the developing world (*CARE: Joel Chiziane*)

because, as one study has stated, "Chronically undernourished people, who commonly also suffer from parasitism and disease, are typically apathetic, listless, and unproductive."[27]

The presence of unhealthy and unproductive people in rural areas probably means that not as much food is being grown as is possible, and the presence of unhealthy and unproductive people in urban areas probably means that fewer manufactured goods and services are being produced. A nation that must spend scarce foreign exchange to buy imported food cannot use those funds to support its development plans. And, more importantly, a nation whose main and most important resource – its people – is weakened by malnutrition is unlikely to generate the kind of economic development that actually does lead to an improved life for the majority of its people. James Grant, former head of the United Nations Children's Fund (UNICEF), has described well the interrelatedness of all key elements of development:

> A cat's cradle of . . . synergisms links almost every aspect of development: female literacy catalyzes family planning programmes; less frequent pregnancies improves maternal and child health; improved health makes the most of pre-school or primary education; education can increase incomes and agricultural productivity; better incomes or better food reduces infant mortality; fewer child deaths tend to lead to fewer births; smaller families improve maternal health; healthy mothers have healthier babies; healthier babies demand more attention; stimulation helps mental

> growth; more alert children do better at school . . . and so it continues in an endless pattern of either mutually reinforcing or mutually retarding relations which can minimize or multiply the benefits of any given input.[28]

How Development Affects Food

The development that took place in Europe and the United States as they industrialized led to an increase in the average family's income, and this meant more money to buy food. As we saw in the preceding section, poverty is the main cause of malnutrition. As incomes rose in the West, hunger disappeared as a concern for the average person. Except for some subgroups in Western countries, malnutrition is no longer a common problem.

Development also affects food in other ways. As a nation develops, major changes start to take place in its agriculture. We will look first at how development affects the amount of food that is produced and how it is produced, and then at the way development affects the types of food people eat.

The production of food

Western agriculture produces an impressive amount of food. The US supermarket, better than any other institution, illustrates the abundance that modern agriculture can produce. The United States produces so much food that huge amounts of important crops such as corn, wheat, and soybeans are exported. Much of this US abundance has come since World War II. By 2000 American farmers were producing twice the output they had in 1930 with only one-third the number of farms.[29] What is the reason for this increase in production? There are many reasons, of course, but basically it is because American agriculture has become mechanized and scientific. By using new seeds, which can benefit from generous amounts of fertilizer, pesticides, heavy machinery, and irrigation, production has soared. But this accomplishment has had its costs, as we shall see below.

Western agriculture basically turns fossil fuel into food. This type of agriculture was developed when oil was inexpensive. Large amounts of energy are needed to build and operate the farm machinery, to build and operate the irrigation systems, to create the pesticides, and to mine and manufacture the fertilizers. Also, huge amounts of energy are needed to process the foods, to transport them to market, to package them, and to display them in retail stores. (Even in this period of increased energy prices, the open freezer in US supermarkets is still common.) It has been estimated that to raise the rest of the world's diet to the American level – especially one featuring its high consumption of beef – would consume nearly all the world's known reserves of oil in 15 years.[30]

Although modern, mechanized agriculture is generally – but not always – much more productive than the traditional agriculture more commonly found in less developed countries, traditional agriculture is generally far more energy efficient than Western agriculture. In traditional agriculture the amount of energy used in the form of farm labor and materials is typically small compared with the yield. Returns up to 50 to 1 are possible, although more common are 15 to 1 returns, whereas in modern industrial agriculture more energy is expended than produced.[31] To produce and deliver to a US consumer one can of corn which has 270 calories in it, a total of about 2,800 calories of energy must be used. To produce about 4 ounces of beefsteak, which also provides about 270 calories, an astounding 22,000 calories of energy must be expended.[32] Anthropologists Peter Farb and George Armelagos give us one perspective we need in order to judge the effects that development, as achieved in the West, is having on agriculture:

> In short, present-day agriculture is much less efficient than traditional irrigation methods that have been used by Asians, among others, in this century and by Mayans, Mesopotamians, Egyptians, and Chinese in antiquity. The primary advantage of a mechanized agriculture is that it requires the participation of fewer farmers, but for that the price paid in machines, fossil fuels, and other expenditures of energy is enormous.[33]

As the late Barbara Ward, a respected British author of many books on development, noted, "the high-energy [US] food system is one reason why the United States, for 5 percent of the world's people, is now consuming nearly 40 percent of its nonrenewable resources."[34] That statement, more than any other, presents the main argument of those who maintain that there is no way the rest of the world can adopt the agricultural methods followed by the United States at present.

Another feature of US agriculture is an increase in the size of farms and a reduction in their number. Table 3.1 shows how farm size and numbers have changed from 1940 to 2000.

Increased demand for farm products, along with government price supports, enabled farmers to replace old sources of power (horses and mules) with new sources (first the steam engine and then the gasoline engine) and to begin using more machinery, improved seeds, fertilizers, and chemicals to control pests. Dramatic increases in farm productivity resulted so that by 1990 only about

Table 3.1 Number and size of US farms, 1940–2000

Year	*Number of farms*	*Average size of farms (acres)*
1940	6,400,000	170
1950	5,600,000	210
1960	4,000,000	300
1970	2,900,000	370
1980	2,400,000	430
1990	2,100,000	460
2000	2,200,000	440

Source: Data from *Statistical Abstract of the United States* (Washington, DC: US Bureau of the Census, 1970, 1992, 2006), p. 582 (1970), p. 644 (1992), p. 548 (2006)

2 percent of US citizens were farmers, down from about 30 percent in 1920. With the increasing financial investment necessary to support the new type of agriculture, and the competition the large farms provide, there has been a noticeable decline in the small, family-owned farm in the United States. In the mid-1990s in the US the large farms, which were only 10 percent of the farms in the country, earned 85 percent of the agricultural income.[35]

Brazil, using Western agricultural methods, has recently become a major exporter of foods. In 2003 it passed the US as the largest exporter of beef. Over the next decade it may pass the US as the world's largest agricultural producer. It has already surpassed the US and Europe in productivity levels in some crops. Using scientific research it has created varieties of crops that can grow in the tropical and savanna soils in its vast interior that were formerly considered poor for crops. According to the director of the Center for International Economic Policy at the University of Minnesota in the US, "(the Brazilians) learned that with modest applications of lime and phosphorus, they can quadruple and quintuple their yields, not just with soybeans, but also with maize, cotton and other commodities."[36]

China, with 1.3 billion people but only about 7 percent of the world's arable land, has had an uneven record in producing grain (corn, rice, and wheat). In the late 1990s China produced a record amount of grain and was an exporter. But rapid urbanization and industrialization has led to the loss of large amounts of farmland. This, along with other factors, led in 2003 to the nation consuming 40 million more tons of grain than it produced. The size of the average farm in China is only about 2 acres, vastly different from the average of 400 acres in the US and the large new farms in Brazil. China, reluctant to become dependent on the US and other countries for its food, is now struggling to find ways to grow enough food for its still growing population.[37]

The growth of what has become known as "agribusiness" – farms run like a big business – has meant an increased concentration of control over the production of food in the United States, although there is still sufficient competition in agriculture so that food in the United States remains relatively inexpensive. The large industrial farms can produce harvests of 100 million tomatoes, but sometimes with less efficiency than small operators can obtain. When committees make decisions instead of the farmer growing the crop, when attention is not given to detail, and when there is a lack of dedication – dedication that usually comes only when someone has a personal stake in the farm – one often finds waste and mismanagement. This happened on large state-owned farms in the Soviet Union, and it is happening on large industrial farms in the United States.

The loss of food

Waste

As national incomes increase, the amount of food wasted tends to increase also. According to rough estimates, the amount of food wasted in low income countries is minimal, less than 10 percent. In middle income countries such as Greece,

Mismanagement on an industrial farm: the case of the oversized carrots

When the author of a book on three different types of farms in the United States saw an entire crop of carrots being plowed under instead of being harvested on a corporation-owned farm in California, he was given the following explanation by a farm supervisor:

> There are enough carrots on [sic] the world right now without these . . . Price isn't so hot, and the warehouses were full when these got to the right size. We were held off harvesting. Someone let time go by and suddenly they were too big. More than eighty acres of them, which comes to sixty million carrots or so. They couldn't fit into plastic carrot sacks they sell carrots in unless they were cut, and that would have cost the processor a bundle. They offered us a hundred and twenty-five dollars an acre for the carrots – and it would have run us two hundred dollars just to have them contract-harvested. So this is the cheapest alternative.

Source: Mark Karma, *Three Farms: Making Milk, Meat and Money from the American Soil* (Boston: Little, Brown, 1980), p. 248

Ireland, and Libya, the amount of food lost or wasted is estimated to be 40 percent. In high income countries such as Canada and the United States, the amount of food wasted could be between 25 and 30 percent. Food loss varies considerably even among countries with similar income levels. For example food loss is estimated to be 60 percent lower in the Netherlands, Finland, Japan, and Sweden, than in Belgium, Switzerland, United States, and Italy, all countries with similar levels of income.[38]

In many developed countries consumers demand that the produce they buy looks cosmetically perfect. This leads to the wasting of much of the food successfully harvested. One study in the late 1990s of food waste in the United States concluded that about 25 percent of the food available for people to consume was lost in the final three stages of the marketing process: retailing, food service, and consumers. Additional amounts of food were lost in the harvesting and distribution stages.[39]

Erosion

Besides the waste of food, there is another waste occurring in the world, which could affect profoundly its ability to produce food in the future: the loss of its farmland by erosion. Although global soil erosion is not expected to seriously hurt world food production, it is a serious problem in a number of locations.[40] Cornell University biologist David Pimentel reports: "Our measuring stations pick up

Chinese soil in the Hawaiian air when ploughing time comes. Every year in Florida we pick up African soils in the wind when they start to plough."[41] Nature makes soil very slowly – under normal agricultural conditions it takes from 200 to 1,000 years to form 2.5 centimeters of topsoil.

A study by the US Department of Agriculture reported in 1999 that 30 percent of the nation's farmland was being eroded at an excessive rate. Almost 2 billion tons of topsoil is lost to erosion.[42] A study sponsored by the United Nations Environment Programme showed that over a 45-year period human activity had led to moderate to severe damage to the land of about 10 percent of the world's vegetated area – an area about the size of China and India combined. Activities involved in the production of food (agriculture and the grazing of livestock) caused most of the damage and most of the land deterioration occurred in Asia and Africa.[43]

Increasingly the world's food supply is relying on irrigation. Irrigation uses more water than any other human activity. Inadequate drainage of irrigated land can lead to waterlogging (an excessive amount of water remaining in the soil) or to salinization (toxic salts are deposited on poorly drained land). It has been estimated that from 10 to 15 percent of all irrigated land at present suffers in some degree from waterlogging or salinization.[44]

Severe soil erosion is expected to seriously affect food production in southeast Nigeria, Haiti, the Himalayan foothills, some parts of southern China, Southeast Asia, and Central America. Salinization is expected to become a major threat in the irrigation systems of the Indus, Tigris, and Euphrates river basins. It is expected also to become a problem in India, Australia, northeastern Thailand, China, the Nile delta, northern Mexico, and the Andean highlands. Nutrient depletion is likely to become a serious problem in large areas of Africa, and numerous other locations from Myanmar to the Caribbean Basin.[45] Desertification, the spreading of deserts, is found in the Sahel region of Africa, in Kazakhstan, and in Uzbekistan. In northern China the desert is growing and leading to massive dust storms that plague Beijing and other cities in China.[46]

Not all the news about erosion is negative. Some encouraging developments are taking place in the world. In the United States, Congress passed a law in 1985 that paid farmers to grow erosion-resistant grasses and trees on the land most susceptible to erosion. By the end of the 1980s about 30 million acres had been placed under this program. The US Department of Agriculture reported in 1999 that soil erosion in the United States had been reduced by about 40 percent since 1982 – but that the reduction had leveled off since 1995.[47] No-till farming was one of the methods used to achieve this.[48] The Worldwatch Institute reports the following encouraging developments throughout the world:

> More soil-friendly farming practices that minimize tilling and reduce the erosive potential of the tilling that is done are coming into wider use, are spreading into countries such as Morocco, the Philippines, and Thailand, and are expanding regionally in parts of sub-Saharan Africa and South America. These methods include contour farming, terracing, vegetative barriers, and improved land use practices at the farm and landscape levels. Better water management practices that control salinization and lower the amount of irrigation water needed per hectare are also spreading.[49]

Urban sprawl

Another situation in many countries that adversely affects their ability to produce enough food for their people is the loss of prime farmland to development. At the beginning of the twenty-first century the United States was losing about 2 acres of farmland every minute because of development; it was being covered over by houses, roads, shopping malls, factories, and by general urban sprawl. While the amount lost was small compared to the amount of actual and potential cropland in the United States, the land lost was often prime farmland, including some of the best fruit orchards, and could be replaced only by marginal land, which was not as fertile, was more open to erosion, and was more costly to use. One-half of the lost farmland was carved into 10-acre lots, many probably for homes for the wealthy.[50]

Street vendors sell food to many urban dwellers in developing-world cities (*Ab Abercrombie*)

The loss of farmland to development in the United States accelerated at the end of the twentieth century. While the population of the country grew about 15 percent from 1982 to 1997, the amount of land turned into urban areas grew nearly 50 percent. This resulted in the loss of 13 million acres of cropland, 14 million acres of pastureland, and 12 million acres of rangeland. The amount of forest land remained unchanged. Much of this loss in the United States came not only because of the sprawl of suburbs near large cities, but also because of the urban sprawl of small and medium-sized cities.[51] The same problem of urban sprawl devouring prime cropland is occurring also in other developed and developing nations.

The type of food

As a nation develops, its diet changes. The wealthier a nation becomes, the more calories and protein its citizens consume. The average citizen of a Western industrialized nation consumes many more calories and much more protein than he or she needs for good health. Much of the excess in protein comes from a large increase in meat consumption. Often the consuming of meat instead of grains in order to get protein, which is needed for human growth and development, is a very inefficient use of food.[52] For every 16 pounds of grain and soybeans fed to beef cattle in the United States, about 1 pound of meat for human consumption is obtained. About three-quarters of the food energy in an Asian's diet comes directly from grain (about 300–400 pounds a year), whereas a US citizen consumes nearly 1 ton of grain per year, but 80 percent of it is first fed to animals.[53] People in North America consume the highest amount of meat per person in the world – about 260 pounds annually. Large amounts of meat are also consumed in Europe and Brazil, both about 160 pounds annually per person. During the last 40 years of the twentieth century, per capita meat consumption in Europe went up about 100 percent, in Brazil about 150 percent, while in North America it increased about 30 percent.[54] It is generally agreed by experts on nutrition that excessive calories and excessive meat consumption can lead to serious health problems. Barbara Ward describes the harmful features of such a diet:

> The car and the television set and the growing volume of office work may well have produced the most literally sedentary population human society has ever known. But at the same time, diets stuffed with the proteins and calories needed for a lumberman or a professional boxer have become prevalent. Everywhere, high meat consumption demands grain-fed animals. Meanwhile, what little grain we do eat through bread usually has little nutritional value and roughage, since these are removed when the flour is refined. Thus, the human bowel is deprived of the fiber it requires to function easily. The eating of fresh vegetables – which also give necessary fiber – has fallen off by between a third and a half in the last half century. Processed, defibered products have taken their place. The results are literally apparent. In all developed nations, obesity and diet-related illnesses are now a major medical problem. . . . Many medical experts are now agreed that with fat, sugar, cholesterol, refined grains, food additives, and the general absence of roughage, modern citizens

Tropical rainforests are being cut down to raise beef cattle for the US fast-food market – the so-called "hamburger connection" (*United Nations*)

are literally – via heart attacks and cancer – eating and drinking themselves into the grave.[55]

A very disturbing development began at the end of the twentieth century, first in the United States, then in other Western developed countries, and finally in some developing countries. More fast-served foods were consumed with a high calorie content. Along with a sedentary lifestyle and lack of exercise, this led to an increasing number of people who were overweight and obese. The food industry contributed to this trend through its extensive advertising, vast expansion of outlets, and its increasing size of the portions of food and beverages served.[56] Table 3.2 gives the number of overweight and obese males and females in a select group of developed and developing countries. Medical personnel warned that this trend would lead to more diabetes, heart problems, high blood pressure, and possibly cancer.

I'd like to end this section with a short explanation of how development has affected the first food North Americans receive after birth. If you are a North American and were born before 1940, the chances are good that the first food you

Table 3.2 Percentage of adults overweight and obese (various countries)

Country	*Male*	*Female*
Greece	79	75
Saudi Arabia	69	76
Germany	76	59
USA	67	62
Mexico	61	65
England	65	56
Spain	58	48
Canada	56	39
Sweden	51	42
Netherlands	54	39
Italy	51	36
Brazil	38	39
China	32	34
Japan	27	21

Data from various years 1990–2003; China data are urban only; "overweight" and "obesity" are labels for ranges of weight that are greater than what is considered healthy for a given height.
Source: International Obesity Task Force (IOTF), "Global Prevalence of Obesity," updated March 16, 2005, at http://www.iotf.org/media/globalprev.htm

received was human milk from your mother's breast, whereas if you were born after 1955,[57] your first food was probably a human-made formula from a bottle (using cow's milk as the basic ingredient). A rapid decline in breast-feeding has taken place in the developing world also. By 2005, according to UNICEF, only about 40 percent of infants in the developing world were being exclusively breast-fed. Partly this was because of urbanization, the increasing number of women in the workforce, and the promotional efforts of formula-making companies (the latter more of a factor in the past than at present). Breast-feeding declined also because of a desire to imitate the United States, to be "modern." As the poor in the less developed nations saw it, if the rich are bottle-feeding their babies, it must be better.

But is it? Except for mothers who are infected with the HIV virus, no, it's not.[58] Nutritionists agree that human milk is the best food for babies. Breast-feeding is also the safest, cheapest, and easiest way to feed babies. Breast-feeding probably improves bonding – a special feeling of closeness – between the mother and the baby, and, as we saw in chapter 2, it can act as a natural birth control. Breast-feeding gives the baby antibodies which enable it to fight off infection; this is especially important since its own immune system is not fully developed during the first year. A 1980 study in Brazil of children of poor parents showed that bottle-fed babies were three to four times more likely to be malnourished than those who were breast-fed. Studies have also shown that in India bottle-fed babies have diarrhea three times more often than breast-fed babies, and in Egypt infant deaths are five times higher among bottle-fed babies than among breast-fed ones.[59] Many of the harmful effects of bottle-feeding in the developing world occur because of the lack of refrigeration and the lack of knowledge about the importance of sterilization. Also, formula is expensive, so poor mothers often dilute it with water – water that is often polluted – and that makes the formula too weak.

There has been a return to breast-feeding in some European countries and partly in the United States. Norway, Denmark, and Sweden lead the way with close to 100 percent of all new mothers breast-feeding their infants at birth. (Globally there was a 15 percent increase in breastfeeding from 1990 to 2000.) After six months (the frequently recommended period for breast-feeding) 80 percent of new mothers in Norway are still nursing their infants. In Britain and the United States about 70 percent of new mothers are now breastfeeding their infants at birth, but only about 30 percent in the US and 20 percent in Britain are breast-feeding six months after birth.[60] Working mothers in the US find bottle-feeding more convenient, and the US culture is still unsettled by the sight of a woman breast-feeding in public. Because US women can no longer turn to their mothers for help or encouragement in breast-feeding (since their mothers didn't do it), a special organization – La Leche League – has been formed by some US women to help others learn about breast-feeding and to aid them with any difficulties they experience. What we find in this case is a modern society turning away from one of the most basic human functions and then having to relearn the advantages of this bodily function and how to practice it.

The general recognition of the harmful effects that were generated by the adoption of bottle-feeding by less developed nations led the World Health Organization in 1981 to adopt, by a vote of 118 to 1 (only the United States voted "no"), a nonbinding code restricting the promotion of infant formula.[61]

The Green Revolution

The bringing of high agricultural technology to the developing world has been called the Green Revolution. The Green Revolution has two basic components: the use of new seeds, especially for wheat, rice, and corn; and the use of various "inputs," such as fertilizer, irrigation, and pesticides. The new seeds, which were developed by over 20 years of cross-pollinating, are highly responsive to fertilizer. If they receive sufficient fertilizer and water, and if pests are kept under control, the seeds produce high yields. The introduction of this new agricultural technology to the less developed nations in the mid-1960s brought greatly increased harvests of wheat and impressive increases in rice production in a number of Asian countries. Over a six-year period, India doubled its wheat production and Pakistan did nearly as well. Significant increases of rice production occurred in the Philippines, Sri Lanka, Indonesia, and Malaysia. Mexico's wheat and corn production tripled in only two decades. From 1950 to 1990 grain harvests around the world nearly tripled. Not only were the harvests much larger, but multiple harvests – in some places up to three – became possible in a year because of the faster maturing of the plants.

Increased food production led to lower food prices globally.[62] These lower prices enabled many people in less developed countries to increase their calorie intake, thus leading to better health and longer life expectancy.

Some negative aspects of the new technology have become apparent. The new highly inbred seeds are often less resistant to diseases than are some of the traditional seeds. Also, the planting of only one variety of a plant – called monoculture – creates an ideal condition for the rapid spreading of disease and for the rapid multiplying of insects that feed on that plant. (The Irish potato blight in the mid-nineteenth century and the US corn blight of 1970 are examples of serious diseases that have attacked monocultures.) The new seeds are also less tolerant of too little or too much water; thus droughts and floods have a more harmful impact on these plants than on the traditional varieties of the grains.

One other negative impact of the new technology was that some farmers and agricultural workers were hurt by it. An evaluation of 40 years of the Green Revolution states: "Those who did not receive the productivity gains of the Green Revolution (largely because they were located in less favorable agro-ecological zones), but who nonetheless experienced price declines, have suffered actual losses of income."[63]

Probably the most serious potential negative aspect of the Green Revolution technology is the question about its sustainability. Critics have raised this question because of its tendency to increase chemical pollution, deplete aquifers, and lead to soil degradation.

Fertilizers

Synthetic fertilizers are usually needed with the new seeds. Fertilizer use has grown dramatically around the globe since 1970, especially in Asia and particularly in China. There is now evidence that the runoff of fertilizers from farmland is a significant source of pollution in rivers and lakes. Excessive nitrogen in the fertilizer and from other sources is overwhelming the natural nitrogen cycle with a variety of ill effects such as a decrease in soil fertility and toxic algae blooms. Excessive nitrogen in lakes greatly stimulates the growth of algae and other aquatic plants, which, when they die and decay, rob the water of its dissolved oxygen, leading to the suffocation of many aquatic organisms. Partially enclosed seas such as the Baltic Sea, the Black Sea, and even the Mediterranean are especially vulnerable to this excessive fertilization – called eutrophication. A large "dead zone" – about the size of the state of New Jersey – of diminished productivity has developed at the mouth of the Mississippi River in the Gulf of Mexico because of the excessive amount of nitrogen from agricultural runoff. A number of measures can drastically reduce this runoff. These include better timing of fertilizer applications, more exact calculation of the amount of fertilizer the crops can absorb, and more accurate delivery of the fertilizer.

Pesticides

There was a large increase in the use of pesticides (insecticides, herbicides, and fungicides) around the world in the 1970s, 1980s, and 1990s, no doubt also connected with the spreading Green Revolution. It is difficult to know how many people are being harmed by pesticides, but it is believed that the number is significant, especially in developing countries. One estimate by the World Health Organization is that perhaps as many as 20,000 deaths occur annually around the world because of pesticide poisoning and 1 million people are made ill.[64] (We will look further at pesticides in chapter 5.)

Irrigation

The Green Revolution also often requires irrigation. The use of fresh water, much of it for irrigation, has increased steadily since the 1960s. Much of this is taking place in the developing nations and in many countries, for example in Africa and the Middle East – according to the World Resources Institute – "water withdrawals appear to be occurring at unsustainable rates."[65] At the height of the Green Revolution in the 1970s irrigated land on the planet increased about 2 percent annually. Since that time irrigation has been growing about 1 percent a year. The growth has slowed partly because of the high costs of installing irrigation and the competition for fresh water. The UN's Food and Agricultural Organization predicts that the 1 percent growth will continue at least until 2010. Egypt, Mexico, and Turkey are planning to greatly expand their irrigation systems. Besides the problems of waterlogging and salinization, which were mentioned earlier, increased use of irrigation can also lead to an increase in infectious diseases such as malaria and schistosomiasis.

Improvements are being made in irrigation. Less water can be used when new irrigation technology is used, such as highly efficient sprinklers and drip feeding. By using these methods, Israel reduced the water it uses for irrigation by about 35 percent between 1951 and 1990 with no loss of productivity.[66]

The future

There is evidence that the positive effects of the Green Revolution are weakening, as the rate of increase in crop yields of those cereals most affected by the Green Revolution has started to slow down. From 1967 to 1982 farmers around the world increased grain yields by 2.2 percent annually. From 1982 to 1994 this increase dropped to 1.5 percent annually.[67] The total increase in food production for all developing countries dropped from 3.2 percent during 1961 to 1980 (called by some the Early Green Revolution) to 2.2 percent from 1981 to 2000 (the Late Green Revolution). Yields dropped from 2.5 percent to 1.8 percent during the same periods.[68] Many plant scientists now believe that they are facing physical limits as they try to get plants to produce more of their weight in grain. They also believe that the increasing scarcity of fresh water, deteriorating soil quality, and limited amounts of unplanted arable land are causing this slowing.[69] Worldwide public funds devoted to agricultural research are decreasing and not many funds are devoted to solving the food problems of the poorest.

Yet there are some positive developments. The Green Revolution technology has yet to be extensively applied to Africa. While Africa does have special problems that are difficult to overcome, such as limited water, poor soils, serious erosion, and much political conflict, there is evidence that if the government is supportive, the technology can be applied. Partly because of the urging by former president of the United States Jimmy Carter that the new president of Ethiopia view a demonstration site using Green Revolution technology, the new government of that country reversed the policy of the former military government, which had focused on heavy industry, and began to focus on agriculture. The government began to lend money to farmers for improved seeds and fertilizer. Grain production in Ethiopia nearly doubled in just two years. Ten years after a major famine, the country began to export grain.[70]

About one-half of the world's people eat rice daily. Chinese rice breeders claim that they are on the verge of producing a super high-yield hybrid that could increase rice yields by 15 to 20 percent. It is not known whether they will be successful or whether the rice produced from the hybrid will be accepted in other Asian countries since its taste may be considered to be inferior to present rice varieties.[71]

Certainly, without the increased production that came with the Green Revolution many developing countries would have already lost the battle to have enough food available for their rapidly growing populations. Dr Norman Borlaug – a US scientist who received the Nobel Peace Prize for his work in developing high-yield wheat, and the person considered to be the "father" of the Green Revolution – has stated the Green Revolution was not meant to be the final solution for the world's food problem: it was designed to give nations a breathing space of 20 or 30 years during which time they could work to bring their population growth under control. Borlaug is as disappointed as many others are that this time has not been used by many nations to take forceful measures to rein in their exploding numbers.

Governmental Food Policies

The availability of food is such a basic need that no government that I know of adopts a "hands off" policy regarding its production, price, and distribution. But many developing nations have given a relatively low priority to agricultural development and to relieving poverty in the rural areas, concentrating on industrial development instead of rural development. Nearly all of the developing nations have scarce public funds, so decisions must be made about where to apply them. It should not be surprising to students of government that public funds usually go to benefit groups with political visibility and power. Political leaders want to stay in power, and it is often the traditional political and economic elites who will influence the leaders' length of stay rather than the scattered and weak – both physically and politically – small farmers and rural poor. In many developing nations the urban masses, who can riot, are much more of a threat to the leaders than the small farmers, and urban people demand plentiful and inexpensive food.

The desire to retain power, of course, is not the only reason why rural development has not been given a high priority in many less developed nations. The desire to achieve the high living standards of the West by following the route taken by the United States and other developed nations – both capitalist and communist in the past – with their emphasis on industrialization was hard to resist; it seemed like a relatively fast way to reduce poverty. US foreign aid in the 1950s and 1960s certainly encouraged developing nations along this route. We who were in the foreign aid program then recognized that this development strategy was a gamble, that maybe benefits would not trickle down to the poor, but the other alternative of trying to work directly with the millions of rural poor did not seem viable. Barbara Ward shows how dominant this strategy of emphasizing industrialization over rural development became: "So far, on average, only 20 percent of the investment of most developing nations has gone to the 70 to 80 percent of the people who are in the rural areas."[72]

How does one respond to the argument that, given limited public funds, it is impossible to give any significant aid to the millions in the rural areas where most of the hunger exists? The response is that there have been a few Asian countries – namely Japan, South Korea, and Taiwan – that have brought significant prosperity to their rural areas by doing certain things. First, they enacted land reform measures – in Japan's case under the US occupation force's direction after World War II – which ended absentee landlordism and exploitative tenancy arrangements. The land was basically turned over to those who farmed it. Second, cooperatives were established to help small farmers purchase needed inputs and market their harvests. The governments also provided information and aid to the farmers through an active agricultural extension service and by supporting agricultural research. Japanese small farmers now have some of the highest yields per acre in the world, and the mechanization they have used on their farms – mainly small machines – has tended to increase rural employment, not decrease it. Double and even triple harvests per year on the same piece of land became possible, and more laborers were required to handle these harvests.

China under Mao Zedong emphasized agriculture instead of industrialization after the disastrous "Great Leap Forward" (a crash program of economic development in the late 1950s). China, with only 7 percent of the world's arable land and about 1.3 billion people, has achieved impressive increases in its agricultural production, but because of its rapid population growth the increased food has mainly gone to feed the increased population. Hunger is certainly less of a problem in China today than it was before the communist takeover – except during the famine in the late 1950s and early 1960s – but the costs have been high. Political opponents have been dealt with harshly and significant damage to the environment came from the efforts to increase the amount of agricultural land. Forests were cut down and marginal pasture land was converted to land for crops. Even though the communist government also made efforts to protect the environment, its actions directed toward increasing agricultural production led to an increased strain on the land. Significant losses of arable land are occurring because of the expansion of cities and industries, soil erosion, desertification, and deforestation. China's goal is to grow 95 percent of its needs of rice, wheat and corn. In the past decade it has had periods of oversupply and decline. Experts are divided on whether China will be able to feed itself in the future without importing large amounts of food, which would affect the world food market.[73]

Another major communist government – the Soviet Union – pursued radically different policies from China. Under Stalin's long rule, the country placed industrialization first, and agriculture was used to support that industrialization. Also, the desire to remove the political opponents of the ruling communists – the prosperous small farmers known as the "kulaks" – and the desire to substitute state-owned and collective farms for privately owned farms, led to what is commonly recognized as the destruction of efficient agriculture in that country. The Soviet Union's inability to grow enough food to feed its people caused it to import large amounts of wheat from the United States and other capitalist countries.

There is space in this chapter to sketch US food policies only briefly. The main point that should be made is that the US government is very active in this area. Up to the 1900s the government's policy was mainly to encourage farm production, but since the 1950s the policy has been directed mainly at coping with an excess of production. The basic policy has been to prop up low farm incomes by using price supports, by purchasing surpluses, and by paying the farmers to grow less food. During the 1950s and 1960s, the policy of the US (and Canadian) governments was to buy up farm surpluses, a process that led to huge public reserves. Food from this reserve often went to poor nations under the Public Law 480 program, whereby surplus food was given or sold to developing nations. World food prices were generally stable during this period since, during bad harvest years, food from the public reserve was released. Now it is no longer the policy of either the United States or Canada to encourage large public food reserves, which means that reserves can no longer can act as a cushion during periods of poor harvests. More recently, the US government has encouraged and supported the export of US farm products to other nations. The United States has become the world's leading exporter of food. The government supports this because exports help correct the large trade deficits that the country often experiences. Subsidy payments in the US were designed also to protect the small family farm by boosting low agricultural prices, but the largest farms have benefited the most with the top 10 percent of agricultural producers receiving 60 percent of the subsidies.[74]

Developing nations have complained that large subsidy payments by the US government and the European Union to their farmers make it difficult for farmers in the poorer nations to compete with the Western farmers. Many less developed nations depend on agricultural exports to earn needed foreign exchange.

Future Food Supplies

How much food can be grown in the world? How many can be fed? Like most of the questions raised in this book, there are no simple answers. Also, it is not hard to find experts who give very different answers to these questions. In this final section we will look at seven topics which are directly related to these important questions: the effect of climate, the amount of arable land, energy costs, alternative/sustainable agriculture, biotechnology, fishing and aquaculture, and, finally, expected future food production.

Climate

Experts are in general agreement that the earth is probably going to have a warmer climate in the future. It is very difficult to predict how this will affect the world's agriculture. It could make conditions worse for the growing of food in some countries and better in others. (This subject will be discussed more fully in the section on climate change in chapter 4.) The experts are also in general agreement that there will probably be more variability in the climate than there has been in the recent past. The climate over the past several decades in the United States and Canada has been unusually good for agriculture, but a good climate cannot be taken for granted. In fact, variability is the hallmark of the earth's climate when it is examined over long periods; one sees long-term cycles of hundreds of years and shorter cycles of 15 to 20 years. A greater variability of climate (higher and lower extremes of temperature and higher and lower amounts of rainfall) will probably lower agricultural production around the world because of the large amount of marginal land which is now being used for agriculture. On this land, such as parts of the American West, the Canadian west, and the Russian east, a slight reduction in rainfall or a slightly shorter growing season can spell the difference between a good harvest and little or no harvest.

A warmer world is apt to have less organic material in its soils as vital nutrients decompose. One study in 2003 indicated that yields could drop by about 17 percent for each degree the growing season warms.[75] In the early years of the twenty-first century there was still much uncertainty regarding this subject.

Arable land

About one-quarter of the earth's land free of ice can be cultivated, the arable land, and experts estimates that about one-half of the arable land is presently being used for agriculture. Large amounts of potential farmland exist in Latin America and Africa. The Brazilian *cerrado* and the grasslands of sub-Saharan Africa have the largest reserves of arable land. Yet most of the good farmland in the world is already being used. Much of the remaining potential arable land is far from population centers and a lot of it is marginal land, which is costly to bring into production and to maintain. Large amounts of energy would be needed to develop it – to build roads to it and to transport its products to market, to irrigate it, and to fertilize it.

Because of these problems, plus the social and political obstacles which must be overcome to develop such areas, it is difficult to estimate the potential

Much of the food in Africa is grown and prepared by women (*World Bank*)

for increasing the amount of farmland. These estimates also must take into consideration the large amount of present farmland which is being lost to agriculture through urbanization, through erosion caused by the cutting down of forests and overcropping, through the spreading of desert-like conditions (desertification) because of overgrazing and farming on the edge of deserts, and through the loss of irrigated lands (salinization and waterlogging) because of poor drainage. The earth's growing population and the type of diet its people choose will also greatly affect the amount of land needed to feed them.

A respected Canadian geographer, Vaclav Smil, writing at the beginning of the twenty-first century, believes that globally the amount of arable land will be sufficient to provide decent nutrition for at least the next two generations. Yet he does not believe there will be enough arable land for everyone. He sees the near future as follows:

> Undoubtedly, the total area of potential farmland is quite large, but its . . . distribution is highly uneven and its initial quality will be generally inferior to the existing

> cropland. . . . [T]he affluent countries should not experience any weakening of their food production capacity because of the declining availability of farmland [but] . . . low income societies tell a different story. Combination of continuing population growth and uneven . . . distribution of potentially available farmland will only increase substantial differences in per capita availability of arable land in those countries. . . . [P]er capita land availability remains high in Latin America, and more than adequate in sub-Saharan Africa. The greatest concerns exist, and will intensify, in the Middle East and in South and East Asia.[76]

Energy costs

The dramatic increase in energy costs in the 1970s had a profound influence on agriculture, and expected rising energy costs in the future will strongly affect food production and the cost of food. As we have seen, modern, Western agriculture is energy intensive, and the spreading of that type of agriculture to the developing world via the Green Revolution also entailed a commitment to using large amounts of energy. In the past, a doubling of agricultural output required a tenfold increase in the amount of energy used.[77] Some people hope for a breakthrough in nuclear fusion research which could lead to vast amounts of electrical energy becoming available; fossil fuels could then be designated for use in agriculture. Others hope that the South will somehow develop an agriculture that does not depend on the high use of energy and energy-related inputs which is common in the developed countries. To do this, it would also have to reject the Western diet as an ideal to strive for.

Alternative/sustainable/organic agriculture

One way to produce food without contaminating the water and air and decreasing the natural fertility of the soil is called alternative, sustainable, or organic agriculture. This resource-conserving and environmentally benign agriculture utilizes a number of old, proven techniques and a new understanding of natural nutrient cycles and ecological relationships. According to the World Resources Institute, it includes "practices such as crop rotation, reduced tillage or no-till, mechanical/biological weed control, integration of livestock with crops, reduced use or no use of chemical fertilizers and pesticides, integrated pest management, and provision of nutrients from various organic sources (animal manures, legumes)."[78] A demand by consumers for foods that were free of possible contamination by chemicals led to a large increase in organic farming in the United States in the 1990s. By 2004 organic foods (which are grown by using many alternative agriculture techniques) had become a $12 billion business in the United States, increasing at an average rate of 20 percent annually. Organic farming is supported with governmental subsidies in Europe, but not so in the US. In 2002, for the first time, the US government began certifying which foods were truly organic.

Whether organic farming ever becomes widespread in the $500 billion food industry in the US is unknown. In 2004 organic food sales in the US were only 2 percent of all food sales. The main criticism of organic farming is that it is less productive than conventional farming, and thus the prices of its foods are relatively high. There is still some debate on this question. A Swiss study published in 2002 comparing organic and conventional farms – the most comprehensive study of its time – found that organic farming leaves the soils healthier and is more energy efficient, but average crop yields are about 20 percent lower.[79] A long-term US study found opposite results with yields about equal in corn and soybean production using organic and conventional methods, and in drought years organic corn yields were significantly higher than conventional farm yields.[80]

Impressive evidence of the worth of alternative agriculture techniques came in China in the summer of 2000. At that time the results of one of the largest agricultural experiments ever undertaken were announced. Under the direction of an international team of scientists tens of thousands of rice farmers in one province participated in a simple experiment that didn't cost them any money, didn't involve the use of any chemicals, and that resulted in them gaining a nearly 20 percent increase in their yields of rice. What the farmers did was plant two varieties of rice in their fields rather than just one. The result of changing from a monoculture to using diversity was to nearly wipe out the most devastating disease that affects rice. The disease destroys millions of tons of rice each year, causing farmers losses of several billion dollars. Scientists involved in the study believe the startling results will have application beyond rice. An ecologist at the University of Washington stated that "what's really neat about this paper [that announced the results of the experiment] is that it shows how we've lost sight of the fact that there are some really simple things we can do in the field to manage crops."[81]

Biotechnology

Biotechnology has been called a technology that will transform modern agriculture. Genetic engineering, the transferring of desirable genes, or traits, from one organism to another, is the best-known part of this technology. New animals and plants are being created today with this technology. Plant and animal species have changed naturally throughout the evolution of life on this planet and human beings have, for thousands of years, influenced that evolution by encouraging the growth of those plants and animals which have traits that benefit humans. But now, as one scientist has stated, "we can do all at once what evolution has taken millions of years to do."[82]

Biotechnology is still controversial. Its defenders point out that food crops can be developed that are resistant to insects and viruses, thus reducing the need for pesticides. On the other hand, plants can be developed that can tolerate herbicides, thus allowing herbicides, which would normally harm the plant, to be used to control the weeds threatening the plant. Fruits can be developed that are resistant to spoilage. A tomato has been developed in the United States that has a natural resistance to becoming overripe, which means the tomato does not have to be picked while it is still green and relatively tasteless. Plants that are more nutritious are being developed such as a new variety of rice that will contain provitamin A, an essential nutrient that is missing in present rice. In Southeast Asia 70 percent of the children under the age of five suffer from a vitamin A deficiency, which leads to vision impairment and increased susceptibility to disease.[83] Plants that can grow under harsh conditions – for example, during droughts, or in salty soils, or in temperature extremes of heat and cold – are also being developed. With this technology, animals, such as pigs, can be developed to have more lean meat, and dairy cows can be developed to produce more milk. In the United States about 35 percent of the corn, 45 percent of the cotton, and 55 percent of soybeans are from seeds which have been genetically modified. It has been estimated that about two-thirds of the foods on supermarket shelves in the US contain genetically altered ingredients.[84] By 2002 genetically altered seeds were used widely in the US, Argentina, and Canada.[85]

In 2002 a team of Chinese government scientists and scientists from a private Swiss biotech company jointly announced, and made public, a draft of the genetic code of two common varieties of rice, one that is commonly eaten in Japan and one widely eaten in China and India. Dr Ronald Cantrell, director of the International Rice Research Institute in the Philippines, said the decoding of the rice genome would "have a tremendous impact on the poor" by enabling researchers to improve the nutrient value and growing characteristics of rice.[86] Rice is the most important food for about 3 billion of the world's people, including many of its poorest. In 2005 the final, accurate version of the genetic makeup of rice was published by an international team from 11 nations, led by Japanese scientists. Rice has been called the Rosetta stone of the cereals, since much of its genetic makeup is a part of the other main cereals such as corn, wheat, and barley, and can be used to study those plants.

The critics of this new technology, who tend to be more numerous in Europe than in the United States, claim that there is a possibility that genetic engineering will alter organisms in detrimental ways that will not be fully known for years. Herbicide-resistant crops might pollinate closely related plants that are now weeds, thus creating a new weed that is also resistant to herbicides. Much negative publicity for bioengineering was generated in the United States in 1999 when a study showed that the pollen from corn that had been genetically altered to produce a natural pesticide can kill caterpillars of the monarch butterfly. Since most of the research today in biotechnology is being performed by private corporations that see it as a way to increase their profits, it is not surprising that most of the present genetic engineering concerns crops and animals that can be profitably sold in the rich nations, not in the poor nations. The critics point to several large corporations that produce herbicides and other farm chemicals as being leaders in efforts to develop herbicide-resistant crops. Instead of encouraging the development of less reliance on chemicals in the growing of foods, this research will increase such reliance.

The European Union requires labels on food identifying it as genetically altered if 1 percent or more of its ingredients have been genetically altered. A crack developed in European opposition in 2004 when the European Union ended its six-year moratorium on the approval of biotech foods.

A report by the US National Academy of Sciences in 2002 called for a more rigorous approval process by the government of biotech foods, although it had already given in 2000 cautious approval of the safety of genetically altered foods then on the market. With biotech companies developing plants with either a combination of genes or with an individual gene which enables the plant to produce pharmaceutical or industrial chemicals, stronger regulations are needed. The Academy warned that genetically altered crops have the potential to pose food safety risks and environmental harm. In a 2004 report the Academy stated it will be difficult to contain all altered genes in plants and animals or prevent any of them from having unintended environmental and public health effects.[87]

Like many technologies, biotechnology appears to have a positive and a negative potential. It is impossible to predict at this point which potential will dominate. Being aware of the negative possibilities and taking steps to counter them may be the best we can do at this time. Government regulations need to be regularly updated to reflect the latest research and the plants and animals need to be monitored while being grown. (The negative side of technology will be discussed further in chapter 6.) Biotechnology could lead to major advances in agriculture in the poorer nations. Some universities, such as the University of Ghent in Belgium, private foundations, such as the Rockefeller Foundation in the United States, and governmental agencies, such as the Swiss Federal Institute of Technology and the European Community Biotech Program, are supporting research in biotechnology that is directed toward that purpose.

Fishing and aquaculture

Not too long ago many people hoped that the world food problem would be solved by harvesting fish from the oceans, but it is now generally recognized that, as one marine biologist has put it, most of the ocean is a biological desert. Nearly all the fish in the world are harvested in coastal waters and in a relatively few places further from land where there is a strong up-welling of water that brings nutrients to the surface.

Over the past 50 years there has been increasing pressure on the world's fish. About two-thirds of the world's major varieties of fish are now fished at or above their capacity to renew themselves, and another 10 percent have been fished so heavily that it would take many years for their numbers to recover. Marine biologists estimate that in the past half century about 90 percent of the large ocean predators such as sharks, tuna, marlin, swordfish, cod, halibut, skates, and flounder have been caught.[88]

According to the World Resources Institute, "substantial potential exists for increasing the ocean fish harvests with better management of fish stocks, although sound management is neither easy nor obvious."[89] The Institute cites the examples of Cyprus and the Philippines where better management of fishing in their waters led to substantial increases in fish harvests in as little as 18 months. The Institute also reported that Canada, the European Union, and the United States had recently adopted tougher controls over ocean fishing and reduced the size of their fishing fleets.

One type of fishing that does hold promise for an increase in catch is aquaculture, the farming of fish inland and in coastal waters. By 2005 about 30 percent of the fish eaten in the world came from fish farms.[90] Nearly nonexistent in the United States a generation ago, aquaculture had developed into a $900 million industry by the end of the twentieth century. Worldwide, aquaculture was a $45 billion industry. More than half of the salmon eaten in the United States, about one-third of the shrimp, most of the clams and oysters, and nearly all the trout and catfish come from fish farms.

Fish farming is popular in developing countries, especially in Asia, which is the home for about 80 percent of the industry. China has about two-thirds of worldwide production.[91] Aquaculture was developed in China several thousand years ago. It is now becoming more popular in the developed nations because people there – partly for health reasons (because fish are low in fat, and fish oil is reported to have beneficial properties) – are consuming more fish and demanding that the fish they buy come from nonpolluted waters. Genetic engineering is also being used to create new species of fish. Here is the way one newspaper described the new techniques being used in aquaculture in the United States:

> Scientists are growing fish twice as fast as they grow naturally, cutting their feed requirement by nearly half, and raising them on a diet of ground chicken feathers and soybeans. Fish are now vaccinated against disease, sterilized so that their energy is spent growing not reproducing, and given hormones to turn females into males and males into females, changes that can be used to improve growth, taste and control of selective breeding.[92]

There are environmental concerns with this rapidly growing industry. Thailand has cleared a large part of its mangrove swamps to make way for shrimp ponds, thus losing a critical habitat for many aquatic species and opening its coasts to erosion and flooding. A 20-acre salmon operation can produce as much organic waste as a city of 10,000 people. There is a fear that fish that escape from the farms can mate with their wild relatives and harm the natural gene pool. It has been estimated that in 1995 between 200,000 to 650,000 salmon escaped from fish farms in Norway, where half the world's salmon production takes place.[93] On the east and west coasts of the US from the mid-1980s to 2000, over 500,000 salmon were estimated to have escaped from their pens.[94]

Future food production

Will the world be able to produce enough food for its rapidly expanding population? This is a hard question to answer. Many experts failed to predict the progress that has been made in food production in the past few decades, so it would be easy to discount the warnings by some of them now. Yet some disturbing signs exist. A statement by the World Resources Institute at the end of the twentieth century seeks to achieve a balance between opposing indicators:

> Prospects for global food and agriculture are at once promising and troubling. On the one hand, global food production has increased since 1970 and has generally been able to meet the demands of a growing world population. . . . It is unclear whether production increases can continue indefinitely. Some factors augur well for global production – for example, improvements in the emerging market economies of Central Europe and possibly a multilateral agreement to liberalize agricultural trade. . . . Better control of diseases (human and animal) could also open up large areas of potentially productive farming and grazing land in Africa.

On the other hand, most agricultural production in the world uses farming practices that are environmentally unsustainable. New efforts are underway in the industrialized countries to encourage more sustainable practices, but these efforts are as yet quite modest. . . . In developing countries, however, population growth and poverty subvert efforts to introduce sustainable practices and encourage agriculture to expand in ways detrimental to the environment. Population growth causes marginal land to be cultivated and contributes to environmental problems such as soil erosion and deforestation . . . it is far from certain that farmers will be able to adopt sustainable practices and still grow enough food to feed a projected world population of 10 billion or more people in the next century.[95]

Conclusions

One of the most fundamental problems many less developed nations face is how to end hunger in their lands. The rapid growth of their populations and the past neglect of agricultural development have resulted in increased suffering in rural areas. Advances in technology have helped to keep the overall production of food in many poor countries ahead of their increased needs, but widespread poverty in the rural districts as well as in some urban areas has meant that many people cannot afford to purchase the food that is available in the market. An emphasis on agricultural development and on increasing employment in both rural and urban areas is needed in order to provide increased income to larger numbers of the poor.

The developed nations face major food problems also. Here the problems are quite different from those faced by the developing nations. The rich nations need to learn how to produce healthful food and to retain a prosperous agricultural sector. There are indications that among some people in the richer nations a new concern does exist with the types of food people eat. Whether this desire for more healthful foods and the awareness of the connection between food and health will spread from a minority to the majority of the people is not yet clear. It is clear, though, that in economic systems where consumers can freely exercise their preferences, the potential exists for important changes to occur fairly rapidly. For example, in the United States the relatively recent awareness of the connection between fatty foods and heart attacks has led to the production of a wide variety of low fat foods.

The picture regarding the health of the farm economy in some developed countries does not look bright. The United States has not yet learned how to maintain a sustainable, prosperous agricultural sector. Its productive capabilities are impressive but, as this chapter has pointed out, its high dependency on uncertain, polluting, and potentially very costly energy supplies, and its tendency to undermine the land upon which it rests make its future uncertain.

Notes

1 UN Development Programme, UN Environment Programme, World Bank, and World Resources Institute, *A Guide to World Resources 2000–2001* (Washington, DC: World Resources Institute, 2000), p. 10.
2 UN Development Programme, UN Environment Programme, World Bank, and World Resources Institute, *World Resources 2005* (Washington, DC: World Resources Institute, 2005), p. 221, and William Bender and Margaret Smith, "Population, Food, and Nutrition," *Population Bulletin*, 51 (Population Reference Bureau, Washington, DC) (February 1997), p. 18.
3 Brian Halweil, "Grain Harvest and Hunger Both Grow," in *Vital Signs 2005* (New York: Worldwatch Institute, 2005), p. 22.
4 Ibid.
5 "Hunger" and "undernourishment" refer to the consumption of insufficient calories, whereas "malnutrition" refers to the lack of some necessary nutrients, usually protein. For the sake of simplicity, I am equating hunger with undernourishment and malnutrition.
6 As cited in World Resources Institute, UN Environment Programme, UN Development Programme, and World Bank, *World Resources 1998–1999* (New York: Oxford University Press, 1998), p. 154.
7 UN Food and Agriculture Organization (FAO), *The State of Food and Agriculture 2005: World and Regional Review*, at www.fao.org, pp. 117–18.
8 According to the World Bank as reported in Celia Dugger, "Report Warns Malnutrition Begins in Cradle," *New York Times* (March 3, 2006), p. A6.
9 Pedro Sanchez and M. S. Swaminathan, "Cutting World Hunger in Half," *Science*, 307 (January 21, 2005), p. 357.
10 UN Development Programme, *Human Development Report 2003* (New York: UN Development Programme, 2003), p. 6.
11 Sanchez and Swaminathan, "Cutting World Hunger in Half," p. 357.
12 UN Development Programme, *Human Development Report 2004* (New York: UN Development Programme, 2004), p. 130.
13 Roy L. Prosterman, *The Decline in Hunger-Related Deaths*, Hunger Project Papers, no. 1 (San Francisco: Hunger Project, 1984), p. ii.
14 Marc Cohen, "Crop Circles", *Natural History*, 112 (October 2003), p. 64.
15 Halweil, "Grain Harvest and Hunger Both Grow," p. 22.
16 John R. Tarrant, *Food Policies* (New York: John Wiley, 1980), p. 12.
17 Sanchez and Swaminathan, "Cutting World Hunger in Half," p. 357.
18 Dugger, "Report Warns Malnutrition Begins in Cradle."
19 Robert Conquest, *The Harvest of Sorrow: Soviet Collectivization and the Terror-Famine* (New York: Oxford University Press, 1986).
20 Jasper Becker, *Hungry Ghosts: Mao's Secret Famine* (New York: Free Press, 1997), and Penny Kane, *Famine in China, 1959–1961: Demographic and Social Implications* (New York: St Martin's Press, 1988).
21 *New York Times*, late city edn (June 7, 1983), p. 1; John Mellor and Sarah Gavian, "Famine: Causes, Prevention, and Relief," *Science*, 235 (January 1987), p. 539.
22 For a fuller discussion of the causes of the African famines see Carl K. Eicher, "Facing Up to Africa's Food Crisis," *Foreign Affairs*, 61 (Fall 1982), pp. 151–74; a series of articles on Africa in the *Bulletin of the Atomic Scientists*, 41 (September 1985), pp. 21–52; and Michael H. Glantz, "Drought in Africa," *Scientific American*, 256 (June 1987), pp. 34–40.
23 Kavita Pillay, "The Politics of Famine in North Korea," *Ninth Annual Report on the State of World Hunger 1999* (Silver Springs, MD: Bread for the World Institute, 1998), p. 24; and James Brooke, "Food Emergency in North Korea Worsens as Donations Dwindle," *New York Times*, national edn (December 5, 2002), p. A16, and "North Korea, Facing Food Shortages, Mobilizes Millions from the Cities to Help Rice Farmers," *New York Times*, national edn (June 1, 2005), p. A6.
24 Sanchez and Swaminathan, "Cutting World Hunger in Half," p. 357.
25 World Resources Institute et al., *World Resources 1998–1999*, p. 154.
26 Presidential Commission on World Hunger, *Overcoming World Hunger: The Challenge Ahead* (Washington, DC: US Government Printing Office, 1980), p. 16.
27 Paul R. Ehrlich, Anne H. Ehrlich, and John P. Holdren, *Ecoscience: Population, Resources,*

Environment (San Francisco: W. H. Freeman, 1977), p. 303.

28 As quoted in "A Shift in the Wind," Hunger Project Papers No. 15, The Hunger Project, San Francisco, p. 4.

29 C. Peter Timmer, "Unbalanced Bounty from America's Farms," *Science*, 298 (November 15, 2002), p. 1339.

30 Raymond Hopkins, Robert Paarlberg, and Michael Wallerstein, *Food in the Global Arena* (New York: Holt, Rinehart & Winston, 1982), p. 102.

31 William Ophuls, *Ecology and the Politics of Scarcity* (San Francisco: W. H. Freeman, 1977), pp. 42–3. The energy expended in modern agriculture is mainly nonhuman energy, of course, and most people consider that to be one of modern agriculture's most attractive features.

32 Peter Farb and George Armelagos, *Consuming Passions: The Anthropology of Eating* (Boston: Houghton Mifflin, 1980), p. 69.

33 Ibid., pp. 69–70.

34 Barbara Ward, *Progress for a Small Planet* (New York: W. W. Norton, 1979), p. 92.

35 Timmer, "Unbalanced Bounty from America's Farms," p. 1339.

36 Larry Rohter, "South America Seeks to Fill the World's Table," *New York Times*, national edn (December 12, 2004), p. 1.

37 Jim Yardley, "China Races to Reverse Its Falling Production of Grain," *New York Times*, national edn (May 2, 2004), p. 6.

38 Bender and Smith, "Population, Food, and Nutrition," pp. 17–18.

39 World Resources Institute et al., *World Resources 1998–1999*, p. 156.

40 Jocelyn Kaiser, "Wounding Earth's Fragile Skin," *Science*, 304 (June 11, 2004), p. 1616.

41 Bill McKibben, "A Special Moment in History," *Atlantic Monthly*, 281 (May 1998), p. 60.

42 William Stevens, "Sprawl Quickens Its Attack on Forests," *New York Times*, national edn (December 7, 1999).

43 World Resources Institute, *World Resources 1992–1993* (New York: Oxford University Press, 1992), pp. 111–16.

44 World Resources Institute et al., *World Resources 1998–1999*, p. 157.

45 Ibid., p. 158.

46 Kaiser, "Wounding Earth's Fragile Skin," pp. 1616–17.

47 Stevens, "Sprawl Quickens Its Attack on Forests."

48 Kaiser, "Wounding Earth's Fragile Skin," p. 1618.

49 World Resources Institute et al., *World Resources 1998–1999*, p. 158.

50 Elizabeth Becker, "Two Acres of Farm Lost to Sprawl Each Minute, New Study Says", *New York Times*, national edn (October 4, 2002), p. A19.

51 Stevens, "Sprawl Quickens Its Attack on Forests."

52 The consumption of meat (and, also, milk from cows and goats) can make nutritional sense. Cows and sheep, for example, can consume grasses, which people are unable to digest, in places where the climate or the condition of the land makes the growing of crops impossible.

53 Ehrlich et al., *Ecoscience*, p. 315.

54 Bill Marsh, "The Evolving World Diet," *New York Times*, national edn (August 20, 2002), p. D4.

55 Ward, *Progress for a Small Planet*, pp. 93–4.

56 Erica Goode, "The Gorge-Yourself Environment," *New York Times*, national edn (July 22, 2003), pp. D1, D7, and Anahad O'Connor, "Study Details 30-Year Increase in Calorie Consumption," *New York Times*, national edn (February 6, 2004), p. A19.

57 Bottle-feeding became more common in the United States than breast-feeding sometime between 1940 and 1955. A lack of good data makes it difficult to pin this down any further.

58 Because the HIV virus that causes AIDS can be transmitted from the mother to the baby through breast milk, it is recommended that mothers infected with the virus bottle-feed their babies. For a discussion of the advantages of breast-feeding for the baby and the mother see Meredith F. Small, "Our Babies, Ourselves," *Natural History*, 106 (October 1997), pp. 42–51.

59 *New York Times*, national edn (December 17, 1982), p. 8.

60 Lizette Alvarez, "Norway Leads Industrial Nations Back to Breast-Feeding," *New York Times*, national edn (October 21, 2003).

61 For a discussion of the controversy over the use of infant formula see Stephen Solomon, "The Controversy over Infant Formula," *New York Times Magazine* (December 6, 1981), p. 100.

62 R. E. Evenson and D. Gollin, "Assessing the Impact of the Green Revolution, 1960 to 2000," *Science*, 300 (May 2, 2003), p. 761.

63 Ibid., p. 762.

64 World Health Organization (WHO), *Public Health Impact of Pesticides Used in Agriculture* (Geneva: WHO, 1990), p. 86.
65 World Resources Institute, *World Resources 1992–1993*, p. 97.
66 Bender and Smith, "Population, Food, and Nutrition," p. 33.
67 "Crop Scientists Seek a New Revolution," *Science*, 283 (January 15, 1999), p. 310.
68 Evenson and Gollin, "Assessing the Impact of the Green Revolution," p. 760. The contribution of one part of the Green Revolution, the new hybrid seeds, to the growth of the yield actually increased during this period.
69 "Reseeding the Green Revolution," *Science*, 277 (August 22, 1997), p. 1038.
70 Ibid., p. 1041.
71 "Crop Scientists Seek a New Revolution," p. 313.
72 Ward, Progress for a Small Planet, p. 178.
73 Nancy E. Riley, "China's Population: New Trends and Challenges," *Population Bulletin*, 59 (June 2004), pp. 25–6, and Yardley, "China Races to Reverse Its Falling Production of Grain," p. 6.
74 Elizabeth Becker, "Far from Dead, Subsidies Fuel Big Farms," *New York Times*, national edn (May 14, 2001), p. A12; see also Alexei Barrionuevo, "Mountains of Corn and a Sea of Farm Subsidies," *New York Times*, national edn (November 9, 2005), p. A1.
75 Erik Stokstad, "Study Shows Richer Harvests Owe Much to Climate," *Science*, 299 (February 14, 2003), p. 997, and Kaiser, "Wounding Earth's Fragile Skin," p. 1618.
76 Vaclav Smil, *Feeding the World: A Challenge for the Twenty-First Century* (Cambridge, MA: MIT Press, 2000), pp. 36–9.
77 Ophuls, *Ecology and the Politics of Scarcity*, p. 54.
78 World Resources Institute, *World Resources 1992–1993*, p. 100. See also John P. Reganold, Robert I. Papendick, and James F. Parr, "Sustainable Agriculture," *Scientific American*, 262 (June 1990), p. 112. A fuller description of alternative agriculture is contained in National Resource Council, *Alternative Agriculture* (Washington, DC: National Academy Press, 1989).
79 Erik Stokstad, "Organic Farms Reap Many Benefits," *Science*, 296 (May 31, 2002), p. 1589.
80 David Pimentel, "Changing Genes to Feed the World," *Science*, 306 (October 29, 2004), p. 815.
81 Carol Kaesuk Yoon, "Simple Method Found to Increase Crop Yields Vastly," *New York Times*, national edn (August 22, 2000), p. D1. See also Dennis Normile, "Variety Spices up Chinese Rice Yields," *Science*, 289 (August 18, 2000), pp. 1122–3.
82 *New York Times*, national edn (December 27, 1988), p. 17.
83 Mary Lou Guerinot, "The Green Revolution Strikes Gold," *Science*, 287 (January 14, 2000), pp. 241, 243.
84 Donald G. McNeil, "Protests on New Genes and Seeds Grow More Passionate in Europe," *New York Times*, national edn (March 14, 2000), p. A1.
85 David Barboza, "Development of Biotech Crops is Booming in Asia," *New York Times*, national edn (February 21, 2003), p. A3.
86 Nicholas Wade, "Experts Say They Have Key to Rice Genes," *New York Times*, national edn (April 5, 2002), p. A19. For more on the scientific significance of this event see Ronald Cantrell and Timothy Reeves, "The Cereal of the World's Poor Takes Center Stage," and Dennis Normile and Elizabeth Pennisi, "Rice: Boiled Down to Bare Essentials," both from *Science*, 296 (April 5, 2002), p. 53 and pp. 32–6.
87 See Carol K. Yoon and Melody Peterson, "Cautious Support on Biotech Foods by Science Panel," *New York Times*, national edn (April 6, 2000), p. A1; Andrew Revkin, "Panel Urges US to Tighten Approval of Gene-Altered Crops," *New York Times*, national edn (February 22, 2002), p. A18; and Andrew Pollack, "No Foolproof Way Is Seen to Contain Altered Genes," *New York Times*, national edn (January 21, 2004), p. A10.
88 Worldwatch Institute, *Vital Signs 2005* (New York: W. W. Norton, 2005), p. 26.
89 World Resources Institute et al., *World Resources 1998–1999*, p. 196.
90 FAO, *The State of Food and Agriculture 2005*, p. 134.
91 Ibid.
92 William Greer, "Public Taste and US Aid Spur Fish Farming," *New York Times*, national edn (October 29, 1986), p. 1.
93 Walter Gibbs, "Fish Farm Escapees Threaten Wild Salmon," *New York Times*, national edn (October 1, 1996), p. B7.
94 Erik Stokstad, "Engineered Fish: Friend or Foe of the Environment?" *Science*, 297 (September 13, 2002), p. 1797.
95 World Resources Institute, *World Resources 1992–1993*, p. 94.

Further Reading

Boucher, Douglas H. (ed.), *The Paradox of Plenty: Hunger in a Bountiful World* (Oakland, CA: Food First, Institute for Food and Development Policy, 1999). Excerpts from Food First's 24 years of research of the world's food system.

Brown, J. Larry and Ernesto Pollitt, "Malnutrition, Poverty and Intellectual Development," *Scientific American*, 274 (February 1996), pp. 38–43. Research into childhood nutrition reveals that poor diet influences mental development in more ways than expected.

Brown, Lester R., *Outgrowing the Earth: The Food Security Challenge in an Age of Falling Water Tables and Rising Temperatures* (New York: W. W. Norton and Earth Policy Institute, 2004). Brown believes that present policies and human demands are undermining the earth's capacity to provide enough food for an expanding population. He recommends a number of steps that could be taken to change this situation.

Gardner, Bruce L., *American Agriculture in the Twentieth Century: How It Flourished and What It Cost* (Cambridge, MA: Harvard University Press, 2002). This book has been called a definitive history. There is a tension in the book between the story of the amazing growth of productivity in American farms and the story of how that growth came about and who benefits from it.

Golkin, Arlene T., *Famine: A Heritage of Hunger* (Claremont, CA: Regina Books, 1987). Visual elements combine with text to examine the root causes of famine and to explore the devastating effect on famine's favorite victim – the world's poor.

The Hunger Project, *Ending Hunger: An Idea Whose Time Has Come* (New York: Praeger, 1985). A visual exploration and discussion of hunger, including population, education, economics, and government policy.

Jensen, Bernard, *Empty Harvest: Understanding the Link between Our Food, Our Immunity and Our Planet* (New York: Avery, 1990). An interesting approach to the topic of food discusses the issue from the perspective of world health, disease and future population.

Lappe, Anna, and Frances Moore, *Hope's Edge: The Next Diet for a Small Planet* (New York: Penguin Putnam, 2002). As in the best-selling book published 30 years ago – *Diet for a Small Planet* – the author and her daughter report on the declining status of diet and agriculture around the world and recommend ways people can improve their diets and protect the environment.

Mather, Robin, *A Garden of Unearthly Delights: Bioengineering and the Future of Food* (New York: Dutton, 1995). The book is about two different visions of agriculture. One is that of agribusiness with its emphasis on mass production, dependence on energy and chemicals, assembly-line processing, long-distance transportation, and profit. The other is that of small-scale organic farming with its labor-intensive methods, seasonal crops, local markets, and stewardship of the land.

Nestle, Marion, *Food Politics: How the Food Industry Influences Nutrition and Health* (Berkeley: University of California Press, 2002). Nestle, a professor in the Department of Nutrition and Food Science at New York University, seeks to show how the food industry, in order to increase its profits, works to encourage people to eat more, thus contributing to the obesity epidemic.

Newman, Jack, "How Breast Milk Protects Newborns," *Scientific American*, 273 (December 1995), pp. 76–9. The author shows how human milk helps infants avoid disease.

Postel, Sandra, "Growing More Food with Less Water," *Scientific American*, 284 (February 2001), pp. 46–51. With the population expanding and the climate getting warmer we will need to grow more food with less water. Postel shows how it can be done.

Pringle, Peter, *Food, Inc.: Mendel to Monsanto – the Promises and Perils of the Biotech Harvest* (New York: Simon & Schuster, 2003). A balanced account of a controversial subject, Pringle criticizes the extreme claims of both the biotech companies and the environmental critics.

Safina, Carl, *Song for the Blue Ocean: Encounters along the World's Coasts and beneath the Seas* (New York: Henry Holt, 1998). The author takes us on a tour to the most threatened fishing grounds, Congressional hearing rooms, expensive Asian restaurants, poor coastal villages, and inland spawning grounds. Although he ends the book on a cautiously optimistic note, this oceanographer warns of the dangers facing the world's oceans.

Smil, Vaclav, *Feeding the World: A Challenge for the Twenty-First Century* (Cambridge, MA: MIT Press, 2000). Can the world feed an expected 10 billion people in this century? After looking at many factors such as arable land, water, top soil loss, diets, and others, Smil believes it can, if efficiency improvements are made. He warns of regional shortages and the unexpected effects of climate change.

CHAPTER 4

Energy

A human being, a skyscraper, an automobile, and a blade of grass all represent energy that has been transformed from one state to another.

Jeremy Rifkin, *Entropy* (1980)

The Energy Crisis

Are we running out of energy? Of course not. Everything is made out of energy, and, as college students learn when they study the laws of thermodynamics in their introductory physics courses, energy cannot be destroyed. These laws also state that energy cannot be created: all we can do is to transform it from one state to another. And when energy is transformed, or in other words, when it is used for some work, the energy is changed from a more useful to a less useful form. All types of energy eventually end up as low grade heat. A "law" in the physical sciences means that there are no exceptions to it, and there are none to the laws of thermodynamics.[1]

So if everything is energy and energy cannot be destroyed, why is there an energy crisis? The crisis has come because of the other laws, the laws that tell us that energy cannot be created, and that, once used, it is transformed into a less useable form. At present, the industrialized world relies on a very versatile, although polluting, fuel – oil. Oil is being consumed at prodigious rates – about 1,000 barrels a second in 2005[2] – its supply is limited, and its price has fluctuated greatly. The developed nations are facing an energy crisis because the era of cheap, and supposedly clean, energy from reliable sources is over. Table 4.1 shows this fact as well as any set of figures can, as it focuses on the changes in the price of gasoline in the United States from 1950 to 2004. The table also helps us understand another important feature of the energy crisis, especially as it has affected the United

Table 4.1 US gasoline prices, 1950–2004

Year	*Retail price per gallon of regular gas ($)*
1950	0.27
1960	0.31
1970	0.36
1980	1.21
1990	1.16
2000	1.69
2004	2.07

Sources: "Dollars and Sense, July–August 1980," in Kenneth Dolbeare, *American Public Policy* (New York: McGraw-Hill, 1982), p. 113; *The World Almanac and Book of Facts 2006* (New York: World Almanac Books, 2006), p. 138

States. The period of cheap gasoline was a relatively long one, and people in the United States got used to having inexpensive petroleum products.[3] Unprecedented economic growth and material prosperity took place in the United States during the 1950s and 1960s, and this was made possible, in part, by cheap energy. Individual lifestyles and modes of industrial production were based on plentiful, inexpensive energy, and when oil prices skyrocketed in the 1970s, the shock to the US economy, and to the economies of many other countries, was profound.

The first oil shock took place in 1973–4. The 1973 Arab–Israeli war led a number of Arab oil-producing countries to stop shipping oil to the United States and other countries allied with Israel. American motorists lined up at gas stations, vying for limited supplies. The Organization of Petroleum Exporting Countries (OPEC), of which most oil-exporting nations are members, seized the opportunity to raise oil prices significantly: they quadrupled.

The second oil shock came in 1979–80. The event which prompted this shock was the Iranian Revolution and the ousting of the Shah as the head of the Iranian government. Iranian oil shipments to the United States stopped, but the real shock came when OPEC doubled its prices. Many North Americans had refused to believe there was a real energy crisis after the first oil shock and had returned to their normal high consumption of petroleum products after the Arab embargo was lifted; but the second oil shock convinced most people that there was indeed an energy crisis. While many had blamed either the US oil companies or the US government for creating the first oil crisis, the second shock clearly demonstrated that something had fundamentally changed in the world. What became apparent to many now was that the United States, and most other developed nations, were dependent on one section of the world for a significant part of their energy, and that they could no longer control events in that part of the world.

The third oil shock came in 1990–1. Iraq invaded Kuwait and threatened Saudi Arabia. In order to prevent Iraq from becoming the dominant power in the Middle East and having significant influence on the production and pricing of oil from that region, the United States led a coalition of forces in forcing Iraq out of Kuwait. The war, which lasted just six weeks, involved a half-million US soldiers and token troops from other nations. A huge, sustained air attack on Iraqi forces in Kuwait and Iraq and on military facilities in Iraq (including on plants for poison gas and nuclear weapons) preceded the ground attack. The allied forces had few casualties, but the retreating Iraqi forces, which suffered large casualties, sabotaged more than 700 oil wells in Kuwait, setting about 600 on fire.

The United States persuaded other Western nations, including Japan, to contribute about $50 billion to help pay for the war. The United States spent about $10 billion for short-term costs. The war and its subsequent damage to their lands and economies cost all the Arab states an estimated $600 billion.[4] The price of oil increased dramatically right after the Iraqi invasion of Kuwait, but by the end of the war the price had dropped back to the prewar level. That price did not reflect the real cost of oil, which should have included the cost of the war. (It has been estimated that by the mid-1980s the United States was spending seven times as much keeping the shipping lanes open to the Middle East oil fields as it was for the oil itself.)[5]

In 2003 the United States invaded Iraq and conquered the country after a relatively short war. Its main public reason for doing so was to destroy weapons of mass destruction which the United States claimed Iraq possessed. No weapons of mass destruction were ever found and the United States later stated that faulty intelligence led it to believe Iraq had such weapons. A few allies of the United States supported the invasion – mainly the United Kingdom – with relatively small military forces, but the invasion was opposed by most nations of the world. It created intense hostility toward America in the Arab world. Most nations believed the main reason for the war was to help the United States secure its sources of oil, although some believed that President George W. Bush had personal reasons for wanting to depose the Iraqi president Saddam Hussein. Three years after the invasion, US forces were still in Iraq battling a fierce insurgency with little prospect of it ending.

The Middle East, which supplies much of the oil imported into the United States and Western Europe, is a highly unstable area. It is torn by regional conflicts (the Arabs against Israel, Iran against Iraq, Syria against Iraq, Egypt against Libya); by religious conflicts (Muslim against Jew, Christian against Muslim, Shi'ite Muslim against Sunni Muslim, fundamentalist Muslims against secular governments); by social and ideological conflicts (traditionalists against radicals); and, in the past, by East–West competition (the United States against the Soviet Union). A large amount of the oil that is involved in international trade is carried on ships which must pass through a single strait in the Persian Gulf – the Strait of Hormuz.

The United States is the largest buyer of oil in the world, with much of it coming from a single country, Saudi Arabia. But many Western European countries are even more dependent on imported oil than is the United States, as is Japan, the industrialized country most dependent on imported oil, producing virtually no oil itself and having few other domestic sources of energy. China is importing more oil as its economy booms.

The large increases in the price of oil made by OPEC in the 1970s led to a massive transfer of wealth from the developed nations to part of the developing world. In the words of one commentator, "It may represent the quickest massive transfer of wealth among societies since the Spanish Conquistadores seized the Incan gold stores some four centuries ago."[6] Higher oil prices led to low economic growth, higher inflation, big trade deficits, and increased unemployment in the United States and other developed nations. Although developing nations use much less oil than do the developed nations, the cost of their imported oil also went up and caused some of them to acquire huge debts to pay for the oil they needed. Daniel Yergin, the coeditor of an important report on energy by Harvard Business School, assessed the potential consequences of the oil shocks in the following terms:

> The unhappy set of economic circumstances set in motion by the oil shocks contains the potential for far-reaching crises. In the industrial nations, high inflation, low growth, and high unemployment can erode the national consensus and undermine the stability and legitimacy of the political system. In the developing world, zero growth leads to misery and upheavals. Protectionism and accumulation of debt threaten the international trade and payments system. And, of course, there is the tinder of international politics, particularly involving the Middle East, where political and social upheavals can cause major oil disruptions and where fears about and threats to energy supplies can lead to war.[7]

Shortage of wood is a part of the energy crisis, since many urban dwellers in developing nations rely on wood as their major source of fuel (*Ab Abercrombie*)

I have focused in this section on the oil crisis, which has affected mainly the industrialized nations. But that is not the full story of the energy crisis that the world faces. Millions of people in developing nations use no fossil fuels at all, relying mainly on wood, charcoal, cow dung, and crop residues for cooking fuel and for heat. The UN estimates that while traditional fuels made up only about 5 percent of energy consumed worldwide in the mid-1990s, they were about 35 percent of the energy consumed in Africa, 20 percent in South America, and 10 percent in Asia. In some of the least developed nations, such as Laos, Nepal, and Uganda, these fuels provided about 85 percent of the energy.[8] The shortage of firewood in the South is increasing as population growth has caused consumption of wood to exceed the growth of new supplies in many areas. Forests are being cut down and are not being replanted. The dependency of the poor in the South on wood is in some ways like the dependency of the rich on oil. Both dependencies can be dangerous and their reduction will require forceful public and private measures.

Responses by Governments to the Energy Crisis

Let us look at a few key countries and regions to see how their governments have responded to the energy crisis.

The United States

The US response to the energy crisis has been rather feeble. No coherent policy for dealing with the crisis has been adopted, although a number of laws dealing with the crisis have been passed. In 1971 President Richard Nixon called for "Project Independence" to make the United States self-sufficient in energy by 1980, and in the late 1970s President Jimmy Carter stated that the energy crisis should be considered the "moral equivalent of war." But in fact, the US response did not seriously reduce the country's dependency on oil. At the beginning of the twenty-first century 50 percent of the oil consumed in the United States was imported and some analysts projected that this could increase to 60 percent by 2010.[9] Why is the United States having difficulty enacting an effective policy to deal with the crisis? Part of the reason is that the inertia of an oil-intensive society is hard to overcome. The nation is used to abundance, in energy as well as in material goods, and the creation of a new outlook and new values is not easy.

The cost of oil in the United States, in one sense, remains very low. In 2004 Americans were spending less of their income on energy, including gasoline, than they had over most of the previous 25 years.[10] What this means is that the "real," or true, cost of oil was not indicated by its price and thus consumers in the United States felt no urgency in demanding – or the government in producing – an energy policy that would break the dominance of oil in their society. As shown by some energy analysts, the real cost of oil would have to reflect not only the military costs necessary to secure it, but also the costs of the environmental degradation it causes – such as the effect it is having on the earth's climate, a subject that will be discussed in a following section. The real cost of oil would have to reflect also the increased healthcare costs that come with its use, and the subsidies by the government to the oil industry.

It was estimated by the International Center for Technology Assessment in 1998 that if the price of gasoline reflected all the environmental, military, and health costs of using it and subsidies to the oil industry, its price would be at least $14 a gallon.[11] Gasoline sales taxes can be used to cover some of these hidden costs – which otherwise are borne by the whole society in their general taxes and in healthcare costs – but the tax on gasoline in the United States has remained much lower than that in other major industrialized nations. In 2000 only about one-quarter of the price of gasoline was taxes in the United States, while in Britain taxes were about three-quarters of the price, and two-thirds in Germany, France, and Italy.[12] In 2005 the typical price of a gallon of gasoline in the United States was $2.26 while in Norway it was $6.66, in the UK $6.17, in Italy $5.94, and in France $5.68.[13]

The most recent effort to address the energy issue in the United States came in 2005 under the leadership of President George W. Bush. The US Congress passed an energy law which still focused on securing more fossil fuels: oil, coal, and natural gas. It also encouraged the construction of more nuclear power plants, and provided more funds for research in renewable energy sources. No provision directly addressed the need to reduce CO_2 emissions from fossil fuels which were causing much of the change in the earth's climate. Nor was there any provision to raise fuel efficiency standards of vehicles, which were actually lower than they were in the late 1980s.[14] Vehicles in the United States consume about 40 percent of its oil consumption.

In 2003, 39 percent of US energy came from oil, 24 percent from natural gas, 23 percent from coal (thus 86 percent from fossil fuels), 8 percent from nuclear, and 6 percent from renewables.[15]

Western Europe

Most Western European countries are more dependent on imported oil than is the United States. Traditionally, European governments have let the prices for imported fuel go up as determined by the world market and have tried to encourage energy conservation through the use of high taxes. France has emphasized nuclear power as its response to the energy crisis, and by 2005 it was producing about 80 percent of its electricity from that source – one of the highest rates in the world. The discovery of oil and natural gas under the North Sea aided mainly Norway and Britain. This large deposit allowed Britain to be self-sufficient in oil for several decades, but production peaked in 1999 and has been declining since then. In 2004 Britain was importing more oil than it exported. In 2003, 41 percent of Britain's energy came from natural gas, 32 percent from petroleum, 18 percent from coal, and 9 percent from nuclear. Denmark now receives about 15 percent of its electricity from wind. And in 2005 Finland began constructing the world's largest nuclear reactor, reversing the trend by most European countries to phase out nuclear power. The Fins were concerned with the growing threat of global warming, their increasing dependence on unreliable areas for oil and natural gas, and high and volatile energy prices.

Japan

Japan has no significant oil, natural gas, or coal deposits and, as stated above, in the mid-1970s it was the most vulnerable of all industrialized countries to OPEC's actions. A consensus quickly developed, after the first oil shock in 1973, that Japan's dependency on oil must be reduced. The government encouraged conservation and increased efficiency in using energy and the people responded. By the early twenty-first century Japan had reduced its dependency on oil from about three-quarters of its energy consumed in the mid-1970s to about one-half.[16]

It is interesting to note some of the differences between Japanese and US societies that have undoubtedly affected their different responses to the energy crisis. Because of their history and their limited land and resources, the Japanese have always assumed scarcity and insecurity of resources such as fuel, whereas the Americans have been accustomed to abundance and have assumed it will continue. Japanese industries have been traditionally more willing than their US counterparts to make long-term investments, the American companies often being more concerned with making short-term profits. The Japanese know that their goods must compete well in international trade if they are to maintain their high living standards. Japan is used to change and adaptation. The consensus that developed in Japan after 1973 emphasized a shift from consumption to restraint. It included a belief that the economy had to shift to "knowledge intensive" industries which use relatively little energy, and that energy efficiency was the key element in the adjustment the country needed to make to this new situation. It moved quickly into knowledge-based and electronic and computer-based industries.

Japan made significant progress in the period between the oil shocks in the 1970s and the third one in 1990–1. By 1990 the energy efficiency of the Japanese economy had improved to such an extent that the production of goods and services took only one-half the energy it took in the late 1970s.[17] The increased efficiency in the automobile and steel industries came after the government set ambitious goals for them to reach.

Another action taken by the government after the early crises was to build large oil storage facilities. By the early twenty-first century, Japan had nearly six months' supply of oil in storage tanks, more than any other nation. The country also sought to diversify its sources of oil and was successful for a while, but by 2002 it was importing 87 percent from the Middle East, about as much as it did before the first oil shock.[18] In 2000 the mix of Japan's energy sources was as follows: 52 percent oil, 18 percent coal, 13 percent natural gas, 12 percent nuclear, and 5 percent water, thermal and other alternatives.

The Japanese government has made nuclear power one of the key parts of its plans to reduce its dependency on imported oil. In 2002 Japan had plans to build about a dozen more nuclear power plants by 2010.[19] Japan also plans to build a number of fast-breeder reactors, to reduce its dependency on imported uranium, and in the early 1990s began importing plutonium from France (recycled from spent uranium fuel from Japanese power plants) for those reactors. By the early 2000s Japan had spent tens of billions on developing fast-breeder reactors which use plutonium as a fuel and, in theory, produce more nuclear fuel than they burn.

In 1999, because of human error, an uncontrolled chain reaction occurred in a nuclear fuel plant. Two people were killed and thousands of people were exposed to moderate levels of radiation. Safety concerns were again raised in 2002 when allegations were made that the world's largest privately owned electric utility in Japan was guilty of numerous serious safety violations in operating its nuclear power plants. Safety concerns, which include concerns with the aging of many of Japan's 53 nuclear plants, were beginning to erode the public's confidence in the safety of nuclear power.[20]

China

Although China's energy situation is not typical for a less developed country because of its vast reserves of coal, it does have a typical developing-world problem: how to provide a growing population with enough fuel in a manner that does not seriously harm the environment. China's population is so large that its use of energy could have a significant effect on the world's environment, as the following quotation from *Scientific American* makes clear:

> The path of industrial development in China . . . could have a greater effect on the atmospheric accumulation of carbon dioxide than that of any other nation. China's critical role stems from its large and growing population, its tendency toward energy-intensive processes, its poor energy efficiency and its massive reliance on coal.[21]

At the beginning of the twenty-first century about 80 percent of China's electricity came from coal, with hydroelectric dams providing much of the rest. This extensive use of coal is creating major air pollution and other environmental problems. China's air pollution levels are now among the world's highest. This extensive air pollution, and serious water pollution, were major contributing factors in the World Bank's rating of 16 Chinese cities among the 20 most polluted cities in the world.[22]

Most of China's coal is situated in the northwestern part of the country, far from the eastern coastal provinces where much of the new economic growth is taking place. Tens of thousands of factories in the eastern provinces are experiencing serious energy shortages and must either shut down or limit their production at times. In 2003 about two-thirds of China had shortages of electricity, while seven provinces had serious shortages.[23] As we can see in the box, energy shortages have

Cooling things off

"With the hottest days of summer fast approaching, Shanghai is making preparations to seed clouds over the city to make it rain, in the hope that a couple of degrees of reduced temperatures will help ward off brownouts or worse, here in China's commercial capital."

Source: Howard French, "China's Boom Brings Fear of an Electricity Breakdown," *New York Times*, national edn (July 5, 2004), p. A4

even led some cities to try to change the weather in an effort to reduce the demand for electricity.

China, now the second largest user of energy in the world, after the United States, is in a very energy-intensive period of its development as it focuses on manufacturing and exporting material goods. As China continues to industrialize and living standards rise, there is concern about where the new energy will come from. At present China's level of energy per person is only 10 percent that of the United States. Here are some of the steps China is taking to try to meet the growing demand for energy:

1 At present it has the world's largest program to create methane gas for use as fuel in rural areas. The gas is produced by fermenting animal and human wastes in simple generators; after the gas is produced, a rich organic fertilizer remains which can safely be used on crops.
2 China was self-sufficient in oil until the mid-1990s. Its use of oil is now growing rapidly. It is making large investments in foreign oil-producing nations from Sudan to Venezuela, and even tried, unsuccessfully, to buy an independent American oil company. The decision by the government in the mid-1990s to produce a so-called "people's car," an affordable compact sedan for the masses, and to double the annual production of vehicles by the end of the decade was popular with the growing middle class. In the opinion of Vaclav Smil, who has been studying the challenges to Chinese economic growth in the twenty-first century, this was not a good decision: "That is an insane route. There is not a single Chinese city that does not suffer from gridlock already."[24] By 2004 car sales were increasing 80 percent per year. China established higher fuel efficiency standards on its cars compared with the United States, but to save money China was importing so-called "sour or dirty" crude oil with a much higher sulfur content than was allowed in Europe or the United States.[25]
3 China plans to increase its nuclear power plants from the present eight to possibly 20 by 2020. These will be located mainly in the industrial coastal areas. Even with this projected increase in nuclear power, nuclear power will not provide more than 4 percent of China's electricity from the present 2 percent.[26] There are problems which could come with this expansion of nuclear power.

One nuclear energy expert at the Chinese Academy of Sciences said in 2005: "We don't have a very good plan of dealing with spent fuel, and we don't have very good emergency plans for dealing with catastrophe."[27]

4 China plans to significantly expand its renewable energy sources, such as solar and wind power, small hydroelectric dams, and biomass using plant and animal wastes. Its emphasis at present is on wind, which it is significantly expanding in numerous windy areas from Inner Mongolia to the eastern coast.[28]

The Effect of the Energy Crisis on the Development Plans of Less Developed Nations

The early stages of industrialization are energy intensive. Modern transportation systems, upon which industrialization rests, utilize large amounts of energy, as does the construction industry. The huge increase in oil prices in the 1970s cast a cloud over the development plans of many developing nations. Most of these plans were based upon an assumption that reasonably cheap oil would be available, as it had been for the West, to support their industrialization. Most of the developing countries have little or no coal or oil themselves. The development plans called for these countries to export natural commodities, nonfuel resources, and light manufactured goods; it was assumed that the earnings from these exports would be sufficient to pay for the fuel they would need to import. The success of the development plans also depended upon the countries being able to generate enough capital locally so that funds for investment in businesses would be available.

When OPEC increased fuel prices, no exceptions were made for the poorer countries; they were required to pay the same high prices for their oil imports as the rich nations had to pay. Added to that burden was the one created by the global recession which the higher oil prices had helped to create. As the recession deepened in the West, the industrialized countries cut back on their imports from the developing nations. Many of these countries borrowed heavily from commercial banks to pay for their higher oil bills and accumulated staggering debt. The World Bank estimated the foreign debt of the less developed countries in the mid-1990s to be about $1.9 trillion. Brazil had the largest foreign debt of all the developing nations, over $150 billion, in the mid-1990s. (One way Brazil has drastically reduced the amount of oil it imports is by using alcohol to fuel its cars. It is using its huge sugar-cane wastes to produce alcohol (ethanol) which can be mixed with or substituted for gasoline, and produces less pollution than gasoline. In 2006 Brazil expects to become energy self-sufficient.)[29]

The new situation created by high oil prices has led some experts to talk about a "Fourth World." This term refers to some of the developing nations, such as Bangladesh, which have few natural resources of their own and little ability to purchase the now expensive oil to promote industrialization. The countries in the Fourth World, the "poor poor," have little chance of developing along the path followed by the West, with its dependency on fossil fuels. If these nations are to

The replacing of human-powered vehicles with oil-fueled vehicles in poor and crowded countries, such as Bangladesh, will be difficult (*World Bank*)

improve their living standards, they will have to follow a development path radically different from the one followed by the developed nations.

Many experts predict that the largest increase in demand for oil in the early twenty-first century will come from the industrializing developing nations with large populations – such as China and India – and not from the developed nations, which have low population growth and have become relatively energy efficient.

Population pressure and the high cost of oil are increasing the demand for traditional fuel in the South, which is mainly wood. This problem has been mentioned above and will be discussed further in chapter 5 on the environment. As firewood becomes expensive or unavailable in rural areas, people switch to burning dried cow dung and crop residues, thus preventing important nutrients and organic material from returning to the soil.

The Relationship between Energy Use and Development

A shift in types of energy

One way to study the progress of the human race is to focus on the way humans have used energy to help them produce goods and services. People have constantly sought ways to lighten the physical work they must do to produce the things that they need – or feel they need – to live decently. The harnessing of fire was a crucial step in human evolution as it provided early humans with heat, enabled them to cook their foods, and helped them to protect themselves against carnivorous animals. Next came the domestication of animals. Animal power was an important supplement to human muscles, enabling people to grow food on a larger scale than ever before. Wood was an important energy source for much of human history, as it still is for a large part of the world's population. The replacement of wood by coal to make steam in Britain in the eighteenth century enabled the industrial revolution to begin. In the late nineteenth century oil, and in the early twentieth century natural gas, began to replace coal since they were cleaner and more convenient to use. Oil had overtaken coal as the principal commercial energy source in the world by 1970. In the 1970s nuclear power was introduced and was producing about 15 percent of the world's electricity by the early 2000s.

Increased use

The use of energy in the world has increased dramatically in the years since the end of World War II in 1945, a period of rapid development in the industrialized countries and one marking the beginning of industrialization in a number of developing countries. Figure 4.1 shows this well. Up through the end of the twentieth century, most of the increased energy use took place in the developed nations. Figure 4.1 shows the world's consumption of energy by type from 1850 to 2000. In 2000 fossil fuels made up about 86 percent of the energy used, with oil about 40 percent, coal about 24 percent, and natural gas about 22 percent. Nonfossil fuels – mainly hydroelectric, nuclear, geothermal, biomass, wind, and solar – accounted for about 14 percent of energy production in 2000.

Figure 4.2 shows per capita energy consumption by region of the world. North Americans consume more energy than any other region. Per capita energy use in North America (United States and Canada) is about twice that of Europeans, eight times that of Asians, and about ten times that of Africans. But per capita energy consumption does not tell the whole story. For example, per capita, the Chinese burn less coal (the most polluting fossil fuel) than North Americans do, but the total amount of coal used in China is twice that used in the United States.[30]

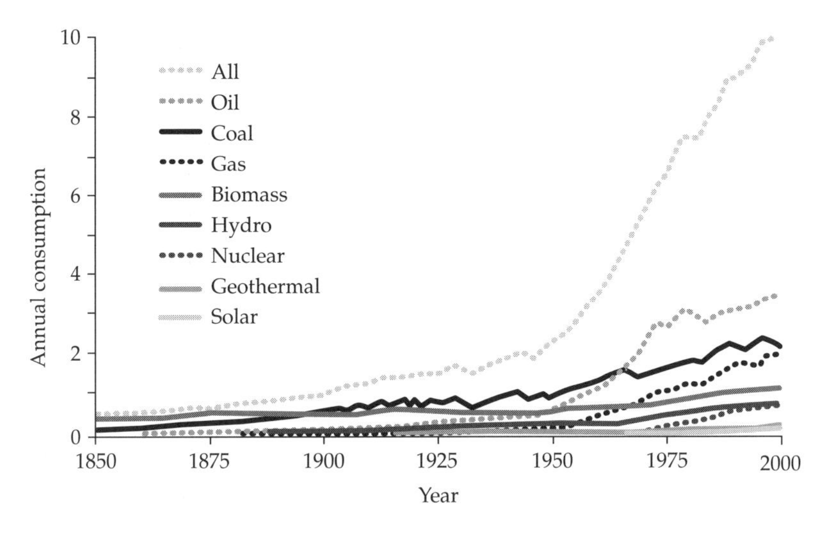

Figure 4.1 Global energy consumption, 1850–2000
Source: Robert Service, "Is It Time to Shoot for the Sun?" *Science*, 309 (July 22, 2005), p. 550; reprinted with permission from American Association for the Advancement of Science

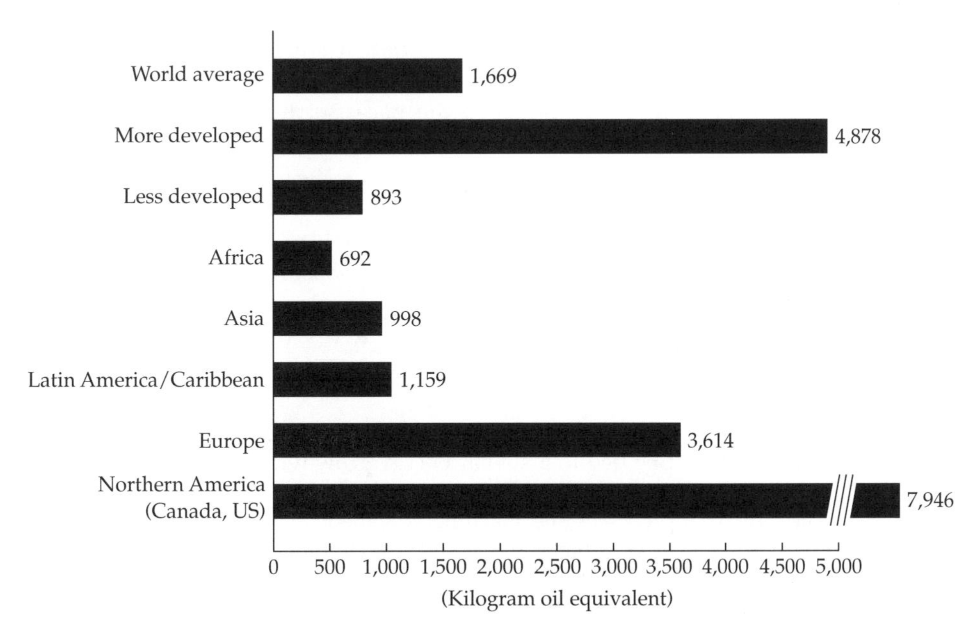

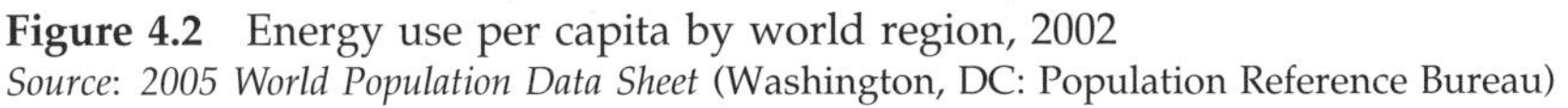

Figure 4.2 Energy use per capita by world region, 2002
Source: *2005 World Population Data Sheet* (Washington, DC: Population Reference Bureau)

The decoupling of energy consumption and economic growth

Historically, there appeared to be a one-to-one relationship in the United States between economic growth and energy growth; for example, a 10 percent increase in the amount of goods and services produced in the country was accompanied by an approximately 10 percent increase in the amount of energy consumed. But the oil shock of 1973 seems to have broken this relationship. Between 1977 and 1985 the US economy grew about 30 percent but the amount of oil used dropped nearly 20 percent.[31] What happened was that the United States began to use energy much more efficiently than it had before 1977, no doubt in response to higher oil prices. But then in the mid-1980s the price of oil fell dramatically and remained relatively low for over a decade. Probably in large part because of that fact, the efforts to further conserve energy in the United States slowed down.

This partial decoupling of energy use and economic growth is not surprising once one realizes that there are a number of countries with high levels of economic prosperity that have traditionally used much less energy than does the United States. In 2002 the United Kingdom, Germany, Austria, and Denmark, countries with high living standards, used about one-half of the energy per person than the United States used. And Japan, which even before the oil shocks used about half the energy the United States used to produce goods and services, so improved its energy efficiency after the first two oil shocks that by the early 1990s it was using half the energy to produce $1 worth of goods and services than it used in the late 1970s.[32] By the mid-2000s the United States was now using nearly 50 percent less energy per dollar of economic output than it had 30 years before.[33] But part of this improvement in the US came because there was a decrease in energy-intensive heavy industries and an increase in service industries, which use less energy.[34]

A number of studies of the US energy situation after the second oil shock concluded that a more efficient use of energy can actually lead to economic growth.[35] A study prepared for the US President in the early 1980s, recommending actions the country should take to deal with the situation presented in the government's *Global 2000 Report*, described some of the energy inefficiencies in the US economy:

> Evidence is mounting that US economic growth, as measured by Gross National Product (GNP), need not be tied to a similar energy growth rate. The most important reason is that the US economy, including much of its building and transportation stock, its industrial processes and machinery, is inefficient in its use of energy, compared both with other economies and with the technological and cost-effective options that already exist. The opportunity is enormous for improving the energy efficiency of US capital stock – in effect creating "conservation energy" – to get the same desirable end result of warmth, comfort, jobs, and mobility that fossil fuel energy provides.[36]

Energy conservation can promote economic growth because the cost of saving energy through such measures as improving the fuel efficiency of cars, improving the efficiency of industrial processes, insulating houses, etc., is lower than the cost of most energy today. Also, investments in improving the energy efficiency of US autos, homes, and factories create many new jobs and businesses throughout the country, thus spurring the growth of the economy in contrast to draining funds out of the economy by purchasing foreign oil.

Part of the reason many European countries use much less energy per person than does the United States is that they are smaller countries with populations not nearly as dispersed. One study has shown that the long distances people and goods move in the United States, in contrast with Europe and Japan, and the US preference for large, single-family homes account for about 40 percent of the difference between high US energy use and lower foreign use. The other 60 percent of the difference is accounted for by the fact that the fuel economy of US automobiles has historically been much poorer than that of many foreign cars, and the energy consumption per unit of output of many American manufacturing firms is higher than that of the foreign companies.[37]

The United States obviously cannot do anything about its size, but there are things that can be done to improve the energy efficiency of its transportation equipment. As mentioned above, the federal government passed a law in 1975, over the strong opposition of the automobile industry, requiring the fuel efficiency of American automobiles to be gradually improved.[38] Most of the long-distance hauling of freight in the United States is by truck, and a truck uses much more energy to move a ton of freight than does a freight train. The US government, by its vast expenditure of funds on the interstate highway system (reported to be the largest public works project in history), its much lower tax on gasoline than in Europe and Japan, and its relatively small amount of expenditures that benefit the railroads, has done much to promote the use of trucks over trains in the country. This policy could be reversed.

During the 1990s no improvement was made in auto fuel efficiency in the United States. There was, in fact, some backsliding. US auto makers, exploiting a loophole in the fuel economy law that allowed them to classify minivans, sports utility vehicles (SUVs), and pickup trucks as "light trucks" (a category of vehicles that had a lower fuel economy requirement than did automobiles), produced many of these vehicles and heavily advertised them. They became very popular. Their fuel efficiency was low, an average of 21 miles per gallon whereas automobiles were averaging 28 miles per gallon. Some of the so-called "light trucks" (which were really passenger vehicles) averaged 15 miles per gallon or less, which was similar to the 14 miles per gallon that US autos averaged 25 years earlier, before the first oil shock. With relatively low gasoline prices, and fading memories of the energy crisis, US auto makers and consumers put fashion and performance ahead of fuel economy.[39]

Japanese-made hybrid automobiles, which use a gasoline engine and an electric motor and were highly energy efficient and had low emissions, appeared on the market in 1999. US companies followed suit in producing such cars in the early twenty-first century, but they were expensive.

In the two decades from the early 1980s to the early 2000s, the US automobile industry used its improved technology to produce vehicles which had faster acceleration, were larger and heavier, and had slightly lower fuel economy. The average vehicle in 2002 had nearly 100 percent more horsepower, and was nearly 30 percent faster in going from 0 to 60 miles an hour than in 1981. It was also about 25 percent heavier.[40] In 2006 the Bush administration instructed the auto industry to improve fuel efficiency by two miles per gallon over a five-year period. This decision was criticized by many US environmentalists for being way short of the bold steps needed to combat a very important, if not the most important, issue the world faces today – climate change.

Climate Change

Most scientists who specialize in the study of the earth's climate believe that the human race is now involved in an experiment of unprecedented importance to the future life on this planet, involving nothing less than the global climate. A change in the global climate is now taking place, mainly because of the burning, by humans, of large amounts of fossil fuels – coal, oil, and natural gas. When these fuels are consumed, carbon, which accumulated in them over millions of years, is released into the atmosphere as a gas, carbon dioxide (CO_2). CO_2 in the earth's atmosphere has increased significantly since the industrial revolution: by about 30 percent between the mid-1700s and the present. This increase is causing a warming of the earth's surface – called "global warming" or the "greenhouse effect" – since CO_2 in the atmosphere allows sunlight to reach the earth, but traps some of the earth's heat, preventing it from radiating back into space. While CO_2 is the largest contributor to global warming, other gases – such as methane, which comes from both natural and human causes; nitrous oxide, which comes from fertilizers and other sources; chlorofluorocarbons (CFCs), widely used in the past in air conditioning and refrigeration; and other halocarbons – can also cause global warming. Many of these gases are increasing significantly in the atmosphere.

Under the sponsorship of the United Nations about 400 scientists from 25 countries contributed material to the most authoritative report on global warming in the early 1990s, and they concluded that because of the release of "greenhouse gases" by humans, an increase in the earth's temperature would occur.[41] Numerous models of the earth's climate have been made by climatologists and nearly all of these predict a warming of the earth because of the increasing CO_2 and other greenhouse gases.

The evidence

Warmer temperatures

There is evidence, as can be seen in figure 4.3, that over the past century the temperature of the earth has increased about one degree Fahrenheit (F) (about one-half degree Celsius (C)). And as of 2005, eight of the ten warmest years in recorded history were in the previous ten years. In the past 1,000 years, the warmest

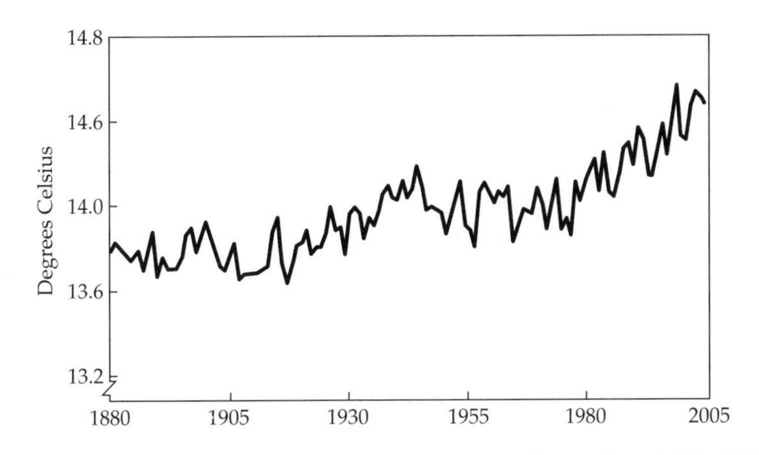

Figure 4.3 Global average land-ocean temperature at earth's surface, 1880–2004
Source: Goddard Institute for Space Studies; as presented in Worldwatch Institute, *Vital Signs 2005* (New York: W. W. Norton, 2005), p. 41

decades have been the most recent ones.[42] Hot, dry weather in recent summers was one of the main causes of many wildfires burning out of control in the western United States. Average temperatures of the Arctic winters have risen about 10 degrees over the past 30 years and the years of the late twentieth century in the Arctic zone were the warmest in the past four centuries. Data from submarines and satellites have shown that the ice sheet over the whole Arctic Ocean has been shrinking in area and thickness. The permafrost is thawing across the Arctic, leading to the damage of buildings, pipelines, and roads in Alaska and Siberia as the land sinks. Temperatures have risen in the Arctic nearly twice as fast in recent decades as in the rest of the world.[43] On average, summer sea ice in the Arctic has shrunk 15 to 20 percent over the past 30 years.[44] Most global warming models predict that the Arctic region will be among the first regions to show effects caused by global warming. Other evidence supporting the assertion that global warming has begun is the fact that most mountain glaciers in the world have been retreating since the late nineteenth century.[45] Scientists have discovered the movements of some glaciers draining the great ice sheets in Antarctica and Greenland have recently accelerated.[46] Studies have shown that since the 1960s spring is coming earlier and winter later to the higher latitude areas in the northern hemisphere. The growing season has been advanced by about seven days in the spring and extended by about two to four days in the fall. The longer growing season and the higher CO_2 levels have led to more vegetation over large parts of northern Europe, northern Russia, and northern China, and in Alaska, Canada and Scandinavia.[47] Recent studies have also shown that the changes in temperature taking place in the earth's atmosphere are consistent with what many computer models predict will happen as CO_2 levels increase. Scientists have found that since 1950, when CO_2 emissions started to increase sharply, there has been a trend for warmer temperatures in the lower atmosphere and cooler temperatures in the high stratosphere.[48]

There is strong evidence that in the past CO_2 and methane in the atmosphere were closely connected in some way with the earth's temperature. European scientists in the Antarctic drilled a hole about two miles deep in the ice and withdrew a core of ice. Like the rings of a tree, the core indicated changing conditions in the past – in fact, back about 650,000 years. The scientists measured the amount of CO_2, methane, and nitrous oxide (all greenhouse gasses) in the air bubbles in the ice and found two amazing facts. First, the amount of CO_2 and methane present in the earth's atmosphere today is higher than in any previous time during those 650,000 years, and secondly CO_2, methane, and the earth's temperature went up and down closely together during that period.[49] The close relationship between the earth's temperature and the carbon dioxide and methane levels is consistent with the global warming theory.

Europe had a record-breaking heat wave in 2003 that killed thousands of people.

Violent weather

More violent weather is expected as the temperature rises. An indication that this may be occurring is that the payout by insurance companies for weather-related damage rose from $16 billion for the decade of the 1980s to $44 billion just for 2004.[50] The 2005 Atlantic hurricane season was the most devastating one in modern times, with seven major hurricanes, including the giant storm that led to the flooding and evacuation of the US city of New Orleans.

Sea level rise

Warmer temperatures should cause the levels of the oceans to rise because melting glaciers and ice caps will add water to the oceans, and water expands when its temperature increases (thermal expansion of the oceans contributes about 25 percent of sea level rise).[51] Sea levels have risen about 8 inches over the past century and since the 1990s the pace has increased markedly.[52]

Coral reefs

Coral reefs are in serious decline around the world because of warmer seas, pollution, disease, and overfishing. An estimated 30 percent have already been severely damaged.[53]

Probable effects

Warmer temperatures

The most common forecast of the computer models is that – based on present trends – the amount of CO_2 in the atmosphere is expected to be double the pre-industrial level before 2100, and that will lead to an increase of from 1.5 degrees C to 4.5 degrees C (3 degrees C equals about 5 degrees F) before the end of the present century, with a consensus forming that 3 degrees C is the most likely figure.[54] (The amount of CO_2 in the atmosphere will not stop increasing after it has doubled, of course. Depending on how much carbon continues to be released on earth, the CO_2 level in the atmosphere could keep rising after the doubling.) Three degrees C would be a more rapid change of temperature than has occurred in the past 10,000 years. While 3 degrees C does not sound like very much, it would be a significant change. According to scientists of the US National Aeronautics and Space Administration, the temperature on earth "would approach the warmth of the Mesozoic, the age of dinosaurs."[55] Scientists have concluded that, because of various "feedbacks," it is likely that the warming will be even more than they are predicting. They also warn that because of our incomplete knowledge about the processes involved in the earth's climate, it is possible we will be confronted with "surprises" in the future.[56]

Heat waves are expected to become more common and more severe. Cities trap heat and their very young, elderly, and poor are especially vulnerable to heat stress. Mid-latitude cities such as Washington, DC, Athens, and Shanghai are more vulnerable than tropical and subtropical cities because their residents are less used to high temperatures. The death toll in cities during extreme temperatures can be surprisingly high, as was seen in Chicago where more than 700 people died during a four-day heat wave in the summer of 1995.

On the positive side, with warmer temperatures we can expect fewer cold weather deaths as winters are milder. Unfortunately there are generally more deaths caused by hot spells than cold spells.

Violent weather

The number of what is called "extreme weather events" (hurricanes (cyclones), violent thunderstorms, and windstorms) is expected to increase.

Sea level rise

A probable effect of a warming of the earth's climate is that the level of the oceans will rise. Such a gradual rising of waters could lead to the evacuation of some coastal cities around the world. Sixteen of the largest cities with populations of over 10 million are located in the coastal regions. The rich countries will probably be able to build dikes to protect their cities, but poor countries such as Bangladesh probably cannot afford to do so. Also much coastal lowland around the world will be threatened. These lands are heavily populated at present, especially in the developing nations. Regions such as the Ganges–Bramaputra delta in Bangladesh, the Nile delta in Egypt, and the Niger delta in Nigeria are especially vulnerable. Island nations such as the Maldives and the Marshall Islands could be inundated.

Air pollution

Higher temperatures tend to favor the formation of pollutants such as ground-level ozone, which is the main component in smog. An increase in winds could disperse pollution whereas a decrease could help pollution levels to rise. An increase in rainfall can wash out pollutants, while a decrease can have the opposite effect.

Infectious diseases

A change in temperature and rainfall can affect the range of many infectious diseases. One obvious example is that the range of mosquitoes that spread malaria, yellow fever and dengue fever could expand with increased temperatures and rainfall. The range of the black fly that carries river blindness is likely to expand, as also is the snail that carries schistosomiasis. People living on the edges of where these diseases are prevalent now are especially vulnerable because many have little resistance built up.

Agriculture

It is very hard to predict how climate change might affect agriculture because some regions could be benefited and some harmed. In a warmer world there would be major changes in the amount of rainfall and its location, with some areas getting more rainfall than at present and some less. Higher CO_2 levels can act as a fertilizer on some plants and reduce their need for water. Higher temperatures can extend the growing season in some regions and more rainfall can benefit crops in some areas. More droughts and extreme temperatures can have the opposite effect. The effect of climate change on plant pests and diseases is similar to that of infectious diseases.

An effort by the US government in 2000 to predict changes in food production in the country found that it was likely that crops would increase in the northern plains, where much of the country's wheat and corn is grown, but decrease in the southern states because of droughts and floods caused by heavy rains. The authors of the study admitted many unknowns exist, such as their inability to calculate the possible effects of flourishing weeds or migrating insect pests.[57]

Disruption of natural ecosystems

Natural ecosystems such as forests, rangeland, and aquatic environments provide a host of services to human and nonhuman life. Many of these services are still relatively unknown. Any disruption of these ecosystems because of climate change could have serious effects. Rough estimates now are that a doubling of CO_2 levels could cause from one-third to one-half of all plant communities and the animals that depend on them to shift their locations.[58] Some animals will not be able to do so because roads and urban sprawl will block their way. The shrinking sea ice in the Arctic is likely to make it much more difficult for polar bears to hunt for seals, which is one of their chief foods, thus leading to the bears' possible extinction.

Uncertainties

Abrupt climate change

Scientists examining the ice cores mentioned above have found evidence that at times in the past the climate of earth changed abruptly to a new level which persisted for hundreds or thousands of years.[59] A threshold was crossed that caused this change, but scientists do not understand what these thresholds were. One scientist described our situation today as similar to when people in a canoe start to rock the boat. Nothing happens for a while until a threshold is crossed and the canoe suddenly tips over and the canoeists find themselves in the water. As strange as it may seem, the next section shows that one of these abrupt changes could be from warm weather to a rapid cooling for parts of our planet.

Slower Atlantic currents

Scientists have discovered some evidence that the currents that bring warm water from the tropics to North America and Northern Europe may be slowing.[60] The melting of the sea and land ice in the Arctic – especially in Greenland – may be diluting the ocean's salty water which is essential to keep the so-called "Atlantic heat conveyor" moving. While the evidence supporting this theory is still too incomplete to convince many scientists that this is happening, there is wide agreement that if this did happen not only will North America and Northern Europe face colder climates, but the monsoons that billions of people in Asia and Africa depend on to support their agriculture could be disrupted.

More intense storms

Some scientists claim to have discovered evidence that storms over the past 35 years have increased in intensity.[61] Especially the number of category "4" and "5" hurricanes (the most intense) have increased, although the total number of storms has not changed.

Clouds and soot

Scientists admit ignorance of the effects global warming will have on clouds. Some types of clouds could cool the earth, while other types could heat it up more. So also is there ignorance of the effect aerosols (soot) will have on weather. It is likely that large amounts of black carbon particles in the air over parts of India and China from the burning of coal and biofuels by millions of villagers are affecting the climate in various ways.[62]

Other positive and negative feedbacks

There are uncertainties over possible "positive feedbacks," those things that might occur as the warming takes place which will make it worse, such as a melting of the permafrost releasing more methane, and "negative feedbacks," those things which could make it cooler, such as an exploding algae population in a warmer ocean absorbing more carbon dioxide.

What is being done at present?

One hundred and ninety-three nations ratified the 1992 Framework Convention on Climate Change which was presented at the Rio Earth Summit. The Convention called on nations to voluntarily reduce their emissions of greenhouse gases to 1990 levels. European nations, Japan, and about 40 small island and coastal states favored putting specific targets and timetables for reaching the targets in the treaty, but the United States opposed this and they were not put in the treaty. The industrial nations pledged to meet the goal of reducing greenhouse gases to 1990 levels by 2000. For the most part the industrialized nations did not meet this goal.

In 1997 many nations met at Kyoto, Japan, and agreed to a proposed treaty that did place legally binding limits on developed nations. No limits were placed on developing nations because they were still producing relatively few greenhouse gases and were making efforts to reduce their widespread poverty. The target set in this proposed treaty (called the Kyoto Protocol) was that developed nations would reduce their greenhouse gas emissions by about 5 percent from their 1990 levels by 2008 to 2012. As of 2006 the United States – the largest producer of greenhouse gasses – had still not ratified this treaty and there was no effort being made in the country to do so. The European Union did ratify it and, as figure 4.4 shows, by 2002 the UK and Germany had reduced their emissions below their Kyoto targets. By 2005 enough nations had ratified the Kyoto Pact to bring it into force.

What more can be done?

As can be seen in figure 4.5, during the second half of the twentieth century CO_2 emissions per capita were much higher in the more developed countries than in the less developed countries. But as development spreads to some of the large less developed nations – such as China and India – and as their population grows, they will produce a relatively larger percentage of the gas, especially as China relies mainly on coal, the fossil fuel which emits the most CO_2. This can be seen in figure 4.4. The developing world is expected to increase its amount of CO_2 emissions to 45 percent of the world's total by 2010.[63] Obviously the developing world must eventually be brought into the climate change treaty and targets put in place for reducing its emissions.

The United States should, obviously, sign the treaty. The US Department of Energy estimates that if no change in energy policies takes place in the country the United States will *increase* its carbon emissions about 35 percent by 2010 from 1990 levels.[64] Eventually all nations will have to agree to reductions in greenhouse gas emissions well beyond those indicated in the Kyoto treaty if there is any

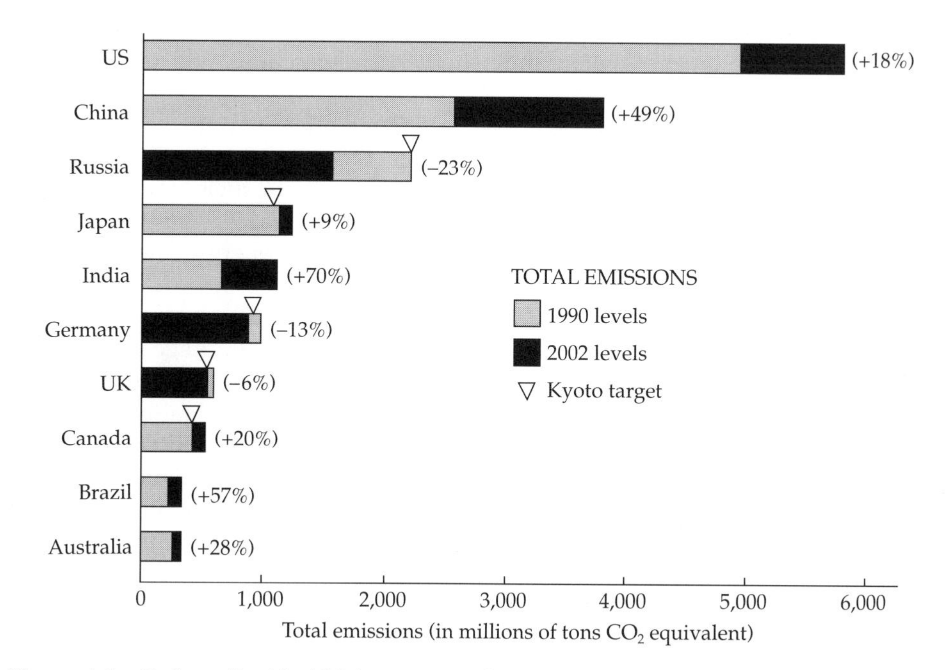

Figure 4.4 Carbon dioxide (CO_2) emissions for selected countries, and the Kyoto Pact target

Note: Percentages in parentheses reflect changes in emissions from 1990 to 2002.

Source: *Science*, 311 (March 24, 2006), p. 1703; reprinted with permission from American Association for the Advancement of Science

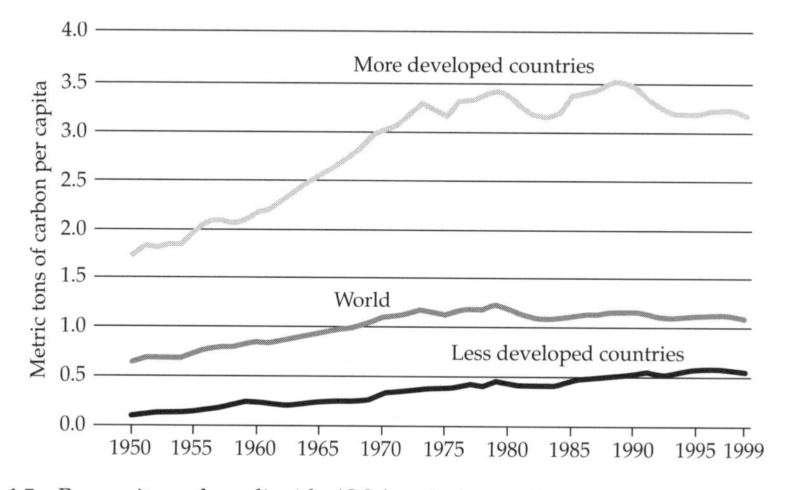

Figure 4.5 Per capita carbon dioxide (CO_2) emissions, 1950–1999

Source: Updated and adapted from F. A. B. Meyerson, "Population, Carbon Emissions, and Global Warming: The Forgotten Relationships at Kyoto," *Population and Development Review*, 24, no. 1 (1998), pp. 115–30; as presented in Roger-Mark De Souza et al., "Critical Links: Population, Health and the Environment," *Population Bulletin*, 58 (Population Reference Bureau, Washington, DC) (September 2003), p. 24

possibility of stabilizing atmospheric CO_2 concentrations at even two or three times their preindustrial level.[65] When, and whether, these actions will take place is unknown.

There are policies which the United States and other nations could pursue that would alleviate the threat of climate change. One would be to de-emphasize programs to promote the increased use of coal and synthetic fuels made from coal and oil, and to encourage the more efficient use of energy and its conservation. The development of renewable energy sources, such as solar energy, and nonfossil-fuel energy, such as nuclear energy, can be promoted.

A carbon tax – a tax on fuel according to the amount of CO_2 released – would give a big boost to the development of renewable energy and encourage technologies that make the use of energy more efficient. Because the prices of fossil fuels today are unrealistically low, not reflecting the health and environmental costs society will have to pay because that fuel is being used, there is little incentive to reduce the use of fossil fuels. By 1998 five countries – Denmark, Finland, Norway, Sweden, and the Netherlands – had carbon taxes. A raising of the fuel economy standards in the United States for cars and trucks would end the present trend toward vehicles with embarrassingly low fuel efficiency.

In the short term, the United States and other high energy users could switch from oil to natural gas as natural gas releases 30 percent less CO_2 per equivalent amount of energy than petroleum (compared to coal, natural gas releases about 40 percent less CO_2). Another policy would be to combat deforestation, since trees, along with other vegetation, absorb large amounts of CO_2.[66] (The increasing destruction of the great tropical rainforests in Latin America and Asia is seen by some experts as representing a real threat to the global climate.)

How the less developed nations can be encouraged to develop without increasing their CO_2 emissions is not clear, but it is clearly in the interests of the industrialized nations to help them do so. It is also in the interests of both rich and poor nations to support population control efforts in the developing world, as more people will release more CO_2 and other greenhouse gases.

The Energy Transition

The world is entering a period of transition from one main energy source – oil – to a new principal source or a variety of sources. This is the third energy transition the world has passed through: the first was from wood to coal, and the second from coal to oil. Many people, although not all by any means, now recognize that the industrialized world must shift from its reliance on nonrenewable and dangerously polluting fossil fuels to an energy source or sources that are renewable and less polluting. Many in the industrialized countries, but fewer in the United States and Australia, which are the only industrialized countries that have refused to sign the Kyoto Pact, understand that their dependence on imported oil must end since it is not a clean fuel, and neither is it cheap, abundant, or secure. But what will be the new principal energy source for the industries of the developed world and the new industries of the developing nations? As in many transitions, the end to be reached is not clear. The only clear thing now is that the present state of affairs is no longer viable. We are now in the beginning years of the energy transition.

For the rest of this section, we shall examine some of the potentialities of the most often discussed energy sources. Energy sources can be divided into those that are nonrenewable (that is, it took millions of years to create them and they are being used up) and those that are renewable, in the sense that most of them currently gain their energy from the sun, which is expected to continue to shine at its present brightness for at least one billion years more.

Nonrenewable energy sources

Oil, natural gas, coal, and uranium are the main sources of nonrenewable energy. According to many analysts the world is not about to run out of **oil**, but within a few decades shortages will become prevalent.[67] The world's demand for oil is now growing 1 or 2 percent each year.[68] Rapidly economically growing China and India are already competing with the West for oil. The production of oil in the United States peaked around 1970 and has been decreasing since then.

Canada has large deposits of tar sands from which oil can be extracted. In the mid-2000s about 1 million barrels of oil a day were being extracted from the oil sands. It is estimated that the tar sands hold as much as 175 billion barrels of oil, but it is relatively expensive to extract.

Proven reserves of **natural gas** are estimated to be larger than oil reserves. Large reserves exist in Russia, Iran, Qatar, and Saudi Arabia. Natural gas is the cleanest fossil fuel, emitting 40 percent of the CO_2 emitted by coal. Europe now uses natural gas for 20 percent of its energy, much of it coming from Russia. A quarter of the energy in the United States now comes from natural gas and there are plans to increase this. According to a well-known energy analyst, natural gas has become the "fuel of choice" for meeting the needs for more electricity in both the developed and developing countries.[69] Some energy analysts expect it to overtake coal and oil as the most important fossil fuel in the world by 2025.[70]

Coal is a much more abundant resource than oil or natural gas, and the United States has very large deposits of it, as do Russia, China, and Europe. It is estimated that the earth has 1 trillion tons of recoverable coal, with one-quarter of it in the United States. China has about one-half as much as the US but burns twice as much as the US at present. It is coal that is fueling China's present economic boom.

Coal, partly because it is relatively cheap and abundant, has made a resurgence in the US under the George W. Bush administration. But its low price is deceptive. Among the many serious pollutants emitted when it is burned, those contributing to climate change are undoubtedly the most dangerous. About 40 percent of CO_2 emissions around the world come from the burning of coal. According to the International Energy Agency (IEA), the pollution from coal will probably increase:

> the energy equivalent of some 1,350 thousand-megawatt coal-fired power plants will be built by 2030. Forty percent of them will be in China . . . India will add another 10 percent or so and most of the remaining half will be added in the West. In the United States the IEA predicts, about a third of the new electric-generating capacity built by 2025 will be coal-fired.[71]

It is possible that uranium, the basic fuel for **nuclear energy**, is widely distributed around the world, but the bulk of positively identified deposits are located in a relatively few countries, one of which is the United States. The mining of uranium can, and has, led to cancer, and the waste products from the mining are radioactive. The United States has fairly abundant supplies of uranium, but, like coal, they will eventually run out.

Nuclear power provides about 15 percent of the world's energy through about 440 nuclear plants. After it remained relatively dormant throughout the world for several decades – except for a few countries such as France, Japan, and China – new interest is now being shown in this source of electricity both in some European countries and in the United States. Increasing demand for energy around the world, insecure oil supplies, and concerns with fossil fuels' contribution to changes in the earth's climate, are fueling this new interest. Nuclear power releases no CO_2 or other greenhouse gasses and no other air pollutants. Partly because of safety concerns caused by the partial meltdown of the core of a reactor at Three Mile Island in the US in 1979, and the explosion of a reactor at Chernobyl in the USSR in 1986, which spread radioactive particles over parts of Europe, new nuclear power plants will be expensive to build. No permanent repository for the tons of nuclear waste from the 100-odd nuclear power plants in the US, waste that must be kept secure for thousands of years, has been found. Most of this radioactive waste still sits in temporary holding containers at the plants and is not safe from an attack by a terrorist from the air.[72]

Renewable energy sources

The energy from the sun can be obtained in a variety of ways: from wood, falling water, wind, wastes, hydrogen, and, of course, from direct sunlight. We will briefly examine each of these. In 2003 only 6 percent of the energy used in the United States came from renewable sources.[73]

First, **wood, agricultural/forestry residues, and animal dung** are still the principal fuels in many developing countries. Rural peoples in sub-Saharan Africa, as in the South Asian countries of India, Pakistan, and Bangladesh, use these traditional fuels to cook their food and to provide heat and light. In fact, except for their own muscle power and the aid of a few domestic animals, the majority of the villagers in many developing nations have no other source of energy. Rapidly expanding populations in poorer countries are placing high demands on the use of wood; at the same time, modern agricultural requirements and development in general are leading to the clearing of vast acres of forests. Acute shortages of firewood already exist in wide areas of Africa, Asia, and Latin America.

The destruction of forests for development purposes in the developing nations is occurring at the same time as the growing demand for wood as fuel (*Caterpillar Company*)

Second, **hydroelectric power**, which is generated from falling water, is a clean source of energy, causing little pollution. A large potential for developing this type of energy still exists in Africa, Latin America, and Asia, although many of the rivers that could be used are located far from centers of population. Large dams, which are often necessary to store the water for the electric generators, usually seriously disturb the local environment, sometimes require the displacement of large numbers of people, and cause silting behind the dam, which limits its life. While most of the best sites for large dams in the industrialized countries have already been developed, a potential exists for constructing some small dams and for installing electric generators at existing dams that do not have them.

Third, **wind** is an energy source that was commonly used in the past for power as well as for the cooling of houses. It is still used for these purposes in some less developed countries and has recently gained respect around the world. In California in the United States 16,000 wind turbines were constructed in just three mountain passes, areas that have fairly steady wind. Actually, the midwestern states in the United States – from North Dakota to Texas – have better wind conditions than California and have a great potential for generating more of their power

Wind turbines in Altamont Pass, California (*US Department of Energy*)

from this source. The dominance of California in producing wind power probably had more to do with the tax incentives that the state gave in order to promote this form of energy rather than wind conditions. The California wind farms began going up in 1981 after the federal government passed a law that encouraged small energy producers, and after both the federal government and the state of California gave tax credits to the wind producers.

A potential exists for the wider exploitation of this source of energy in the US. In 2002 wind provided about 1 percent of the energy in the country.[74] Modern technology is producing more efficient wind collectors. The cost of producing electricity from wind has fallen to such a degree – because of the advances in wind turbine technology – that wind power is now competitive with power produced from fossil fuels.

Attracted by the success of wind farms in California, a number of European countries such as Germany, Denmark, Spain, Italy, Britain, and the Netherlands, have greatly increased their wind power. About 20 percent of Denmark's electricity now comes from wind power. Germany gets 10 percent of its electricity from wind. The European Union has set a goal of getting 10 percent of its electricity from wind by 2010. (Europe's overall goal is to have about 20 percent of its electricity – and about 10 percent of all its energy – come from renewable sources by 2010.) By the early part of the twenty-first century Europe had become the largest user of wind power with 75 percent of the world's wind power residing there.[75] China has begun installing wind turbines from Inner Mongolia to offshore of its eastern coasts as part of its goal to secure 10 percent of its energy from renewable sources by 2020. Wind power produced 3 percent of India's electricity in the mid-2000s.

The main problem with wind, of course, is that it is usually not steady, and thus the energy it creates must be stored in some way so it can be used when the wind dies down. There is not yet any easy and inexpensive way to do this. Another problem with wind is that the choice windy places in the world are relatively few and unevenly distributed. They are also often in remote locations, far from population centers, and in areas of great natural beauty, which the windmills spoil. In an effort to defuse public opposition to the windmills' location and to benefit from strong and steady winds in coastal areas, many offshore wind farms are now in the planning stage. Past problems such as the noise the wind turbines make as the blades whirl (some blades are as large as the wingspan of a 747 aircraft) and the killing of birds have been partly solved by improvements in turbine design and more care given to their location.

Fourth, **biomass conversion** is the name given to the production of liquid and gaseous fuel from crop, animal, and human wastes; from garbage from cities; and from crops especially grown for energy production. Millions of generators that create methane gas from animal and human wastes are producing fuel for villages in India and China. Brazil is using its large sugar-cane production to produce low pollution alcohol for fuel for automobiles. In 2006 Brazil expects to become energy self-sufficient, relying on its own oil production and large use of alcohol (called ethanol in the United States).[76] An important part of Brazil's success came when the automobile industry in Brazil developed new technology that permitted it to produce an engine that can use either gasoline, alcohol, or a combination of both. This allows drivers to select the cheapest fuel, which at present is alcohol.

St Louis and some other US cities are burning their garbage mixed with coal and/or natural gas to produce electricity. It is difficult to estimate how widespread this form of energy generation will become in the future. Some see good potential while others mention its negative aspects, such as the emission of harmful gases and of foul odors from burning garbage, and the use of land to grow energy crops instead of food crops in a hungry world.

Fifth, the use of pollution-free **direct sunlight** probably has the greatest potential of all the forms of solar energy for becoming a major source of energy in the future. Each year the earth receives from the sun about ten times the energy that is stored in all of its fossil fuel and uranium reserves. Direct sunlight can be used to heat space and water, and to produce electricity, indirectly in solar thermal systems, or directly by using photovoltaic or solar cells. Solar thermal systems collect sunlight through mirrors or lenses and use it to heat a fluid to extremely high temperatures. The fluid heats water to produce steam, which is then used to drive turbines to generate electricity.

Japan produces the most solar cells in the world (about 50 percent in 2004) and has the largest solar power capacity. Germany is second in the use of solar energy, and the United States third. (The United States was once the world's leader in solar energy but the ending of governmental incentives after the 1980s, and low

Solar thermal power plant, California (*US Department of Energy*)

natural gas prices, ended its leadership.) In 2005, 98 per cent of solar systems in the United States were used to heat swimming pools, a use that hardly existed in the rest of the world.[77] China is the world's leader in the production and use of solar thermal plants.

Although dropping in cost, solar cells are still relatively expensive. Their high cost is probably the most serious hindrance to their wider use. A major reduction in their cost would probably come about if they were mass-produced, but without a large demand for solar cells, which their high cost prevents, mass production facilities will not be built by private enterprise. A way out of this vicious cycle could come as the costs of oil and natural gas continue to rise, or if the world finally unites in an effort to combat climate change. Solar energy could be used well in moderately or intensely sunny places. Much of the developing world fits this criterion. The developing world is, in fact, often mentioned as a vast potential market for solar energy because many of its rural areas still lack electricity, and solar energy is collected about as efficiently by small, decentralized collectors as it is by larger, centralized units.

The cost of solar energy from solar thermal plants has been dropping rather rapidly. If one includes the hidden costs of fossil fuels – i.e. the costs society bears

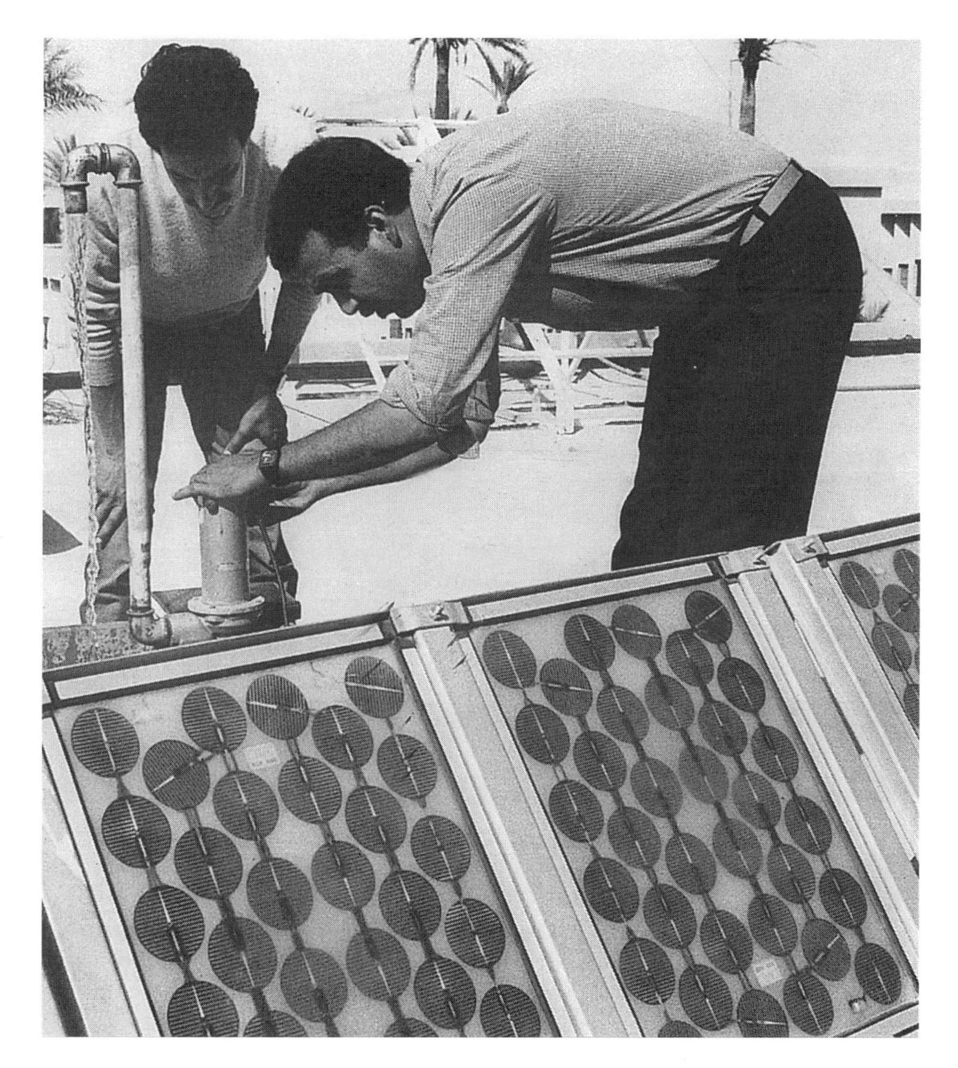

Solar energy provides power for a water pump in Morocco (*USAID Photo Agency for International Development*)

now and will bear in the future because of the pollution they produce and the costs of military forces to ensure access to them – solar energy is probably less expensive than fossil fuels right now.

Sixth, **geothermal energy**, heat that is produced within the earth's interior and stored often in pools of water or in rock, or as steam under the earth's cool crust, is not a form of solar energy but it is a renewable form of energy. Iceland uses this form of energy to heat many of its homes, and Russia and Hungary heat extensive greenhouses with it. Two US cities, one in Oregon and one in Idaho, use geothermal energy, and a geothermal power plant that produces electricity has been built in northern California. In the early 1990s geothermal energy

Geothermal power plant, California (*US Department of Energy*)

provided New Zealand and Kenya with nearly 10 percent of their electricity, the Philippines with nearly 20 percent, and E1 Salvador with nearly 40 percent. For a few favorable locations in the world, geothermal energy can be utilized, but it is not expected to have a wider potential.

Finally, **hydrogen-powered fuel cells** have the potential to become a major nonpolluting and efficient source of energy for vehicles. In fuel cells hydrogen is combined at low temperatures with oxygen supplied from the air to produce electricity, which is used to run an electric motor. Vehicles powered by the electric motor would be clean, quiet, highly efficient, and relatively easy to maintain. No battery is required and basically the only substance coming from the exhaust is water. Hydrogen can be obtained from water by a process that itself uses electricity. If the electricity used to make hydrogen comes from renewable and nonpolluting sources such as solar power, wind power, or hydroelectric power, hydrogen fuel cells are a renewable and clean source of energy. If a polluting fuel such as coal is used to make hydrogen, the fuel cell would be neither clean nor renewable. In the mid-2000s the Bush administration in the United States was focusing governmental research on coal and nuclear power as possible sources of energy to make hydrogen.[78]

By 2000 nearly all automobile companies were putting a major effort into developing cars using fuel cells. But major hurdles exist before fuel cell cars are mass-produced. In the mid-2000s hydrogen fuel cells cost about a hundred times as much per unit of power as the internal combustion engine powered by gasoline.[79] The US National Academy of Sciences in 2004 estimated that the transition to a hydrogen economy would take decades because of the serious challenges involved.[80] One major problem is the need to create thousands of hydrogen fueling stations. One industrialist in the United States put it this way:

> "It's the classic chicken-and-egg dilemma. There's no demand for cars and trucks with limited fueling options, but no one wants to make the huge investment to create a fueling infrastructure unless there are fleets of vehicles on the road. So the question is: How do we create demand?"[81]

Conservation/energy efficiency

Conservation is not commonly thought of as an energy source, but according to an influential study of the US energy situation by the Harvard University Business School in 1980, it should properly be regarded as a major untapped source of energy. "But is conservation really a 'source' of energy?" asked one of my bright students. "Good question," I responded. "Think of something that makes it unnecessary for you to use a product. Isn't it, in a sense, the same as the product?"

How much energy could the United States save by adopting conservation measures? The Harvard study found that the savings could be impressive:

> If the United States were to make a serious commitment to conservation, it might well consume 30 to 40 percent less energy than it now does, and still enjoy the same or an even higher standard of living. That saving would not hinge on a major technological breakthrough, and it would require only modest adjustments in the way people live.[82]

To many people, the term "conservation" means deprivation, a doing without something; but the Harvard study, and many others since, have shown that much energy conservation can take place without causing any real hardship. There are three ways to save energy: by performing some activity in a more energy-efficient manner (for example, designing a more efficient motor); by not wasting energy (turning off lights in empty rooms); and by changing behavior (walking to work or to school).

Many businesses now recognize that making their operations more energy efficient is a good way to increase profits. (This subject was first discussed in an earlier section of this chapter: "The decoupling of energy consumption and economic growth.") The investments the companies make to redesign their business operations so they reduce their energy usage are soon repaid by lower energy bills. Dow Chemical discovered after the 1973 oil crisis that the company's standard practice up to then was never to turn off its de-icing equipment during the year, which meant that its sidewalks and service areas were being warmed even on the Fourth of July. Over the most recent decade DuPont increased its production by about 30 percent but cut its energy use nearly 10 percent, saving more than $2 billion. Five other companies – IBM, British Telecom, Alcan, Norske Canada, and Bayer – collectively saved another $2 billion by reducing their CO_2 emissions by about 60 percent. British Petroleum (BP) met its 2010 goal in 2001 of reducing its CO_2 emissions 10 percent below its 1990 level, thus cutting its energy bill by about $650 million over ten years.[83]

One major conservation method US industry could adopt is called "cogeneration," which is the combined production of both electricity and heat in the same installation. Electricity is currently produced by private and public utilities, and the heat from the generation of the electricity is passed off into the air or into lakes and rivers as waste. In cogeneration plants, the heat from the production of electricity – often in the form of steam – is used for industrial processes or for heating homes and offices. The production of electricity and steam together uses about one-half the amount of fuel as does their production separately. Cogeneration is fairly common in Europe but not in the United States, where electric utilities often give cheaper rates to their big industrial customers, thus reducing the incentive to adopt the process.

If the United States ever does reach the goal of energy savings that the Harvard report believes is possible, it will be because of a combination of governmental policies encouraging conservation and of action by millions of individuals. The United States is a country where people respond well to incentives to promote conservation practices, but such governmental incentives have so far been rather weak. In contrast to weak efforts by the central government, some of the US states have done more to encourage the conservation and use of renewable energy. For example, the state of California allowed homeowners to deduct 55 percent of the cost of solar devices from their state taxes. (This law no doubt partly explains why California leads the nation in the number of solar devices installed in homes.) The city of Davis, California changed its building code so that all new homes in the city must meet certain energy performance standards.

American homes are not designed to use energy efficiently. If houses with large window surfaces were positioned to face the south, they could gain much heat from the low winter sun, and these windows could be shaded by deciduous trees or an overhang to keep out the high summer sun. The popular all-glass American skyscrapers built during the 1960s are huge energy wasters, since their large areas of glass absorb the hot summer rays. Since their windows cannot be opened, at times the buildings' air conditioners must work at high levels just to cool their interiors to the same temperature as the outside air. Simple measures like planting trees to obtain shade can have a significant cooling effect on a house, a city street, or a parking lot, reducing temperatures by as much as 10 to 20 degrees over unshaded areas. Townhouses, the modern name for the old row houses, are again becoming popular in many cities; they are much more energy efficient than the common, single-family house because of their shared walls.

Saving energy often takes an initial investment, as the box "Conservation: the case of the inexpensive water heater," illustrates. Knowing this fact helps one

Conservation: the case of the inexpensive water heater

A personal blunder illustrates well several conservation principles. A few years ago my hot water heater stopped working. I was greatly relieved when the plumber assured me that he could replace it right away so we could soon have hot water again. I remembered that the government had a policy of rating appliances for their energy efficiency, and asked the plumber if the water heater he would install was energy efficient. "Oh yes," he replied, "it has a sticker on it." I was also pleased to hear that the price of the new heater was lower than I feared it might be. After the new heater was installed, I read the label on it, which rated its use of energy. I found to my dismay, and embarrassment, that the heater had the lowest rating for energy efficiency that it was possible to give on that model. A water heater lasts about 15 years, so I will have to pay for my error for some time. (Wrapping an insulation blanket around the water heater later helped to reduce the heater's deficiencies.)

In addition to teaching me that conservation requires careful planning, this little episode illustrated well to me a key aspect of energy conservation and explained why many people don't do it. Conservation often requires an initial investment – the more efficient water heaters are more expensive than the least efficient – and the decision to spend more now in order to save in the future is not always easy to make. People naturally look at the purchase price of the appliance – or home – and often follow the rule, "the cheaper the better," as long as the appliance or home is adequate. What that price does not tell you – but what the government's sticker on my water heater did tell me – is how much it will cost to run that appliance, or heat and air-condition that home, over its many years of use.

understand why the decontrol of prices of oil and natural gas, which will lead to higher prices of those fuels, is probably not enough by itself to cause many people to use less energy. The better educated and more affluent might recognize that an investment in insulation or a more expensive water heater makes good sense and will save them money over the long run, but those with lower incomes do not have the extra money to make the initial investment. Some of the poor spend a higher portion of their income on energy than do those on higher incomes, and thus could benefit greatly from the better insulated house or the more fuel-efficient car, but they usually end up with a poorly insulated house and a gas-guzzling car. Higher prices for fuel will probably help to reduce energy consumption, but stronger governmental incentives and regulations, such as substantially higher tax credits for installing insulation and substantially higher fuel efficiency standards for automobiles, could produce a significant movement toward conservation.

Some real progress is being made in conservation/energy efficiency efforts around the world, but much more can be done. Here is how Amory Lovins, an authority on the subject, sees the positive features:

> Many energy-efficient products, once costly and exotic, are now inexpensive and commonplace. Electronic speed controls, for example, are mass-produced so cheaply that some suppliers give them away as a free bonus with each motor. Compact fluorescent lamps cost more than $20 two decades ago but only $2 to $5 today; they use 75 to 80 percent less electricity than incandescent bulbs and last 10 to 13 times longer. Window coatings that transmit light but reflect heat cost one fourth of what they did five years ago.[84]

Lovins believes that Europe and Japan, although up to twice as energy efficient as the United States, can still make significant improvements in conserving energy. Even more opportunities to conserve energy exist in the developing countries, Lovins believes, where, on average, countries are three times less efficient than the United States. And finally Lovins is encouraged by what he sees in China, which has what he calls "ambitious but achievable goals" to shift from coal production to decentralized renewable energy and natural gas.[85]

Conclusions

The energy transition the earth is passing through is possibly the most important one human beings have encountered during their long evolution on the planet. The very suitability of the planet for high civilization is threatened by the fossil fuels that they rely on to power the machinery that makes their products, runs their transportation systems, fuels their high-tech agricultural systems, and heats and cools their homes. The burning of these fossil fuels is causing the climate to change in potentially very destructive ways. Wars are already being fought over the control of oil, the main fossil fuel the earth depends on at present. As long as that dependence remains, more conflicts seem likely.

Other energy sources are available but will take major efforts by governments and individuals to make them prominent. The careful reader of this book is learning about these renewable and nonpolluting sources of energy and of some of the difficulties standing in the way of their wider use.

Time is limited for this energy transition to occur. If too much time is taken, the population of the earth is large enough and its industrialization great enough – and both are still growing – that the changing climate could bring widespread suffering and destruction. The suffering and destruction will be caused, in large part, by the very energy that fuels our civilization.

Specifically, the efforts of the leading industrial nation, the one that produces more goods and services than any other and releases more pollutants that affect the climate than any other, have been very disappointing. American scientists have been leaders in gathering the evidence that the climate is changing because of human actions, but so far the US national government has been unresponsive. Has this been because of the political power of the fossil fuel and automotive industries which have opposed taking action, or is it because the American public has a lack of understanding that new energy initiatives are urgently needed for the long-term health of their country and of the planet itself? Or is it both?

Many European countries, along with Japan and others, are taking actions to address this issue. China is starting to address it but because of its heavy reliance on coal, it has become the second largest contributor to the problem of climate change – the energy/environmental issue with the potential to change our world.

Notes

1 An interesting discussion of the laws of thermodynamics and what they tell us about energy is contained in Jeremy Rifkin, *Entropy: A New World View* (New York: Viking Press, 1980), pp. 33–43.
2 Richard Kerr and Robert Service, "What Can Replace Cheap Oil – and When?" *Science*, 309 (July 1, 2005), p. 101.
3 The era of relatively cheap energy in the United States extended into the twenty-first century. Prices in table 4.1 are not adjusted for inflation. When inflation is considered, in the early twenty-first century gasoline was still relatively inexpensive in the United States.
4 *New York Times*, national edn (January 16, 1992), p. A7, and Youssef Ibrahim, "Gulf War's Cost to the Arabs Estimated at $620 Billion," *New York Times*, national edn (September 8, 1992), p. A4.
5 Joseph Romm, "Needed – a No-Regrets Energy Policy," *Bulletin of the Atomic Scientists*, 47 (July/August 1991), p. 31.
6 Lester R. Brown, *The Twenty-Ninth Day* (New York: W. W. Norton, 1978), pp. 205–6.
7 Daniel Yergin and Martin Hillenbrand (eds), *Global Insecurity: A Strategy for Energy and Economic Renewal* (Boston: Houghton Mifflin, 1982), p. 7.
8 Robert Livernash and Eric Rodenburg, "Population Change, Resources, and the Environment," *Population Bulletin*, 53 (Population Reference Bureau, Washington, DC) (March 1998), p. 27.
9 Timothy Wirth, C. Boyden Gray, and John Podesta, "The Future of Energy Policy," *Foreign Affairs*, 82 (July/August 2003), p. 134.
10 Neela Banerjee, "The SUV Is Still King, Even as Gas Prices Soar," *New York Times*, national edn (May 4, 2004), p. C6.
11 "The Hidden Cost of Gas," *Sierra*, 87 (March/April 2002), p. 15. See also Harold M. Hubbard, "The Real Cost of Energy," *Scientific American*, 264 (April 1991), p. 36.
12 Sarah Lyall, "Britons Running out of Patience as Costly Gas Runs Out," *New York Times*, national edn (September 12, 2000).
13 Simon Romero, "The $6.66-a-Gallon Solution," *New York Times*, national edn (April 30, 2005), p. B1.
14 Danny Hakim, "EPA Holds Back Report on Car Fuel Efficiency," *New York Times*, national edn (July 28, 2005), p. C1.
15 Bill Marsh, "One Recipe for a (Mostly) Emissions-Free Economy," *New York Times*, national edn (November 4, 2003), p. D3.
16 Ken Belson, "Why Japan Steps Gingerly in the Middle East," *New York Times*, national edn (September 17, 2002), p. W1.
17 David Sanger, "Japan Joins in Embargo against Iraq," *New York Times*, national edn (August 6, 1990), p. C7.
18 Belson, "Why Japan Steps Gingerly in the Middle East," p. W1.
19 Ibid., p. W7.
20 Howard French, "Safety Problems at Japanese Reactors Begin to Erode Public's Faith in Nuclear Power," *New York Times*, national edn (September 16, 2002), p. A8.
21 John H. Gibbons, Peter D. Blair, and Holly L. Gwin, "Strategies for Energy Use," *Scientific American*, 261 (September 1989), p. 142.
22 Howard French, "China's Boom Brings Fear of an Electricity Breakdown," *New York Times*, national edn (July 5, 2004), p. A4.
23 Chris Buckley, "China, as Summer Nears, Braces for Power Shortages," *New York Times*, national edn (April 8, 2004), p. W1, and Chris Buckley, "Chance to Revive Sales Draws Nuclear Industry to China," *New York Times*, national edn (March 10, 2004), p. W1.
24 Patrick Tyler, "China Planning People's Car to Put Masses behind Wheel," *New York Times*, national edn (September 22, 1994), p. A1.
25 Keith Bradsher, "China Pays a Price for Cheaper Oil," *New York Times*, national edn (June 26, 2004), p. B1.

26 Buckley, "Chance to Revive Sales Draws Nuclear Industry to China," p. W1.
27 Howard French, "As China Races to Build Nuclear Power Plants, Some Experts Still See Danger," *Spartanburg Herald-Journal* (January 16, 2005), p. A16.
28 Howard French, "In Search of a New Energy Source, China Rides the Wind," *New York Times*, national edn (July 26, 2005), p. A4.
29 Larry Rohter, "With Big Boost from Sugar Cane, Brazil Is Satisfying Its Fuel Needs," *New York Times*, national edn (April 10, 2006), p. A1.
30 Jeff Goodell, "Cooking the Climate with Coal," *Natural History*, 115 (May 2006), pp. 40–1.
31 Amory Lovins, "More Profit with Less Carbon," *Scientific American*, 293 (September 2005), p. 81.
32 Sanger, "Japan Joins in Embargo against Iraq." In the early 1990s Japan's energy intensity (the amount of energy required to produce a unit of gross domestic product) was the lowest of all market industrialized countries. These countries, in the previous two decades, had reduced their energy intensity by 25 percent, with the sharpest decline coming after 1979. World Resources Institute, *World Resources 1992–1993* (New York: Oxford University Press, 1992), p. 145.
33 Lovins, "More Profit with Less Carbon," p. 74.
34 Neela Banerjee, "Pushing Energy Conservation into the Back Seat of the SUV," *New York Times*, national edn (November 22, 2003), p. B2.
35 See, for example, Marc H. Ross and Daniel Steinmeyer, "Energy for Industry," *Scientific American*, 263 (September 1990), pp. 88–98; Gibbons et al., "Strategies for Energy Use," pp. 136–43; National Research Council, *Energy in Transition 1985–2010* (San Francisco: W. H. Freeman, 1980); Robert Stobaugh and Daniel Yergin (eds), *Energy Future: Report of the Energy Project at the Harvard Business School* (New York: Ballantine Books, 1980); Roger W. Sant, Steven C. Carhard, et al., *Eight Great Energy Myths: The Least-Cost Energy Strategy, 1978–2000* (Arlington, VA: Energy Productivity Center of the Mellon Institute, 1981).
36 Council on Environmental Quality and the Department of State, *Global Future: Time to Act* (Washington, DC: Government Printing Office, 1981), p. 61.
37 Joel Darmstadter, "Economic Growth and Energy Conservation: Historical and International Lessons," Reprint No. 154, Resources for the Future, Washington, DC, 1978, p. 18.
38 In 1974 the average fuel efficiency of all American cars was 14 miles per gallon. The law required that this be increased to 27.5 miles per gallon by 1985.
39 For an explanation of how US auto companies were still able to build fuel-inefficient dinosaurs, and some of the tactics they used to keep from being fined for violating the fuel economy law, see James Bennet, "Trucks' Popularity Undermining Gains in US Fuel Savings," *New York Times*, national edn (September 5, 1995), pp. A1, C3; and Keith Bradsher, "Auto Makers Seek to Avoid Mileage Fines," *New York Times*, national edn (April 3, 1998), p. A14.
40 Danny Hakim, "Fuel Economy Hit 22-Year Low in 2002," *New York Times*, national edn (May 3, 2003), p. B1. See also Matthew Wald, "Automakers Use New Technology to Beef up Muscle, Not Mileage," *New York Times*, national edn (March 30, 2006), p. C4.
41 J. T. Houghton, G. J. Jenkins, and J. J. Ephraums (eds), *Climate Change: The IPCC Scientific Assessment* (Cambridge, UK: Cambridge University Press, 1990).
42 Richard Kerr, "Millennium's Hottest Decade Retains Its Title, for Now," *Science*, 307 (February 11, 2005), p. 828.
43 Rodger Doyle, "Melting at the Top," *Scientific American*, 292 (February 2005), p. 31.
44 Worldwatch Institute, *Vital Signs 2005* (New York: W. W. Norton, 2005), p. 40.
45 Houghton et al., *Climate Change*, p. xxix.
46 Richard Kerr, "A Worrying Trend of Less Ice, Higher Seas," and Ian Joughin, "Greenland Rumbles Louder as Glaciers Accelerate," both in *Science*, 311 (March 24, 2006), pp. 1698–721.
47 World Resources Institute, UN Environment Programme, UN Development Programme, and World Bank, *World Resources 1998–1999* (New York: Oxford University Press, 1998), p. 174.
48 Ibid., p. 173.
49 Edward Brook, "Tiny Bubbles Tell All," *Science*, 310 (November 25, 2005), pp. 1285–7.
50 "Signs of a Changing Planet," *Sierra*, 90 and 91 (July/August 2005 and March/April 2006), pp. 17 and 13.
51 Gabriele Hegeri and Nathaniel Bindoff, "Warming the World's Oceans," *Science*, 309 (July 8, 2005), p. 254.

52 Andrew Revkin, "Climate Data Hint at Irreversible Rise in Seas," *New York Times*, national edn (March 24, 2006), p. A12.
53 T. P. Hughes et al., "Climate Change, Human Impacts, and the Resilience of Coral Reefs," *Science*, 301 (August 15, 2003), p. 929.
54 Richard Kerr, "Three Degrees of Consensus," *Science*, 305 (August 13, 2004), p. 932.
55 J. Hansen et al., "Climate Impact of Increasing Atmospheric Carbon Dioxide," *Science*, 213 (August 28, 1981), p. 966.
56 Houghton et al., "Climate Change," pp. xi, xxvii.
57 Andrew Revkin, "Report Forecasts Warming Effects," *New York Times*, national edn (June 12, 2000), pp. A1, A25.
58 World Resources Institute et al., *World Resources 1998–1999*, p. 72.
59 Richard Alley, "Abrupt Climate Change," *Scientific American*, 291 (November 2004), pp. 62–9.
60 Detlef Quadfasel, "The Atlantic Heat Conveyor Slows," *Nature*, 438 (December 1, 2005), pp. 565–6, and Andrew Revkin, "Scientists Say Slower Atlantic Currents Could Mean a Colder Europe," *New York Times*, national edn (December 1, 2005), p. A8.
61 Richard Kerr, "Is Katrina a Harbinger of Still More Powerful Hurricanes?" *Science*, 309 (September 16, 2005), p. 1807, and Claudia Dreifus, "With Findings on Storms, Centrist Recasts Warming Debate," *New York Times*, national edn (January 10, 2006), p. D2.
62 William Chameides and Michael Bergin, "Soot Takes Center Stage," *Science*, 297 (September 27, 2002), pp. 2214–15.
63 World Resources Institute et al., *World Resources 1998–1999*, p. 170.
64 Ibid., p. 176.
65 Ibid., pp. 176–7.
66 Humus, the organic material in topsoil, also stores large amounts of carbon.
67 Richard Kerr, "Bumpy Road Ahead for World's Oil," *Science*, 310 (November 18, 2005), p. 1106.
68 Kerr and Service, "What Can Replace Cheap Oil – and When?" p. 101.
69 Daniel Yergin and Michael Stoppard, "The Next Prize," *Foreign Affairs*, 82 (November/December 2003), p. 108.
70 Simon Romero, "Demand for Natural Gas Brings Big Import Plans, and Objections," *New York Times*, national edn (June 15, 2005), p. C8.
71 Goodell, "Cooking the Climate with Coal," p. 37.
72 James Lake, Ralph Bennett, and John Kotek, "Next-Generation Nuclear Power," *Scientific American*, 286 (January 2002), pp. 73–81.
73 Bill March, "One Receipt for a (Mostly) Emissions-Free Economy," p. D3.
74 Mark Fischetti, "Turn Turn Turn," *Scientific American*, 287 (July 2002), p. 86.
75 Marlise Simons, "Wind Turbines Are Sprouting Off Europe's Shores," *New York Times*, national edn (December 8, 2002), p. 3.
76 Larry Rohter, "With Big Boost from Sugar Cane, Brazil Is Satisfying Its Fuel Needs," *New York Times*, national edn (April 10, 2006), p. A1.
77 Worldwatch Institute, *Vital Signs 2005*, p. 36.
78 Paul Meller, "Europe and US Will Share Research on Hydrogen Fuel," *New York Times*, national edn (June 17, 2003), p. W1, and Matthew Wald, "Will Hydrogen Clear the Air? Maybe Not, Some Say," *New York Times*, national edn (November 12, 2003), p. C1.
79 Matthew Wald, "Questions about a Hydrogen Economy," *Scientific American*, 290 (May 2004), p. 68.
80 Matthew Wald, "Report Questions Bush Plan for Hydrogen-Fueled Cars," *New York Times*, national edn (February 6, 2004), p. A19.
81 Steven Ashley, "On the Road to Fuel-Cell Cars," *Scientific American*, 292 (March 2005), p. 68.
82 Stobaugh and Yergin, *Energy Future*, p. 10.
83 Lovins, "More Profit with Less Carbon," p. 74.
84 Ibid., p. 76.
85 Ibid., p. 83.

Further Reading

Berger, John, *Charging Ahead: The Business of Renewable Energy and What It Means for America* (New York: Henry Holt, 1997). Berger believes that the industrialized nations are on the threshold of the world's next great energy transformation: a vast expansion of clean, renewable energy. He focuses on leading-edge companies, the pioneers of renewable energy, and describes the foot dragging of government and fossil-fuel giants.

Gelbspan, Ross, *The Heat Is On: The High Stakes Battle over Earth's Threatened Climate* (Reading, MA: Addison-Wesley, 1997). Gelbspan, a Pulitzer Prize-winning journalist, argues that government policymakers have not taken action on climate change in the United States because of a clever propaganda campaign by the oil and coal industry and OPEC which, with the aid of uninformed and credulous news media, has convinced policymakers that global warming is a theory and not a fact.

Goodell, Jeff, *Big Coal: The Dirty Secret behind America's Energy Future* (New York: Houghton Mifflin, 2006). The secret is that much (about one-half) of America's electricity is generated from burning coal and this is rarely acknowledged. Coal is used because it is cheap and abundant. Goodell writes: "Our shiny white iPod economy is propped up by dirty black rocks."

Goodstein, David, *Out of Gas: The End of the Age of Oil* (New York: W. W. Norton, 2004). The author is a professor of physics and vice provost of the California Institute of Technology. He predicts that the peak of oil production in the world will come in the present decade or the next and that after that oil supplies will decline forever. Goodstein makes a grim forecast: unless we solve our energy and environmental crises soon, "civilization as we know it will not survive."

Hansen, James, "Defusing the Global Warming Time Bomb," *Scientific American*, 290 (March 2004), pp. 70–7. Hansen is the Director of the NASA Goddard Institute for Space Studies and was one of the first to testify before the US Congress about climate change, thus trying to alert the nation to the dangers of ignoring this subject.

Hoffmann, Peter, *Tomorrow's Energy: Hydrogen, Fuel Cells, and the Prospects for a Cleaner Planet* (Cambridge, MA: MIT Press, 2001). The book covers the main aspects of this subject from the history of its discovery and the various attempts in the past to use hydrogen as a fuel, its safety record, and the difficulties of using it as a widely available fuel.

Klare, Michael, *Blood and Oil: The Dangers and Consequences of America's Growing Petroleum Dependency* (New York: Metropolitan Books, 2004). Since America's own supply of oil is drying up, its demand is increasing, and the sources of oil in the world today are often in conflict-ridden areas with strongly anti-American sentiments, Klare sees conflict as inevitable.

Philander, S. George, *Is the Temperature Rising? The Uncertain Science of Global Warming* (Princeton, NJ: Princeton University Press, 1998). In simple, nontechnical language, this book by a professor of geoscience at Princeton is an introduction to the basics of the earth's climate and weather. Philander shows that because weather and climate are so complex, there will always be uncertainties and scientific dissent. He argues that to continue to defer action on environmental problems until more accurate scientific information is available could lead to a crisis.

"Powering the Next Century," *Science*, 285 (July 30, 1999), pp. 677–711. A special feature on energy focuses on alternative approaches for generating, distributing, and conserving energy. Ten articles are included in this special report.

Romm, Joseph, *The Hype about Hydrogen* (New York: Island Press, 2004). Romm sees a hydrogen-based transportation system in the United States being at least 30 years away and an infrastructure needed to use it costing hundreds of billion dollars to build. He does not believe we can wait that long to solve our global warming crisis.

Schneider, Stephen H., *Laboratory Earth: The Planetary Gamble We Can't Afford to Lose* (New York: HarperCollins, 1997). A leading climatologist from Stanford University agues that the United States is already behind in dealing with the disastrous shift in the planet's climate.

Smil, Vaclav, *Energy at the Crossroads: Global Perspectives and Uncertainties* (Cambridge, MA: MIT Press, 2003). Smil has spent four decades studying energy. Here he presents an introduction to the complexities and uncertainties of the subject. He shows the difficulties of changing from our dependence on carbon-emitting fuels, at the same time explaining the need to do so.

Vaitheeswaran, Vijau, *Power to the People: How the Coming Energy Revolution Will Transform an Industry, Change Our Lives, and Maybe Even Save the Planet* (New York: Farrar, Straus & Giroux, 2003). An optimistic, entertaining book: the author believes that improvements in energy efficiency, the use of renewable, clean fuels, and new technology to capture carbon could enable us to pass through the energy transition safely.

Weart, Spencer, *The Discovery of Global Warming* (Cambridge, MA: Harvard University Press, 2003). The *New York Times* called this book the intellectual journey to belief in global warming.

World Resources Institute, British Petroleum, General Motors, and Monsanto, *Building a Safe Climate: Sound Business Future* (Washington, DC: World Resources Institute, 1998). In an unusual collaboration, these four organizations explored the question whether it is possible to have a healthy economy and a healthy environment. They present their conclusions in this report.

CHAPTER 5

The Environment

We travel together, passengers on a little spaceship, dependent on its vulnerable resources of air and soil; all committed for our safety to its security and peace; preserved from annihilation only by the care, the work, and I will say, the love we give our fragile craft.

Adlai E. Stevenson (1900–65)

The Awakening

The relationship between the environment and development has not been a happy one. Development has often harmed the environment, and the environmental harm has in turn adversely affected development. Industrialization brought with it many forms of pollution, pollution that is undermining the basic biological systems upon which life rests on this planet. It took millions of years for these systems to evolve.

The first world conference on the environment was held in Stockholm, Sweden, in 1972 under the auspices of the United Nations. At that conference the developed nations, led by the United States, pushed for greater efforts to protect the environment, while many less developed nations feared that an effort to create strict antipollution laws in their countries would hurt their chances for economic growth. The developing nations maintained that poverty was the main cause of the deterioration of the environment in their countries. What they needed, they said, was more industry instead of less.

Ten years later, the nations of the world again met together to discuss the state of the global environment, this time in Nairobi, Kenya. The positions of the rich and poor nations had changed dramatically. The developing nations generally showed enthusiasm for further efforts to protect the environment, since in the ten years between the conferences they had seen that environmental deterioration, such as desertification, soil erosion, deforestation, and the silting of rivers and reservoirs, was harming their efforts to develop and to reduce poverty. On the other hand, many of the rich nations at Nairobi, led by the United States, called for a slowing down of environmental initiatives until they had recovered from their economic recessions.

Even though the positions of the developed and developing nations had become somewhat reversed during the ten years between the two environmental conferences, there is no doubt that an awareness of the threat to the environment caused by human activities had by 1982 become worldwide. Only 11 nations had any kind of governmental environmental agency at the time of the first conference, whereas over 100 nations, 70 of them in the developing world, had such agencies at the time of the second. These agencies did much to educate their own governments and people about environmental dangers.

In 1992 the third environmental conference sponsored by the United Nations was held in Rio de Janeiro, Brazil. Popularly called the Earth Summit, and formally the Conference on the Environment and Development, it was attended by the largest number of leaders of nations in history. They were joined by about 10,000 private environmentalists from around the world and 8,000 journalists. Although frequent clashes took place between the representatives of Northern rich countries and relatively poor Southern countries in the preparatory meetings, which took place during the two years preceding the conference, two major treaties were signed by about 150 nations at the conference. One concerned the possible warming of the earth's climate, which was discussed above in the chapter on energy. The treaty called on nations to curb the release of so-called greenhouse gases that may be causing a change in the world's climate. Because of the insistence of the United States, no specific targets or timetables were placed in the treaty, but the treaty did call for nations to eventually reduce the emissions of their greenhouse gases to 1990 levels.

The second treaty – the biodiversity treaty, providing for the protection of plant and animal species – was signed by most nations. The United States did not sign it and stood fairly alone in its opposition to it. The opposition by the first Bush administration in the United States to these environmental initiatives can be explained partly because 1992 was a presidential election year and President Bush was vulnerable to attack because of slow economic growth and a huge governmental deficit. (He was in fact defeated for reelection in large part because of poor economic conditions in the United States.)[1]

The conference made the term "sustainable development" known throughout the world. The term means that economic growth in the present should not take place in such a manner that it reduces the ability of future generations to live well. Economic growth and efforts to improve the living standards of the few or the many should be sustainable; in other words, they should be able to be continued without undermining the conditions that permit life on earth, thus making future development impossible or much more difficult. The term represents an effort to tie economic growth and protection of the environment together, a recognition that future economic growth is possible only if the basic systems that make life possible on earth are not harmed. It also implies a recognition that economics and the environment are *both* important, that economic development and the reduction of poverty are essential to the protection of the environment.

Sustainable development was endorsed by the conference and a new organization – the Sustainable Development Commission – was set up under the United Nations to monitor the progress nations are making to achieve it.

In 2002 the fourth United Nations environmental conference was held in Johannesburg, South Africa, under the title: World Summit on Sustainable Development. One of the main clashes between nations was unusual. The European Union proposed a target which nations would set as their goal for switching to renewable energy from fossil fuels. The goal was to have 15 percent of world energy to come from renewable sources by 2010. European nations felt that in order to hold nations responsible for their actions, specific targets were necessary for without them it is difficult to measure progress or lack of progress. The United States strongly opposed this provision and with the help of oil-producing nations such as Saudi Arabia and Canada, along with Japan, got the provision dropped from the final conference agreement. The final agreement endorsed many of the Millennium Development Goals, discussed in chapter 1. But many environmentalists were disappointed with the results of the conference, many blaming the lack of leadership by the United States, whose president, George W. Bush, was one of the few country leaders who didn't attend the conference. In the words of the UN Secretary-General Kofi Annan, "Obviously, this is not Rio." In order to recognize some positive features of the conference, the head of the United Nations Environment Programme gave this evaluation: "Johannesburg is less visionary and more workmanlike, reflecting perhaps a feeling among many nations that they no longer want to promise the earth and fail, that they would rather step forward than run too fast."[2]

Nations vary greatly on how well they are treating the environment. At the time this third edition of *Global Issues* was written, a recent study by Yale and Columbia universities in the US rated New Zealand and five Northern European countries as the best, while the United States was rated twenty-eighth.[3] Rapidly economically growing China's environmental problems are "grave," according to a high Chinese government official, and are costing the nation more than $200 billion a year, 10 percent of the nation's GDP.[4]

In this chapter we will examine some of the effects development has had on the air, water, and the land of our planet. We will then focus on some of the dangers that have been created in the workplace and in the home. After looking briefly at the use of natural resources in the world, we will learn why the extinction of species is accelerating. Development's role in the extinction of human cultures will be explored also. The chapter will end with an explanation of what makes environmental politics so controversial.

The Air

Smog

Industrialization has brought dirtier air to all parts of the earth. From factories and transportation systems, with their telltale smokestacks and exhaust pipes, toxic fumes are being emitted constantly into the air. A few spectacular instances in the twentieth century resulted in large numbers of people becoming ill or dying because of the toxic gases in the air they breathed: 6,000 became ill and 60 died in the Meuse Valley in Belgium in 1930; 6,000 became ill and 20 died in Donora, Pennsylvania in 1948; and in London tens of thousands became ill and 4,000 died in 1952. (It was this last-mentioned instance that led the United Kingdom to pass various laws to clean up the air, which have proved to be quite successful. By the late 1970s, 80 percent more sunshine reached London than had in 1952.)[5]

A number of industrialized countries, including the United States, have made significant progress in reducing air pollution in their large urban areas. Since the 1970 Clean Air Act was passed in the United States, lead has been reduced by 95 percent, sulfur dioxide by about 30 percent, and particulates (tiny particles in the air) by about 60 percent.[6] And there is even evidence that a sharp decline in air pollution in the Arctic occurred in the 1980s, probably because the Soviet Union had shifted from coal and oil to the cleaner natural gas and Western Europe had reduced its sulfur dioxide emissions in order to combat acid rain.[7]

In spite of this progress, much remains to be done. In the late 1990s about 80 million people in the United States still lived in urban areas where the air was considered poor enough to be called unhealthy during some days of the year.[8] From 2000 through 2002 parts of southern California, the area with the dirtiest air in the country, had over 400 days when the air was unhealthy to breathe.[9] (There has been significant progress made in cleaning the air in Los Angeles. In 1997 there was only one Stage One smog alert – when people with respiratory ailments are advised to stay indoors and others to refrain from vigorous exercise outdoors – whereas there were 122 such warnings in 1977 and 148 in 1970.) But even some of the progress cited above, the reduction of particulates, turned out to be less impressive than first thought. Studies in the 1990s and early 2000s indicated that extremely small particulates spewed by vehicles, factories, and coal power plants, which were not illegal to release until 1997, were the greatest risk to health and estimated to be causing up to 60,000 premature deaths a year in the United States.[10]

Asthma rates among children in the US have been increasing rapidly and may be linked to air pollution. Researchers from Harvard University and the American Public Health Association predicted in 2004 that asthma rates would continue to worsen because of fossil fuel emissions from a growing number of cars, pollen, and from mold increases due to global warming. Poor and minority children in the inner cities had the highest rates of asthma.[11]

Europe has improved its air also, but more needs to be done there as well. A recent study of air pollution in Austria, France, and Switzerland found that car-related air pollution kills more people in the three countries than do car accidents.[12] Studies in the Czech Republic and Mexico City found that the risk of infant death is double when air pollution is worse.[13] And a 2002 study of the effects of air pollution on farming found that ozone, a key component of smog, was costing European farmers about 6 billion euros annually.[14]

Less developed countries that were starting to industrialize and modernize were facing air pollution problems even greater than those of the West. Cities such as Mexico City, Bangkok, Beijing, Delhi, and Jakarta had serious air pollution. In most

Your "friendly" coal power plant

About one-half of the electricity in the United Slates comes from power plants fueled by coal. A typical 500-megawatt coal plant, which can power a city of about 140,000 people, burns about 40 train cars of coal each day and yearly releases into the air the following pollutants: 3.7 million tons of carbon dioxide, 10,200 tons of nitrogen oxide, 10,000 tons of sulfur dioxide, 720 tons of carbon monoxide, 500 tons of small particles, 220 tons of hydrocarbons, and 225 pounds of arsenic, 170 pounds of mercury, 114 pounds of lead, 4 pounds of cadmium and other toxic heavy metals.

Source: "A Typical Coal Plant," *Nucleus* (Spring 2000), p. 5

Vehicles, such as this truck/bus, provide a lot of air pollution in the cities of the developing countries (*Ab Abercrombie*)

of the megacities of the South air pollution is worsening because of increased industry, vehicles, and population. Pollution levels sometimes exceed the air quality standards of the World Health Organization (WHO) by a factor of three or more. The WHO estimated that about 1.4 billion people in urban areas breathe air with pollution exceeding the WHO guidelines.

Mainly because of its heavy use of coal, China has some of the worst air pollution in the world. Levels of particulates in the air in some of its major cities exceed WHO levels by at least six times. In the early 2000s, using looser standards than are used by the UN and Western countries, China estimated that 60 percent of its major cities have serious air pollution.[15] The sale of cars is increasing about 80 percent a year and Chinese refineries were purchasing a cheaper quality of petroleum which had a very high sulfur content, an amount high enough to ruin catalytic converters being installed in China's new cars to reduce air pollution.[16] It is estimated that 25 percent of deaths in China are caused by lung disease at least partly attributed to serious urban and household air pollution as well as to widespread cigarette smoking.[17] By the mid-2000s China's concern with pollution was growing, but efforts to reverse the damage done to the environment and being done by a quarter-century of rapid economic growth were still inadequate to the task.

Indoor air pollution can also be severe in poorer nations, mostly in rural areas, but also sometimes in urban areas, where 3.5 billion people rely on traditional fuels for heating and cooking. Wood, straw and dung are often used as fuels, and women and children especially are exposed to the smoke when these fuels are burned. In many of these dwellings the air pollution indoors is far worse than outdoor pollution. The World Bank has identified indoor pollution in developing nations as one of the four most urgent environmental problems.[18]

With the collapse of communist regimes in Eastern Europe and in the Soviet Union in the late 1980s and early 1990s came evidence of startling amounts of pollution in those countries, pollution that had been kept secret for many years

The world's largest pollution cloud

A two-mile thick cloud of brownish haze, about 4 million square miles large, has been discovered high over the Indian Ocean by scientists. The cloud, about the size of the United States, is composed of pollutants, mainly from the burning of fossil fuels as well as from forest fires and wood-burning stoves in the Indian subcontinent, China, and Southeast Asia. Winds during the winter monsoons bring the pollution out to sea as the prevailing winds are coming down from the Himalayan Mountains. In the late spring and summer the winds reverse and the haze is blown back over the land. The pollutants combine with the monsoon rains and come back to earth as acid rain.

The UN Environment Programme (UNEP) reported in 2002 that the cloud had killed tens of thousands of people in the past ten years, including 52,000 in India in 1995 alone.

Sources: William Stevens, "Enormous Haze Found over Indian Ocean," *New York Times*, national edn (June 10, 1999), p. A23; Worldwatch Institute, "Air Pollution Still a Problem," in *Vital Signs 2005* (New York: W. W. Norton, 2005), pp. 94–5

to prevent it being used to criticize the regimes. Poland had some of the worst air pollution, in part caused by the heavy use of coal in its industry. By 1990 the air was so polluted in some industrial regions of Poland that adults and children suffering from respiratory problems were placed in salt mines, 600 feet below the surface, so they could breathe unpolluted air. The Polish Academy of Science in 1990 reported that one-third of the nation was living in "areas of ecological disaster."[19]

Airborne lead

The story of airborne lead illustrates well the connection between industrialization and air pollution. Scientists are able to estimate the amount of lead there was in the world's air in the past by taking core samples of ice in the Greenland ice cap. The air bubbles in the ice, ice which represents past rainfall, shows that from 800 BC to the beginning of the industrial revolution around 1750, the amount of lead in the air was low. There was a major increase after 1750 and a huge increase after World War II when the use of leaded gasoline rose sharply. In 1965 the lead concentration in the Greenland ice was 400 times higher than the level in 800 BC. Other studies showed that, in 1980, the bones of Americans contained 500 times more lead than those of prehistoric humans.[20]

Children are the ones who are most susceptible to harm from breathing lead. They inhale two to three times as much lead in the air per unit of body weight as do adults because their metabolic rates are higher and they have greater physical activity than adults. *There is no known safe level of lead in the human body.* High levels of lead poisoning can lead to death, but even low levels can cause learning difficulties and behavioral problems. In the early 2000s nearly a million children in the United States were estimated to have unhealthy levels of lead (as defined by the Centers for Disease Control (CDC), the US Government's top health organization) in their blood,[21] and black children were more likely to have high levels of lead than were white children. Many of the black children live in old houses or apartments in the inner cities where they consume lead from lead-based paint flaking off from walls (which young children, who tend to put everything in their mouths, wind up ingesting), and from old lead water pipes. (Disturbing research shows that the lead paint industry in the US actively promoted the use of leaded paint for 40 years, even after studies showed that lead could poison children and its use had been banned or restricted in a number of countries.[22])

Research published in the *New England Journal of Medicine* (one of the most respected health journals in the US) in 2003 found that levels of lead even lower than the CDC acceptable level affected children's brains and that there was no known way to restore intelligence lost because of lead damage. These studies indicate the possibility that 90 percent of children in the US have been harmed by lead poisoning.[23]

There has been a significant improvement in reducing the amount of lead in the blood of Americans. Between 1976 and 1991 the amount dropped by nearly 80 percent.[24] Because of tighter federal government air pollution requirements, new cars were required to use nonleaded gasoline, and many experts believe that

Aggressiveness and delinquency linked to lead in bones

A study of 800 boys in the United States showed a direct link between the amount of lead in the boys' bones and their behavior. Those with a relatively high level of lead in their bones had more aggressiveness and delinquency than those boys with low levels of lead. Other studies have shown that childhood antisocial behavior is a strong predictor of criminal behavior as an adult. The director of the study cautioned that the study did not show that lead was the cause of childhood delinquency, but only that it was probably one cause. That is not surprising, he stated in an interview, because "lead is a brain poison that interferes with the ability to restrain impulses."

Source: Jane E. Brody, "Aggressiveness and Delinquency in Boys Is Linked to Lead in Bones," *New York Times*, national edn (February 7, 1996), p. B6

the reduced use of leaded gasoline was the cause of the lower lead levels in blood. About 90 percent of the lead in air comes from leaded gasoline. After reducing lead in gasoline, a total ban on leaded gasoline in the US came into effect in 1995. The European Union effectively banned all leaded gas in 2000. Some lead experts believe that recent lower levels of violent crime in the US are the result of lower lead blood levels in US children born after 1980 rather than better crime enforcement.[25]

Another indication that the efforts to reduce the lead used in gasoline were having a beneficial effect can be seen by a study of the Greenland ice cap conducted in the early 1990s. This study of the lead concentrations in Greenland snow showed a drop in the lead concentrations to about the levels existing in the early 1900s, before the widespread use of leaded gasoline.[26]

Ninety percent of gasoline used in the world is now lead free and some developing countries are starting to come to grips with this problem. One of the few studies of the lead problem in Africa was made in Egypt in 1995 and updated since. It found that IQ levels of children had been lowed four points because of lead poisoning and that adults exposed to high levels of lead had higher rates of heart attack and other health problems than those who had not been so exposed.[27]

Although significant improvements have been made, lead remains a problem in the US. One-quarter of the homes in the country with children under six contain lead-based paint,[28] and US industry still releases large amounts of lead into the atmosphere – in 2001 it was about 440 million pounds.[29]

Acid rain

When fossil fuels are burned, sulfur dioxide and oxides of nitrogen are released into the air. As these gases react with moisture and oxygen in the atmosphere in the presence of sunlight, the sulfur dioxide becomes sulfuric acid (the same substance used in car batteries) and the oxides of nitrogen become nitric acid. These acids then return to earth in rain, snow, hail, or fog. When they do, they can kill fish in lakes and streams, dissolve limestone statues and gravestones, corrode metal, weaken trees, making them more susceptible to insects and drought, and reduce the growth of some crops. The effects of acid rain on human health are not yet known. Some scientists fear that acid rain could help dissolve toxic metals in water pipes and in the soil, releasing these metals into people's water supplies.

In the United States, acid rain comes mainly from sulfur dioxide produced by coal-burning electricity-generating power plants in the Midwest and from the nitrogen oxides from auto and truck exhausts. Acid rain has caused lakes in the northeastern part of the country to become so acidic that fish and some other forms of life are unable to live in them. Other areas of the country, such as large parts of the South, Northwest, Rocky Mountains, and the northern Midwest, are especially sensitive to acid rain since the land and lakes in these areas contain a low amount of lime. Lime tends to neutralize the falling acid. An international dispute was created between Canada and the United States because a large amount of the acid rain falling on huge sections of Canada comes from industrial emissions in the United States.

Europe is facing a similar problem. Many lakes in Norway and Sweden are now so acidic that fish cannot live in them, and about one-third of the forests in Germany are sick and dying. Much of the acid rain falling in Northern and Central Europe comes from industry in Britain, Germany, and France. The section of Europe with the greatest damage from acid rain lies in Eastern Europe. The efforts of the communist governments in that region to keep up with the West led to industrial growth fueled with lignite coal, which is cheap and abundant in the region but also extremely polluting. In one area where East Germany, Czechoslovakia, and Poland met more than 300,000 acres of forests have disappeared and the ground is poisoned by the huge amount of acid rain that fell there from the coal-fed power plants and numerous steel and chemical plants. Local foresters dubbed the area the "Bermuda Triangle of pollution," as winds carried the sulfur dioxide and other pollutants to other areas of Europe.

Acid rain was first observed in industrial England in the late 1800s, but nothing was done about it. In the 1950s the response to the increasing air pollution in the United States and Europe was to build tall smokestacks on factories so that emissions of toxic gases would be dispersed by the air currents in the atmosphere. These tall smokestacks led to a noticeable improvement in the air around many factories, smelters, power plants, and refineries, but the dispersal of noxious gases in the atmosphere gave more time for these gases to form into acid rain. We now realize that the tall smokestacks violated a fundamental law of ecology, one that biologist Barry Commoner has labeled the "everything must go somewhere" law.[30] Matter is indestructible, and there are no "wastes" in nature. What is excreted by one organism as waste is absorbed by another as food. When the food is toxic, the organism dies. Thus is explained the beautiful clear water in lakes with a high acid content: many forms of plankton, insects, and plants have ceased to exist there.

In 1990 the US Congress passed the Clean Air Act, which calls for a large reduction of sulfur dioxide emissions from power plants. An innovative provision was put into the law that allows polluters to buy and sell their rights to pollute (the total amount of emissions indicated in the law must not be exceeded and the level will be lowered over time). The hope was that this provision would encourage polluters to find the cheapest way to cut their pollution. This hope has generally been realized according to the US Environmental Protection Agency (EPA). During the decade of the 1990s the cost of the scrubbers on coal plants that remove sulfur dioxide fell by about 40 percent, thus making it cheaper for plants to remove the pollutant. Because of this result, the sulfur dioxide "cap and trade" program has been cited as a successful example of free-market environmentalism and has served as a "model" for other programs along the same line.[31]

But the 1990 Clean Air Act or the "cap and trade" system has not solved the acid rain problem in the US. A 2001 study of the problem by ten scientists showed that even after ten years of reducing sulfur dioxide, the acidic level in many lakes in the northeastern US had not decreased. The scientists found that the problem was much more complex than they had assumed and that the lingering effects of acid rain are stronger than they had anticipated. The scientists concluded that the Clean Air Act was much too weak to solve the problem and more drastic reduction of sulfur dioxide was needed, as well as reductions of nitrogen oxides, which are still relatively less regulated.[32]

A study in 1996 of the forests in Europe found that of the trees sampled in more than 30 countries, 25 percent were damaged, that is, they had lost more than a quarter of their leaves. While the study could not prove acid rain was the cause, one-half of the countries participating in the study believed that air pollution was a cause.[33]

Acid rain is now becoming a major problem in Asia with the increased use of fossil fuels as industrialization spreads. High levels of acid rain have been reported in southeast China, northeast India, Thailand, and South Korea, which are near or downwind from major urban and industrial centers. According to a high government official in China, in 2004 acid rain was falling on two-thirds of the country.[34]

Ozone depletion

The ozone layer in the atmosphere protects the earth from harmful ultraviolet rays from the sun. Scientists believe that life on earth did not evolve until the ozone layer was established. That layer is now being reduced by substances produced by humans, mainly in the developed nations. Chlorofluorocarbons (CFCs) – used as a propellant in aerosol spray cans, as a fluid in refrigerators and air conditioners, as an industrial solvent, and in the production of insulating foams – can destroy ozone. Ozone can also be destroyed by halons, which are chemicals used in fire extinguishers, and when nuclear bombs are exploded.

Scientists are agreed that any major depletion of the ozone layer would cause serious harm to humans, other mammals, plants, birds, insects, and some sea life.

Skin cancer would increase, as would eye cataracts. Increased ultraviolet light could also adversely affect the immune system of humans, which protects them from many possible illnesses. As is mentioned in chapter 6, one of the most harmful effects of a nuclear war would be the damage it would do to the ozone layer, which would affect life far beyond the combat area.

By analyzing past data, British scientists in the mid-1980s discovered that, during two months of the year, a hole was occurring in the ozone layer over the South Pole. Almost every year since it was discovered, the hole has continued to get larger. The hole (which is actually a significant reduction in the ozone normally found above that region, not a 100 percent decrease) galvanized the world to act to reduce the danger. About 60 nations met in Montreal, Canada, in 1987, and agreed to cut the production of CFCs by 50 percent by 1998. But further evidence that the depletion of the ozone layer was progressing faster than expected led the nations of the world to meet again – this time in London in 1990. The 90 nations attending that meeting agreed to speed up the phasing out of ozone-destroying chemicals. They agreed to halt the production of CFCs and halons by the year 2000. Less developed nations were given until 2010 to end their production and a fund was set up, mainly contributed to by the industrial nations, to help the poorer nations obtain substitutes for ozone-depleting chemicals.

New disturbing evidence of the ozone depletion danger was made public in the early 1990s. The US Environmental Protection Agency announced in 1991 that data from satellites, which had been collected over the previous 11 years, revealed that the ozone layer over large parts of the globe, including the layer above the United States and Europe, had been depleted by about 5 percent. This loss was occurring twice as fast as scientists had predicted. Based on the new findings, the agency calculated that over the next 50 years about 12 million people in the United States will develop skin cancer and more than 200,000 of them will die from it.[35]

Based on the new US evidence and on new data collected by an international team of scientists, which showed that the depletion was occurring in the dangerous summer months as well as in the winter, 90 nations met in Copenhagen, Denmark, in 1992 and agreed to further accelerate the ending of ozone-destroying chemicals. All production of CFCs was to end by 1996 and halon production was to end by 1994. (Developing nations were again given a ten-year grace period to phase out the production of these two chemicals.)

Chlorine compounds enter the atmosphere mainly as a component of CFCs and it is chlorine that scientists now believe is causing the destruction of the ozone layer. One atom of chlorine can destroy 100,000 atoms of ozone. CFCs will remain in the atmosphere for about 50 to 100 years.

The Montreal Protocol has brought impressive results. The transition away from the widely used CFCs and other ozone-depleting chemicals has been faster than many thought possible. In the ten-year period after the Protocol was signed in 1987, consumption of these chemicals dropped over 70 percent, with most developed nations meeting the Protocol's goal, as amended, to cease CFC production by 1996. Figure 5.1 shows the sharp decline in CFC production from 1989 to 1995. CFC concentrations in the atmosphere have peaked and have started a slow decline.

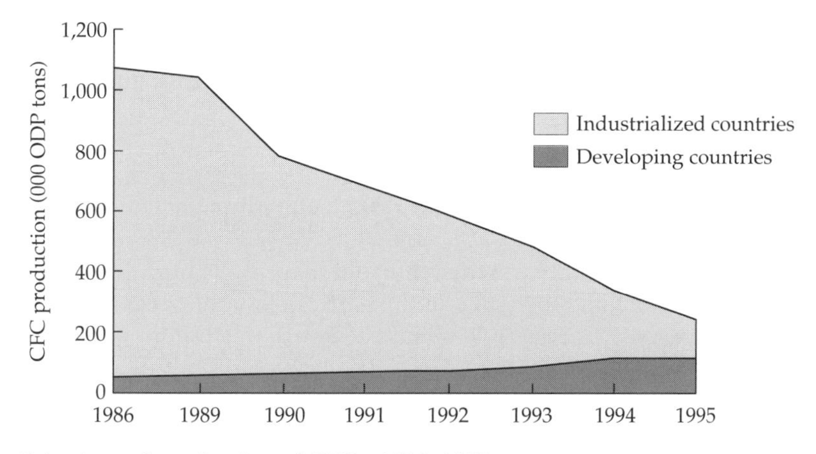

Figure 5.1 Annual production of CFCs, 1986–1995

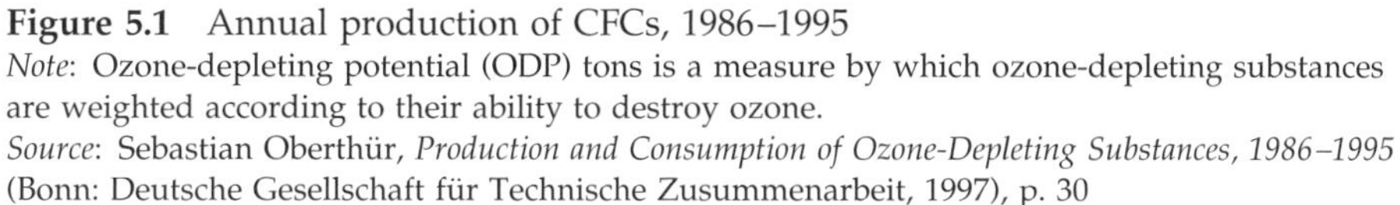

Note: Ozone-depleting potential (ODP) tons is a measure by which ozone-depleting substances are weighted according to their ability to destroy ozone.
Source: Sebastian Oberthür, *Production and Consumption of Ozone-Depleting Substances, 1986–1995* (Bonn: Deutsche Gesellschaft für Technische Zusummenarbeit, 1997), p. 30

The depletion of the ozone layer: how to protect yourself

People should try to keep out of the sun from 11 a.m. to 3 p.m. – when ultraviolet (UV) rays are at their strongest – and they should use hats and sunscreen lotion, which protect against both UV-A and UV-B rays, when exposed to the sun. Sunglasses that block UV rays should also be worn. While it is important to use sunscreen lotions, dermatologists agree that people who are at especially high risk of getting melanoma, the deadly kind of skin cancer, should not rely on sunscreens but should stay out of the sun. At high risk are those with a fair skin who sunburn easily, those with many moles, and those with a family history of skin cancer. Clouds offer little protection, but shade does. Melanoma rates are now on the increase in many countries, including the United States.

Sources: Walter Willett et al., "Strategies for Minimizing Cancer Risk," *Scientific American*, 275 (September 1996), pp. 94–5; Jane Brody, "How to Save Your Skin in the Season of Sun," *New York Times*, national edn (May 24, 2005), p. D7

Ozone loss is expected to gradually diminish until about 2065, when the ozone layer is expected to return to its 1980 condition.[36]

A development that could delay the restoration of the ozone layer is that scientists have found a disturbing connection between ozone depletion and climate change. Different processes produce these, but studies have shown that heat-trapping greenhouse gases, such as carbon dioxide, while warming the lower atmosphere, actually cool the high stratosphere. And evidence now exists that a colder stratosphere can increase the rate of ozone depletion.[37] But the connection between ozone depletion and climate change is complicated as ozone loss can also affect the climate.[38]

Climate change (global warming)

The release of carbon dioxide into the atmosphere from the burning of fossil fuels and other gasses is causing a change in the earth's climate. This subject has been discussed in chapter 4 ("Energy"). It is likely to be the most serious environmental problem of the twenty-first century.

The Water

Development, to date, has tended to turn clean water into dirty water as often as it has turned fresh air into dirty air. In the United States the deterioration of the nation's rivers was dramatized in the late 1960s when the Cuyahoga River, which flows through Cleveland, caught fire because it was so polluted. That event helped lead to the first Earth Day in 1970 and helped prod the US Congress into passing the Clean Water Act of 1972, which set a ten-year goal to return the nation's waterways to a state where they would be "fishable, and swimmable." Ten years later, many US rivers, streams, and lakes were cleaner than they had been when the Act was passed, but many still remained too polluted to allow safe fishing or swimming.

By 1990 the $75 billion that had been spent in the United States on upgrading sewage treatment facilities during the previous two decades had resulted in a significant improvement of the nation's waters. A survey about that time revealed that 80 percent of the nation's rivers and streams were now safe for fishing and 75 percent were safe for swimming. But that survey indicated also that about 130,000 miles of rivers were still unsafe for fishing and 150,000 miles were unsafe for swimming.[39]

Why was there still a significant problem after this large expenditure and 20 years of effort? A large part of the reason was that little progress had been made in reducing the pollution from urban and agricultural runoffs. Especially during storms, huge amounts of polluted water from city streets and the lawns of houses drain directly into rivers and lakes, untreated by local sewage treatment plants, and huge amounts of water drain from farms and golf courses, water laden with pesticides, herbicides, and fertilizers.

The status of freshwater fish in the United States reveals the extent of the problem of water quality in the country. In the early 2000s the EPA reported that about 15 percent of river miles, 30 percent of lake area and 100 percent of the Great Lakes had fish consumption advisories because of contamination by PCBs, mercury, and dioxins.[40] Usually the warnings did not advise that no fish should

Water pollution in the USA is partly caused by large amounts of pesticides, herbicides, and fertilizers which run off from fields during storms (*Vince Winkel*)

be eaten but rather that their consumption should be limited. For example, the state of New York recommends that people eat no more than one meal of fish weekly from any fresh water in the state.[41]

Prospective mothers and children are especially considered to be at risk from contaminated fish. A study of children whose mothers ate contaminated fish from Lake Michigan scored significantly lower on intelligence and achievement tests when they were 11 years old. The researchers attributed these results to exposure to PCBs (polychlorinated biphenyls, an especially persistent and toxic class of industrial chemicals) during fetal development. Because of PCB contamination in the Great Lakes, the EPA recommended that states advise women who might become pregnant, and children to limit their consumption of salmon from the lakes to one meal every month or two. (The state of Michigan, unlike the other states in the area, refused to issue a special warning about eating salmon from the Great Lakes, contending that this was overzealous regulation by the Federal government.)[42] Salmon fishing is a thriving recreational industry in the Great Lakes region.

The Great Lakes are less polluted by PCBs than they were in the past – the use and discharge of the chemical is now tightly controlled and its production is banned – but concentrations of the chemical in fish have continued to rise as the chemical works its way up the food chain. While PCB contamination in other waters in the country is still a problem, contamination of fresh waters also comes from air pollution, such as mercury contamination from coal-burning power plants, industrial sites, and incinerators, and from other chemicals. In 2004 the head of the EPA said that mercury emissions from human causes in the US had declined nearly 50 percent from 1990 to 1999 but admitted that virtually every river and lake in the country had fish contaminated with mercury.[43]

Other developed countries are also experiencing serious water pollution problems. In the 1970s and 1980s the River Rhine was commonly called the sewer of Europe. By the mid-2000s a cooperative effort, some of it begun in the 1950s, by the five nations on the Rhine to clean up the river was having significant results. About 20 to 25 billion euros were spent, much of it on new sewage treatment plants, and salmon have been restocked in the river.[44] Its success was causing some to cite this effort as a model for other international efforts to reduce pollution.

What is causing the polluted water in the developed nations? Industry must take a large part of the blame since traditionally industrial wastes have been dumped into nearby water as often as they have into the air overhead. Many industries are no longer dumping wastes into nearby rivers, but some dumping still goes on. The source of much of the most serious water pollution today in the developed countries is chemicals. The chemical industry has had a huge growth in the industrial world since World War II. Chemicals are now finding their way into waterways, many of which are being used for drinking water. A nationwide study of streams and lakes in the US at the beginning of the twenty-first century found low levels of many chemicals. About half the waterways had trace amounts of insecticides, antibiotics, fire retardants, disinfectants, degraded detergents, insect repellents, some nonprescription drugs, and steroidal compounds. It is not known if these levels are harmful to plants, animals or humans. Water treatment plants are unable to remove these substances from the water. Studies in Europe a decade earlier found similar results in European waterways.[45]

The developing world faces huge water problems. At the beginning of the twenty-first century about one-half of the world's population had no access to clean water.[46] In the developing countries about 90 percent of the waste water was discharged directly into waterways without being treated.[47] In China only 6 of the country's 27 largest cities had safe drinking water,[48] and in Mexico City 7 out of every 8 toilet flushes went untreated.[49] And Bangladesh possibly faced the worst water problem of all. In an effort to help the country get clean water, the government and international aid organizations in the 1970s and 1980s funded the digging of tube wells, about 10 million overall, but no one tested the groundwater for arsenic. It is now recognized that many of the tube wells are contaminated with arsenic, a deadly pollutant, and 20 to 35 million people are drinking this water. The WHO has declared the situation the "largest mass poisoning of a population in history."[50]

A surprising positive development regarding the use of water occurred in the United States at the end of the twentieth century. After the amount of water used by industry and agriculture consistently grew faster than population growth for the first eight decades in the century, an unexpected change occurred. Instead of water use continuing to rise, from 1980 to 1995 it actually declined by about 10 percent even though the size of the population had increased about 15 percent during the same period. The main cause of the decrease was that industry and agriculture had learned to use water more efficiently rather than look for more water. Most of the best dam sites were already being used, the cost of new dams had risen, and the negative environmental effects of dams became well known. Also federal antipollution laws made it cheaper for industry to find ways to use less water and to recycle it rather than clean it before discharging waste water into rivers and lakes. Agriculture, which uses more water than any other human activity – about 70 percent of all water withdrawals – significantly reduced the water needed for irrigation by adopting new methods such as drip irrigation at the roots of plants rather than spraying the water into the air where much of it is lost through evaporation and by the wind. Water use by individuals in the country has not decreased but it has leveled off. (The amount of water used for domestic purposes, in and around the home, averaged about 100 gallons per person per day.)[51] Water use also decreased in Europe and there has been a slowdown worldwide in the expansion of irrigation.

Because of increasing population and rapid industrialization in some developing nations, water use in the developing world is expected to continue to grow. But developing nations might adopt some of the more efficient ways to use water that the West has discovered and thus reduce their need. With expected warmer temperatures coming with global warming, increasing deforestation that makes water supplies more erratic, water scarcity in the South is expected to be a major concern in the twenty-first century. At the beginning of the century it was estimated that about 2.5 billion people lived in river basins where water was scarce and of these about 1.5 billion people lived in areas of high water scarcity.[52] More conflicts between countries over the availability of water, such as those that have taken place in the past between India and Pakistan, Israel and Syria, and Mexico and the United States, may occur. New efforts by nations to cooperate to deal with water scarcity might also occur. (This cooperation could follow the example of the long-term successful efforts by nations bordering the Mediterranean Sea – both Arab and non-Arab – to reduce the pollution in that body of water.) It is not clear whether conflict or cooperation will be the main result of coming water scarcity.

The Land

Whenever development has occurred, its effect on the land has been profound. The economic growth that comes with development increases the amount of goods and services available for human consumption. More natural resources from the land are required for the production of these goods, of course, and their extraction disturbs the land greatly. But even more widespread are the changes to the land that come with the disposal of the goods after they are no longer of use, and of the wastes that are created in the manufacture of the goods. Many of these wastes are artificial substances that never existed before in nature; thus nature has few, if any, ways of breaking them down into harmless substances. Development also affects the vegetation on the land, in some ways reducing it and in some ways helping to preserve it. In this section we will focus on two of the many changes to the land that come with development: the disposal of wastes, and deforestation. These two changes are affecting many human beings in such direct ways today that it is important that we look at them closely.

Solid wastes

It seems to be a common occurrence in a number of developed countries that, as more goods and services become available, more are desired and less value is placed on those already in hand. After the end of World War II, an unprecedented period of economic growth in the industrialized world took place, leading to a huge increase in the consumption of material goods.

As consumption rose so did wastes. "Throwaway" products that were used briefly and then discarded became common, as did items that wore out quickly. Such facts disturbed few people in the United States since they found enjoyment in buying new, "better" products. Many such products were relatively cheap in the 1950s, 1960s, and early 1970s since energy and other raw materials were inexpensive. Between 1960 and 2000 the amount of solid wastes generated in the United States per person annually grew by about 60 percent, until it reached about 1,600 pounds.[53] On average each American generated about 2.7 pounds of waste per day in 1960, whereas in 2000 this figure had grown to 4.5 pounds.[54]

In the early 1990s the EPA published new regulations requiring all cities and towns in the United States to improve their landfills or dumps where their solid wastes were being disposed of, so pollutants would not leak from them into the groundwater or into nearby waterways. The municipalities had two years to comply with the regulations and many of them had to close their dumps. Now there are fewer landfills in the country, but the new ones that have been built to meet the new government regulations are very large and serve a regional area.

One obvious way cities could respond to the new requirements was to get their citizens to produce fewer solid wastes and to start recycling trash, which many communities did to reduce the amount of trash going to their landfills. Another way to reduce trash was to make citizens pay variable costs for the disposal of their solid wastes, based on amount and type. Seattle is an example of a US city that has successfully followed that principle. Seattle began charging its citizens according to the amount of trash they put out for disposal. Yard wastes, such as grass clippings, if separated by the citizens so the city could use them for composting, were charged at a much lower rate than regular trash, and paper, glass, and metal (which could be recycled) were hauled away free. Seattle, which was already more environmentally conscious than most other American cities, found that during the first year it started charging its citizens for the amount of waste they produced, the total tonnage the city needed to haul to the landfills fell by about 20 percent. By the mid-1990s, 90 percent of the residents of Seattle were recycling their waste.

A huge solid waste problem has been created by the growing use of plastics around the world. This modern achievement of the chemical industry takes petroleum and turns it into containers which are very difficult to break and won't decay in any individual's lifetime. This last quality is what is causing the problem. Current estimates are that sunlight might break plastic down over about 500 years but there is no known living organism that can digest even a single molecule of plastic.[55]

Every year about 250 billion pounds of plastic pellets are produced in the world. A research sailing ship in the early twenty-first century found an area in the Pacific Ocean, about the size of the state of Texas – approximately 1,000 miles across – filled with floating plastic debris.[56] A Japanese scientist and his colleagues at Tokyo University have discovered that floating plastic fragments in the sea absorb toxic chemicals such as DDT, PCBs, and other oily pollutants. Here is how this problem has been described:

> The potential scope of the problem is staggering. . . . When those pellets or products degrade, break into fragments, and disperse, the pieces may also become concentrators and transporters of toxic chemicals in the marine environment. Thus an astronomical number of vectors for some of the most toxic pollutants known are being released into an ecosystem dominated by the most efficient natural vacuum cleaners nature ever invented: the jellies and salps living in the ocean. After those organisms ingest the toxins, they are eaten in turn by fish, and so the poisons pass into the food web that leads, in some cases, to human beings.[57]

In 2003 the research vessel that documented the huge floating body of plastic debris in the Pacific took underwater photographs of transparent filter feeding organisms with colored plastic fragments in their bellies.[58]

Toxic wastes

The first warning of the danger of toxic wastes came from Japan. In the 1950s and 1960s hundreds of people were paralyzed, crippled, or killed from eating fish contaminated with mercury that had been discharged into Minamata Bay by a chemical plant. In the late 1970s the warning came to the United States. Many people in a residential district of Niagara Falls, New York, were exposed to a dangerous mixture of chemicals that were seeping into their swimming pools and basements. Most of these people did not know, when they bought their homes, that the Hooker Chemical Company had dumped over 20,000 tons of chemical wastes in the 1940s and 1950s into a nearby abandoned canal, ironically known as Love Canal. News of the Love Canal disaster spread through the country as the story of the contamination slowly came out in spite of the denials of the chemical company and the apathy of the local government. Eventually hundreds of people were evacuated from the area. The state and federal governments bought over 600 of the contaminated homes. After putting a "wall" of clay and plastic around the buried toxic waste, the federal government later declared much of the Love Canal neighborhood fit for resettlement. The name of the neighborhood was changed from Love Canal to Black Creek Village.

Since 1989 the US EPA has been requiring some of the producers of toxic wastes to report annually how much they are releasing into the environment. Two industries that produce the largest amount of toxic wastes – the mining and oil exploration industries – were exempt from the reporting requirement. Half of all wastes were produced by the chemical industry, with other significant amounts produced by the metal, oil refining, paper, and plastics industries.

This reporting law, formally called the Toxic Release Inventory, has been considered to be a very effective piece of legislation. Even the chemical industry, which has worked to weaken it, has admitted that because of it toxic emissions have been reduced by about 50 percent. Nearly immediately after the first report was released, several large corporations announced that they would voluntarily reduce their emissions by 90 percent over the next three years. States and local residents have used the report to put pressure on companies to reduce their pollution.

In 2001 US manufacturers reported a nearly 15 percent decline from the previous year in toxic releases into the air, water, and ground. This was progress, but 6 billion pounds of hazardous materials were still being released into the environment. In 1996 the reporting requirements were expanded to include, for the first time, electric utilities, incinerator operators, recyclers, and many mining companies, an increase of 6,400 new plants over the previous 23,000 that had to report toxic emissions.

The persistence of some toxic chemicals is shown by the fact that byproducts of the chemical used to make stain protectors in carpets and food wrappers are showing up in seals and polar bears in the Arctic and in dolphins in the mid-Atlantic.[59] And a number of studies have shown that indigenous peoples in the Arctic are being exposed to relatively large amounts of pesticides, industrial chemicals, and heavy metals, with uncertain health effects.[60] Why are Arctic peoples being exposed to such poisons? One reason is that many rivers, ocean and air currents flow northward, thus bringing toxins released in other parts of the world. Also many indigenous people have a diet rich in fish and marine mammals; thus they absorb the toxins the fish and mammals have been exposed to. Breast milk and samples of blood in umbilical cords in Arctic women contain moderate to extremely high levels of toxins such as DDT, PCBs, dioxins, mercury, lead, and a flame retardant.[61]

Governmental and industrial responses to the waste problem

In 1980 the US government created a $1.6 billion fund to finance the cleaning up of the worst toxic waste sites. The law which set up this fund (popularly called Superfund) allowed the government to recover the cost of the cleanup from the companies that dumped wastes at the sites. In 1986 $9 billion more for the cleanup was approved by the US government, to come mainly from a broadly based tax on industry and a tax on crude oil. The Congressional Office of Technology Assessment has estimated that it will require about 50 years and $100 billion to clean up toxic waste dumps in the country. By the early 1990s the EPA had identified nearly 40,000 toxic waste sites in the country as being potentially hazardous. Of the 1,400 sites that were identified as the worst and needing priority action to clean them up, only about 500, or 35 percent, had been cleaned up by 1998. In 1995 the taxes on industry to provide money for the Superfund expired and the US Congress refused to extend them. General tax revenues were used thereafter to provide funds, but by 2004 the Superfund had a serious shortage of money.

There are other ways government can help control the waste problem. Barbara Ward, the late British economist, mentions four ways a government can encourage the reduction of wastes and promote the reuse of wastes: (1) it can make manufacturers pay a tax that could cover the cost of handling the eventual disposal of their products; (2) it can stimulate the market for recycled products by purchasing recycled products for some of its own needs; (3) it can give grants and other incentives to cities and industries to help them install equipment that recycles wastes; and (4) it can prohibit the production of nonreturnable containers in some instances.[62]

Inefficient and wasteful technologies and processes to produce goods are still common in the United States and other developed nations, since many of these were adopted when energy was cheap, water plentiful, many raw materials inexpensive, and the disposal of wastes easy. Some industries now realize that they can increase their profits by making their procedures more efficient and producing less waste. One such company is 3M, which, according to one study, has reduced its pollution as well as increased its profits, "not by installing pollution control plants but by reformulating products, redesigning equipment, modifying processes . . . [and] recovering materials for reuse."[63]

Germany is one of the leaders in creating imaginative ways to deal with toxic wastes. Recognizing that the ideal solution to this problem is to concentrate on reducing the production of toxic waste rather than focusing on its disposal or the cleanup, the country is implementing what is called a "closed-cycle economy."

When products are built they are designed with concern with how they will be disposed of when no longer wanted. Parts are marked so they can later be identified electronically to facilitate their recycling. According to the head of Germany's environmental protection agency, as manufacturers are held financially and legally responsible for the safe disposal of their products, it is expected they will support the revolutionary concept of the environmentally friendly closed-cycle economy.[64]

Deforestation

The UN's Food and Agriculture Organization (FAO) reported in 2005 that the destruction of the world's tropical forests continued at a high rate. Most of the

The cutting down of tropical forests is accelerating (*USAID Photo Agency for International Development*)

deforestation is taking place in the developing world. About 50,000 square miles of forest was cut down annually from 2000 to 2005, much of it in areas never previously logged.[65] This loss was partially offset by reforestation efforts (especially in China), new forest plantations, and the gradual expansion of forests in the developed nations.

Over the past 8,000 years about one-half of the world's forests have been cut down to make room for farms, pastures, and for other uses. Only about 20 percent of the original forests remain. These forests are known as "frontier forests," or "old growth forests," and they are very different from the human-modified forests that are prevalent in the world today. The frontier forests contain between 50 and 90 percent of the world's plant and animal species. Seventy-six countries have lost all of their frontier forests and another 11 are close to losing what remains of them. Three countries – Brazil, Canada, and Russia – have nearly 70 percent of the remaining frontier forests. Large sections of these forests remain free from risk at present, but the World Resources Institute estimates that about 40 percent are endangered by human activities.[66] The greatest threat to them in the future is from logging – much of it illegal – from making pasture land for cattle, and in the Brazilian Amazon, cutting down the forest for the growing of soybeans.

In contrast to the situation in the developing countries where the tropical rainforests are located, the forests in the developed nations actually increased in the twentieth century. As marginal farmland was taken out of production and trees were allowed to return to the land, forests grew in the industrialized countries. In the United States over the past 100 years about 25 million acres (10 million hectares) of agricultural land have reverted to forests in the eastern and southern parts of the country. These forests contain much less diversity of life than the frontier forests, of course. Also, many of the forests in the developed nations are not healthy. The FAO reported that from 1980 to 1994 forest cover expanded in Europe by 4 percent, but forest conditions actually worsened. Forests were being damaged by fires, pests, drought, and air pollution. More than 25 percent of the forests surveyed in Europe in 1995 were suffering significant defoliation. The proportion of completely healthy trees in Europe fell from about 70 percent in 1988 to about 40 percent in 1995.[67]

Deforestation is a serious problem because it can lead to erosion of the land, it can cause the soil to harden, and it can make the supply of fresh water erratic. Scientific studies support the hypothesis that deforestation can lead to significant changes in the climate. These changes usually mean less rainfall. Sometimes deforestation leads to too much water in the wrong places. Serious floods are occurring now in India in areas that had never experienced flooding; it is believed that the cutting of forests in the Himalayan mountains, the watershed for many rivers in India, is causing the flooding. Rioting has even been reported among some of the tribal peoples of India who are protesting the cutting of their forests by commercial firms. In just one year – 1998 – government officials in two countries admitted that the cutting of trees had led to disastrous events in which many people died. In China it was recognized that the clear cutting of forests along the upper reaches of the Yangtze River contributed to the unusually severe flooding that year. In Italy mudslides from a deforested mountain covered five villages.

China has started a massive reforestation program hoping to stop the expansion of its deserts. Billions of trees have been planted. But the deserts continue to expand and sandstorms (that can be detected as far away as the western US) have increased from about five a year in the 1960s to about 25 in the 1990s.[68]

Forests absorb a significant amount of CO_2 yearly, thus helping to combat climate change. One hectare of a tropical rainforest can absorb 200 tons of CO_2 annually.[69] Scientists working in the Amazon have estimated that the Amazon rainforest alone could be absorbing over 1 billion tons of CO_2 each year. These scientists have found that in the sections of the forest they are studying, more trees are growing per hectare than in the past and the trees are growing faster and larger than before. They attribute this surprising finding to the increased amount of CO_2 in the atmosphere. They have also concluded that only large, undisturbed sections of the forest absorb large amounts of CO_2. Sections of the forests that have been logged, burned or fragmented actually lose CO_2 to the atmosphere.[70]

Forests in the temperate zones of the earth absorb far less CO_2 than do the tropical rainforests, but they do absorb a significant amount. Studies in high-latitude forests have shown that the soils of the forests actually absorb much more CO_2 than the trees themselves. Scientists estimate that peat and other organic matter in the soils absorb two-thirds of the CO_2, while the trees absorb the remaining one-third.[71]

Deforestation not only destroys a valuable "sink" for CO_2, but it also releases the gas. Trees that are burned after they are cut, which is common when the forest land is cleared for settlements or for farming, release CO_2 into the atmosphere.

Small-scale farmers clearing land for crops are one of the main causes of deforestation, as in this Brazilian rainforest (*Campbell Plowden/Greenpeace*)

The FAO has concluded that deforestation is mainly caused by small-scale, subsistence farmers and by government-backed conversion of the land to other purposes such as ranching and settlements. (Recent research in Brazil indicates that now only 17 percent of the deforestation in Brazil is caused by small subsistence farmers.)[72] Subsistence farmers are cutting down forests because they are poor, unemployed, and landless. A highly inequitable distribution of land in the country, such as in Brazil, results in relatively few people owning a large amount of the land and many owning none at all. Logging leads to roads being built into previously inaccessible forests and landless peasants follow these roads looking for land to farm.

The FAO does not consider logging a cause of deforestation because, at least in theory, the forest can grow back after it has been logged. In reality logging often degrades the forest, leading to serious erosion and making it less suitable as a habitat for a wide variety of plants and animals. China and Malaysia have stopped much of their own deforestation, but are now importing logs from Indonesia and other areas. So many fires are set in Indonesia annually by poor farmers, illegal loggers, and plantation owners to clear forest land that the smoke from the fires covers a huge area, causing disruptions in air and sea travel and health problems (40,000 people were hospitalized in 1997). In 2006 Indonesia agreed to export logs from much of its remaining tropical forests to China and replace the forests with vast palm oil plantations. The oil will be sold to China for use in products such as detergents, soaps, and lipsticks.[73]

An example of a government-supported resettlement effort that led to serious deforestation took place in northern Brazil. In the early 1970s the Brazilian government began a large colonization project in the Amazon basin, moving people in from the poverty-ridden northeastern section of the country. It was hoped that the resettlements would help reduce the poverty in the northeast and provide food for an expanding population. Unfortunately, both hopes faded as colony after colony failed. The main reason for the failure was that tropical forest land is actually not very fertile, in spite of the huge trees growing on it. Such trees get their needed nutrients directly from decaying leaves and wood on the forest floor, not from the topsoil, which in many places is thin and of poor quality. This fact explains why many of the settlers had experiences similar to that of the following Brazilian peasant who described what happened to his new farm in the Amazon: "The bananas were two feet long the first year. They were one foot long the second year. And six inches long the third year. The fourth year? No bananas."[74]

The cutting of the trees in a tropical forest puts a severe strain on the soil since the trees protect the soil from the violent rains that are common in the tropics. And once the soil is washed away, it is not easily recreated. Some studies now estimate that from 100 to 1,000 years are needed for a mature tropical forest to return after human disturbances have taken place.[75]

If only small plots of the forest are cleared, regeneration of the forest is possible. Some peoples have practiced what is known as shifting cultivation in the tropical forests. They clear a piece of land and farm it for a year or two before moving on to a new piece of land. As long as this remains small in scale, the damage to the forest is limited, but any large-scale use of this type of agriculture can lead to irreversible damage to the forest.

Some tropical soils contain a layer known as laterite, which is rich in iron. When these soils are kept moist under a forest they remain soft, but if allowed to dry out, which happens when the forest cover is removed, they become irreversibly hard – so hard that they are sometimes used for making bricks.

In Central America and in Brazil, large areas of forests are being cut down to make pastures for the raising of cattle. The cattle are being raised mainly to supply the fast-food hamburger market in the United States. The growing of cattle on large ranches for export does not, of course, do anything to solve the food problems in the exporting countries, or to provide land to the landless.

Local people can earn income from the forests by what has been called "ecotourism." This type of tourism focuses on the growing number of people from the developed nations who wish to visit tropical forests and other spots that have been left more or less in a natural state. If evidence exists that local people can earn more income by letting the forests remain than by cutting them down, a strong argument can then be made supporting their preservation. Also, local people can be enlisted in the efforts to prevent deforestation since they will have an economic stake in the preservation of the forests.

The landless and the poor around the world today are assaulting the remaining forests for agricultural land and for fuel. As poverty is the root cause of the hunger problem, discussed in chapter 3, and one of the root causes of the population explosion, discussed in chapter 2, so also is it one of the causes of the deforestation problem. Development can reduce poverty, and when it does this for the multitude, it can reduce one threat to the world's forests. Development can also lead to the destruction of the forests as they are cleared for cattle farms, for lumber, for commercial ventures, and for human settlements. As with the population problem, development in its early stages seems to worsen the situation; but development that benefits the many and not just the few can eventually help relieve it.

The Workplace and the Home

Cancer

Cancer is often considered to be a disease of developed countries. It is estimated that one out of four people in the United States alive at present will contract cancer and many of them will die from it. Cancer now kills more children than any other disease, although accidents are still the number one cause of death of children. It is commonly believed by the general public that exposure of workers to cancer-causing substances – carcinogens – in the workplace, and the exposure of the general population to pollution in the air and water and to carcinogens in some of the food they eat, are the main causes of this dreaded disease. There is no question that many workers – such as the millions of people who worked with asbestos – have been exposed to high levels of dangerous substances. But scientists do not now believe that contamination at the workplace is the main cause of cancer; nor do they believe that water pollution or food additives are causing most of the cancer cases. The consensus among leading cancer experts at present is that smoking and diet, mainly diets high in animal fat and low in fiber, are the main causes.[76] Also, a long-term, very large study published in 2003 convinced most experts that being overweight or obese significantly increases the likelihood of a person contracting cancer.[77] Another recent study identifies very small particles in polluted air from burning fossil fuels and wood as increasing the risk of lung cancer.[78]

Some experts fear that while chemicals cannot be blamed for most cancer today, there is a possibility that chemical-related cancers may increase greatly in the future because of the large increase in the production of carcinogenic chemicals since the 1960s. Cancer can occur 15 to 40 years after the initial exposure to a carcinogen, so chemicals may yet prove to be a major culprit.

Pesticides

The story of pesticide use illustrates well the dangers that new substances, which have become so important to modern agriculture, have brought to people at their workplaces as well as in their homes at mealtime. Rachel Carson is credited with making a whole nation aware of the dangers of persistent pesticides such as DDT.

Her book *Silent Spring*, which appeared in 1962, shows how toxic substances are concentrated as they go up the food chain, as big animals eat little animals. Since most of the toxic substances are not excreted by the fish or animal absorbing them, they accumulate and are passed on to the higher animal that eats them. Carson's warning led to a sharp reduction in the use of long-lived pesticides in many developed countries; but if she were alive today (she died of breast cancer in 1964), she would probably be disturbed to learn that short-lived but highly toxic pesticides are now increasing in use in the United States. The use of herbicides has especially increased dramatically as farmers, railroad companies, telephone companies, and others find it cheaper and easier to use these chemicals to get rid of unwanted vegetation than to use labor or machines. These new highly toxic pesticides pose a special risk to the workers who manufacture them and to the farmers who work with them in the fields. Although DDT was banned for use in the United States in 1972, residues of it could still be found in most people in the United States 20 years later.[79]

Pesticide use is increasing in the less developed nations – not just the use of short-lived pesticides but of persistent pesticides such as DDT as well. US law explicitly permits the sale to foreign nations of substances that are banned, highly restricted, or unregistered in the US. US companies, as well as many in Europe, have increasingly turned to the overseas market to sell their products as more restrictions on the use of pesticides occur in the developed nations.

By the mid-1990s the Consumers Union, a respected nonprofit US organization, found that much domestic produce had higher levels of pesticide residues than imported produce. Using US Department of Agriculture statistics from 27,000 food samples taken from 1994 to 1997, it found that although nearly all the pesticide residues were within legal limits, they were frequently well above the levels the US Environmental Protection Agency said were safe for young children. Fresh peaches had by far the highest toxicity level.[80]

The US Department of Agriculture has estimated that 32 percent of US crops were lost to pests in 1945 whereas in 1984, despite a large increase in the use of pesticides, the loss due to pests had risen to 37 percent.[81] This situation may explain why a number of agricultural experts are now advocating a more balanced program for controlling pests. A selective use of pesticides would go along with the use of biological controls, such as natural predators, and other nonchemical means to control pests.

In the early 1990s the US government announced that it was going to try to reduce the amount of pesticides used on US farms. A five-year study by the US National Academy of Sciences on the effect of agricultural chemicals on children was published in 1993. It criticized the method the government had been using to calculate the safe amount of pesticide residue on foods. It found that the risk calculations by the government had not taken into account the fact that people are also exposed to pesticides from sources other than on foods, such as in their drinking water, on their lawns, and on golf courses. It found that infants and children might be especially sensitive to pesticide residues on food. They consume 60 times the amount of fruit adults do, in relation to their weight, so are getting higher doses of the pesticides that are used on fruits. And this is taking place early in their lives. The head of the committee that prepared the report drew the following conclusion: "Pesticides applied in legal amounts on the farm, and present in legal amounts on food, can still lead to unsafe amounts."[82] In 1999 the EPA responded to these concerns and banned most uses of a pesticide widely used on fruit and vegetables, and tightened restrictions on another, because of their possible harm to children. This was the first time the agency had issued regulations specifically designed to protect children.[83] In 2005 the EPA issued new guidelines on the use of many chemicals which recognized that children might be more at risk from the use of these chemicals than adults.[84]

Pesticides have played an important role in the successes of the Green Revolution; it is doubtful that food production would have stayed ahead of population growth in the world without them. What seems to be called for now is a highly selective use of pesticides, not their banishment.

Chemicals

Development has led to the introduction of many chemicals that have never been adequately tested to verify their safety. In the 1980s the National Research Council of the US National Academy of Sciences concluded that tens of thousands of important chemicals had never been fully tested for potential health hazards. The Council found that this included about 90 percent of the chemicals used in commerce, 60 percent of the ingredients in drugs, 65 percent of pesticide ingredients, 85 percent of cosmetic ingredients, and 80 percent of food additives.[85] The general conclusion of the Council was that we are ignorant of the potential harm that might be caused by many products we come in contact with in our lives today.

Not everything causes cancer, of course, but development has brought forth so many new products in such a short time that we cannot be sure which ones do and which do not. Barry Commoner shows that new products often bring large profits to the first industry that introduces them, so there is a strong incentive for industries to be innovative. New products, especially in the United States since World War II, are often made of synthetic materials that pollute the environment, but the pollution usually does not become evident until years after the introduction. Commoner states that "by the time the effects are known, the damage is done and the inertia of the heavy investment in a new productive technology makes a retreat extraordinarily difficult."[86]

This dismal situation may be about to change. In 1998 the EPA announced that it was exploring ways to test chemicals widely used in everyday products for the possible ill effects that they can cause. The EPA realized that new technology that was being used to screen millions of chemical compounds each year in an effort to find potential new medicines could also be used to find potentially dangerous chemicals.

In 1999 the US chemical industry, after being threatened by the government with a new regulation that would make new toxicity testing mandatory, announced that it would voluntarily spend about $1 billion over a six-year period to test thousands of the estimated 87,000 chemicals in use today for possible toxic effects. The industry thus shifted its position from one in which it assumes its products are safe, to one in which it assumes that they can be toxic and tries to show which are and which are not.

Europe is active on this issue also. In the mid-2000s the European Commission of the European Union proposed a sweeping new law that would force chemical companies to prove the safety of a chemical before it is sold, rather than wait for the need for regulations later after problems in its use develop. In some cases Europe was taking the lead in banning dangerous chemicals, which put pressure on the US to follow suit.[87]

An encouraging development in the US in the mid-2000s was the decision by the US chemical giant DuPont to spend about 10 percent of its research budget on trying to find biological-based substances, such as in corn and in sugar, to replace fossil fuels as the building blocks of its chemicals. But for the near future, DuPont, like other chemical companies, is counting on chemicals based on fossil fuels to make most of its profit.[88]

The vital need for major reforms in the chemical industry can be illustrated by one class of modern "miracle" chemicals that have been used to make such popular products as Scotchgard stain protector, Teflon non-stick cookware, and Gore-Tex water resistant clothing. These fluorochemicals have been found in the blood of sea and land animals, birds, and Arctic women. It is possible that these chemicals will never break down into harmless substances, and, according to the American Red Cross, are now found in the blood of nearly all Americans from whom it receives donations.[89] To its credit, the 3M corporation stopped producing the chemical used to make its popular Scotchguard product when some of the above information become known.

The Use of Natural Resources

Since the world's population is growing exponentially, as we learned in chapter 2, it is probably not surprising that the consumption of nonfuel minerals is also growing exponentially. But, unlike petroleum, the supplies of minerals are not becoming exhausted. Another great difference between nonfuel natural resources and energy supplies is that the actual cost of producing most minerals has decreased over the past century.[90] This reduced cost has occurred, even as lower-grade ores are being mined, because of advances in technology – such as better exploration techniques, bigger mechanical shovels to dig with, bigger trucks to haul the ore away, and bigger ships to transport it to processing plants. Whether new technology will continue to keep the cost of minerals low in the future is a subject that is debated by scientists and economists. As ores containing a lower concentration of the desired minerals are mined and less accessible deposits are turned to, processing costs will rise. Also, mineral extraction is highly energy intensive and rising energy costs will directly affect the price of minerals. Some analysts have observed that mineral prices in the past did not reflect the true environmental costs of extracting and processing the minerals, but with new pollution laws in most industrial countries, the mining industry will have to assume more of these costs.

One trend that is apparent is that most industrialized nations are becoming more dependent on foreign countries for their minerals. The United States is a mineral-rich country; in the 1950s it was nearly self-sufficient in the most important industrial minerals. By the late 1970s it was self-sufficient in only 7 of the 36 minerals essential to an industrial society. Western Europe and Japan are even more dependent on imported minerals than is the United States. This increasing dependency on ores from foreign countries, many of which are essential for the advanced technologies common in the West, has strongly influenced the developed nations' policies toward the developing world, where many of the minerals are found.

There are five main steps a country can take to counteract shortages of a needed material if it cannot locate new rich deposits of the ore: (1) it can use resources more efficiently; (2) it can recycle waste products containing the desired material; (3) it can substitute more abundant or renewable resources for the scarce material; (4) it can turn to ore deposits having a lower concentration of the needed mineral; or (5) it can reduce its need for the material.

Resource efficiency

Two concepts have emerged in the developed world that are designed to reduce the use of natural resources by using the resources more efficiently. One is called "eco-efficiency" and the other is called "product stewardship." The use of either or both of these concepts by some industries led to a 2 percent improvement in resource efficiency per year in the developed world from 1970 to 1995.[91]

Eco-efficiency involves redesigning products and the processes that are followed to make them so that fewer natural resources are used and less waste is produced. In the United States a pioneer in the use of this concept is the 3M corporation. 3M established a program it called "Pollution Prevention Pays." The corporation claims that over a 20-year period this program resulted in the company cutting its toxic emissions by about 750,000 metric tons and the company saved about $800 million in expenses.[92]

Proctor & Gamble also followed this concept. In 1989 the corporation introduced concentrated detergent powders called Ultra detergents. These detergents cleaned the same amount of clothes as the old detergents but were half the volume, used 30 percent fewer raw materials, required 30 percent less packaging, and significantly cut the energy required to transport the product to the market.

By following eco-efficiency principles SC Johnson Wax from 1990 to 1997 increased its production by 50 percent while at the same time it cut its manufacturing waste by half, reduced packaging waste by a quarter, and reduced the use of volatile organic compounds by about 15 percent. The company saved more than $20 million annually from these changes.[93]

Some companies are attempting to drastically reduce their use of natural resources and their toxic emissions by adopting a "closed loop" process. In this manufacturing process, wastes are completely recycled or reused.

Product stewardship is being practiced mainly in Europe at present. This is the principle that a company should be held responsible for the environmental impacts of its products throughout their whole life cycles. The trend is for more laws and agreements between government and industry that are based on the Polluter Pays principle. This principle embodies the idea that the manufacturer of a product should be responsible for the harm to the environment that comes from its production, use, and disposal. Some companies following this principle, especially in the field of office equipment, are starting to recondition or rebuild old equipment rather than build new equipment from scratch. Xerox is one company that has followed this principle in the United States. The company encourages product returns so that it can recondition old copiers. It has found that it can even make a profit by recycling toner cartridges.[94]

All of these concepts are leading to a revolutionary change in thinking about the responsibilities of the manufacturer. Instead of having society as a whole pay for the consequences of the manufacturer's actions, the principle is slowing spreading that the manufacturer should accept this responsibility. If the manufacturer is held responsible, it will have an incentive for redesigning products and manufacturing processes to cut their costs. In some instances, as the examples above show, the practicing of this principle can lead to a "win/win" situation as the manufacturer and the environment both benefit from the new way of thinking.

Recycling

It is generally agreed that more recycling of waste material needs to be done in the United States. In the late 1980s recycling became relatively popular in the country because more citizens became aware of environmental problems and because many towns were faced with trash dumps that were becoming filled. (New dumps were becoming very expensive to open because of tighter federal government regulations.)

A demand for recycled material grew in the United States as more industries started using it. By the mid-1990s this was taking place as a growing economy emerged and new plants able to process recycled material began operating. One solution that has been suggested to promote recycling is that government should require the use of more recycled material, such as in newsprint. The US government did take this step in 1993 when President Clinton ordered all federal government agencies, including the military, to purchase paper with a minimum of 20 percent recycled fibers in it. But even with respect to newsprint, there is no simple solution since newspaper cannot be recycled indefinitely because the quality of the fibers degrades.

Even with the new interest in recycling in the United States, the country is still not doing as much of it as other industrial nations do. In the early 1990s the United States was recycling about 15 percent of its trash while Japan was recycling about 50 percent. European countries also do much more than the United States. Since 2002 the EU has required all its members' auto manufacturers to be responsible for the recovery and recycling of all of its new autos. As of 2004 the EU requires all the electronic companies of its members to pay for the collection and recycling of its products. And by July 2006 no electronics sold in Europe can any longer contain some of the most toxic materials such as lead, cadmium, mercury, hexavalent chromium, and the flame retardants PBDE and PBB.[95]

In 2005 the US was recycling about 30 percent of its waste, although the average American was creating more waste.[96] In the early 2000s about 50 percent of the paper in the US was being recycled, very little of the plastic, about 35 percent of the metals, little glass, and only about 5 percent of its wood.[97] To increase recycling in Japan, many cities have increased the number of categories into which items to be recycled must be separated. Yokohama, a city of nearly 4 million people, had ten categories in 2005. Recycling costs more than dumping but about the same as incineration, which land-scarce Japan uses for much of its garbage.[98]

One unfortunate trend is the shipping of electronic waste, including computer monitors and circuit boards, to developing countries, such as China, India, and Pakistan, for recycling. These items contain lead and other toxic material. The people recycling the items, often children or adults with no protective clothing, are being exposed to dangerous substances. A report at the beginning of the twenty-first century by five American environmental groups estimated that from 50 to 80 percent of electronic waste collected in the US for recycling was being placed on container ships for use or recycling by developing countries. The US was the only developed country that had not signed the 1989 United Nations treaty which was designed to limit the exporting of hazardous waste. The producers of hazardous wastes were encouraged to deal with their waste problems within their own borders whenever possible. The European Union at that time was considering passing a law which would require manufacturers of products which contain hazardous material to take responsibility for them from "cradle to grave."[99]

While recycling is desirable, it is only a partial solution to resource shortages and to pollution by the minerals industry. Recycling also creates pollution and uses energy. The move by the US soft drink and beer industry to use aluminum cans that can be recycled is obviously not the final solution to the litter problem. About 40 percent of aluminum cans are never collected for recycling,[100] and the manufacture of aluminum uses a lot of energy. Probably a better solution was the move by some American states to require returnable soft drink and beer containers to be used in their states instead of throwaways. Oregon's experience with its container law has been that highway litter was significantly reduced, recycling was stimulated, the price of beverages remained about the same, and new jobs were created.[101] In the states that have container laws about 80 percent of all containers are returned to the stores.[102]

An examination of the reaction by some companies to the efforts of a few states to require that deposits be collected on beverage containers to encourage their return for recycling or reuse illustrates the economic pressures that discourage real recycling. One investigator found that beverage companies (both soft drink and beer), companies that make glass beverage containers, and retail grocery chains financially support the Keep America Beautiful campaign, which encourages recycling and the picking up of litter by citizens, largely the containers that these companies produce or sell. (Studies have shown that 50 percent of US litter is beverage containers.) The investigator found also that these companies were strong supporters of efforts to defeat attempts by states to pass legislation requiring returnable containers and deposits. A front organization funded by these companies and others spent $2 million in 1987 to defeat an attempt to pass a deposit law in the District of Columbia, the area around the nation's capital city.[103] According to the Public Interest Research Group, a nonprofit watchdog group in Washington, DC, 21 beverage companies spent $4 million from 1989 to 1992 to defeat a national bottle bill.[104] These companies complain that having to reprocess empty containers will increase their expenses.

The "throwaway" economy that developed in the United States after World War II still exists. The new efforts to recycle are a step forward, but much remains to be done. While it is certainly good that citizens become involved in picking up litter along their highways, better yet would be a system that discouraged the litter. Recycling is much better than burying containers, but better yet would be to reuse them. The investigator who explored the hidden motives behind industries' support for efforts to reduce litter in the country commented in an interesting way on the situation in the United States:

> Only in America could custom compel the discarding of a perfectly good vessel simply because someone had quaffed the contents, but that's what we do with 50 billion cans and bottles every year. An additional 50 billion or so are "recycled," a uniquely American interpretation of the word because they too are discarded, then crushed, melted, and remade rather than simply washed and refilled. It's as if we were a nation of dukes and earls, pitching our brandy snifters at the hearth.[105]

Denmark has banned throwaways, but this action is unlikely to be taken in the United States any time soon.

Substitution

When a material becomes scarce, it is sometimes possible to substitute another material for it which is more abundant or to use a renewable resource in place of the scarce item. For example, the more abundant aluminum can be used in place of the scarcer copper for most electrical uses. Difficulties arise at times when the substituted material in turn becomes scarce. Plastic utensils and containers replaced glass products in most US kitchens because of certain advantages plastic has over glass, such as being less breakable and lighter in weight. But plastics are made from petrochemicals, which are now becoming scarcer. Also, the plastics industry produces more dangerous pollutants than does the glass industry. Another limitation to substitution is that some materials have unique qualities that no other materials have. Tungsten's high melting point, for example, is unmatched by any other metal. And substitutions can produce disruptions in the society, causing some industries to close and new ones to open. The last-mentioned point can mean, of course, new opportunities for some people and fewer for others. New ways of doing things can also be substituted for old ways, sometimes resulting in a reduced use of resources. The trend in some businesses to use communications in place of transportation (business meetings with participants on video screens instead of physically being present) might be such a development.

Mining of low-grade ores

Many of the deposits with the highest concentration of the desired minerals have now been mined, but there are large, less rich deposits of many desired minerals scattered around the world. These can be, and in many places are, mined. There are significant costs incurred when such mining takes place. The cost of the mining increases, since more ore must be processed, mines must be bigger, more energy and water must be used. Because more ore must be processed, more wastes are produced. Large strip mines are often used and these have a devastating effect on the land. Even the best attempts to restore the strip-mined land are costly, and very imperfect.

Reducing needs

The fifth way to counteract shortages of a material is to reduce the need for the material. Many consumer goods – such as automobiles and clothes – become obsolete in a few years as styles change. This planned obsolescence leads to a high use of resources. Many products also wear out quickly and must be replaced with new ones. In the US more durable products could be designed by industry, but they would often be more expensive. It is probably because of this reason that US industry generally does not make such products. Higher prices would mean fewer sales, a slower turnover of business inventories, and thus probably lower profits. They could also mean fewer jobs.

Overdevelopment

Perhaps a good way to end this section is to explain the concept of overdevelopment. According to the Australian biologist Charles Birch, "over-development of any country starts when the citizens of that country consume resources and pollute the environment at a rate which is greater than the world could stand indefinitely if all the peoples of the world consumed resources at that rate."[106] From this perspective, it can be seen that the United States could be considered the most overdeveloped country in the world, followed closely by many other industrial countries. People in the United States, who constitute about 5 percent of the world's population, consume about 30 percent of the world's annual use of natural resources, and do so, as this chapter has shown, with devastating effects on the environment. This devastation is being reduced as new environmental laws are enacted and gradually enforced in the developed world, but it has not been reduced to such an extent that the concept of overdevelopment is outdated.

The Extinction of Species

No one knows for sure how many species of living things there are on the earth. Biologists today generally make educated guesses that the number is between 10 million and 100 million. (Scientists have given a name to about 2 million of them, and of those named, only about 10 percent have been studied in any detail.) Throughout the earth's history, new species have evolved and others have become extinct, with the general trend being that more new species are created than die out. It is now believed that because of human actions this trend has been reversed, with extinctions outnumbering the creation of new species. And the trend appears to be increasing.

According to Edward O. Wilson of Harvard University, probably the most respected of all US biologists, the world has experienced five major periods, or "spasms," of extinction of large numbers of species, from which it took millions of years to recover. These extinctions were caused by natural forces, such as a change of climate. Wilson believes that because of the vast growth of the human population and the related widespread deforestation and overuse of grasslands that are now occurring on our planet, the earth is heading into the sixth and worst period of extinction of species. Wilson estimates the present rate of extinction as about 27,000 species per year, or three per hour. (The normal "background" rate is about 10 to 100 per year.) If the present rate continues, Wilson estimates that 20 percent of all the species in the world will be extinct in 30 years.[107] Robert May, zoologist at the University of Oxford, who presides over the Royal Society and until 2000 was chief science advisor to the British government, estimates the present extinction rate as 1,000 times as great as before the arrival of human beings.[108]

Whereas hunting used to be the main way humans caused extinction, it is now generally believed that the destruction of natural habitats is the principal cause of extinctions. As the human population grows, humans exploit new areas of the world for economic gain and often destroy life forms as they do so. Biologists believe that about one-half of all species live in tropical forests, which as we have seen above are being cut down at an increasing rate. There is now a growing recognition that climate change, discussed in chapter 4, could become as dangerous to biodiversity as the loss of habitats.[109]

A dramatic example of how habitats are destroyed can be seen by looking at a large land development scheme known as "Jari" in the Amazon valley. The Amazon is the largest tropical rainforest on earth. The year-round warm temperature, heavy rainfall, and abundant sunlight produce excellent conditions for the evolution of species. Species can be destroyed, however, when the land is cleared to make way for farms and commercial enterprises.

In the late 1960s, the US shipping executive and financier Daniel Ludwig, one of the richest people in the United States, purchased Jari, a parcel of land in the Brazilian Amazon approximately the size of the state of Connecticut. Ludwig invested about $1 billion to construct a paper-pulp factory there. (The factory and a wood-burning power plant were constructed in Japan and towed to the Amazon on huge barges.) Large parts of the forest on Ludwig's estate were cut down and burned to make way for the planting of two or three species of fast-growing trees he brought into the area. As Ludwig said, "I always wanted to plant trees like rows of corn."[110] Ludwig got his rows of trees, but he probably also caused the extinction of an unknown number of plants, insects, and animals. One author described what it was like to walk through one of the new forests at Jari: "no snakes lurked beneath the log, no birds sang in the branches, and no insects buzzed in the still air."[111]

After 14 years building Jari, Ludwig abandoned the project in 1982 and sold it to Brazilians at a loss. In 1999 the Brazilian owners of Jari were trying to sell it to foreign investors for 1 dollar. The Brazilian consortium that bought it from Ludwig has seen only one profitable year since taking it over. Jari has $350 million in debt and the new owners will have to invest several hundred million dollars if it is to continue manufacturing high quality cellulose, the fiber used to make paper.[112]

Jari is unique because of the large size of the undertaking, but other smaller developments are becoming more and more common in the remaining rainforests in Latin America, Africa, and Southeast Asia. Scientists fear that the extinctions these developments are causing could be a direct threat to the well-being of human as well as other life on earth.

Many of the species in the tropics have never been studied by scientists. But based on past experience, it is believed that many of these unknown species contain properties that could directly benefit humans. Nearly one-half of the prescription drugs now sold in the United States have a natural component in them.[113] The importance of some of these drugs can be illustrated by the example of just one plant from the tropical rainforests, the rosy periwinkle. Drugs are now produced from this plant that achieve 80 percent remission in leukemia and Hodgkin's disease patients.

Exotic species are vital to the health of modern agriculture. The wild varieties and locally developed strains of a number of major grains grown today have characteristics that are of vital importance to modern seed producers. Seeds are needed with natural resistance to the diseases and pests that constantly threaten modern agriculture. Many farmers today utilize only a relatively few, highly productive varieties of seeds in any one year. The monocultures that are planted are especially vulnerable to diseases and to pests that have developed resistance to the pesticides being used. An example of how this works was shown in 1970, when 15 percent of the corn crop in the United States was killed by a leaf disease, causing a $2 billion loss to farmers and indirectly to consumers because of higher prices. That year, 70 percent of the corn crop used seeds from only five lines of corn. The disease was finally brought under control with the aid of a new variety of corn that was resistant to the leaf disease. The new corn had genetic materials originating in Mexico.[114]

Insects from tropical forests can at times prove extremely valuable to American farmers. Citrus growers in the United States saved about $25–30 million a year with the one-time introduction from the tropics of three parasitic wasps that reproduced and preyed on the pests attacking the citrus fruit.[115] (The introduction of exotic species by humans for profit, or amusement, or by accident into areas to which they are not native is now recognized as having great potential for harm. Since the new species usually has no natural predators in the new area, it can multiply rapidly, destroying or displacing other desirable animals or plants, as was the case with the introduction of rabbits in Australia, and starlings and the kudzu plant in the United States.)

American biologist Paul Ehrlich of Stanford University does not believe that developing nations can preserve tropical habitats on their own since their financial needs are so great. What is needed in the world, he feels, is a new awareness that the diversity of life forms on earth is a priceless treasure that benefits all humanity and that all share a responsibility for helping to preserve. He states: "Over 95 percent of the organisms capable of competing seriously with humanity for food or of doing us harm by transmitting diseases are now controlled gratis by other species in natural ecosystems."[116]

In 1992 at the UN-sponsored conference on the environment in Rio de Janeiro, a proposed treaty to try to slow down the loss of species was presented. The treaty, formally named the Convention on Biological Diversity, called for the study of each nation's biodiversity and a commitment to preserve the biodiversity that exists on earth. By 2000 most nations of the world had ratified the treaty, but the United States still had not.

A recent suggestion for a practical way to combat this daunting problem of the extinction of species is that conservation efforts could be focused on a relatively few, highly vulnerable "hot spots" where there is a large concentration of species found nowhere else in the world. The 25 hot spots that have been identified contain the last habitats for about 45 percent of the earth's plant species and 35 percent of its land-based vertebrate (fish, amphibian, reptile, bird, mammal) species. The top eight spots are southern coastal India and Sri Lanka, the island of Madagascar, Indonesia, Brazil's Atlantic forest, the Caribbean Islands, Burma and Southeast Asia, the Philippines, and the eastern mountains and coastal forests of Kenya and Tanzania.[117] At the beginning of the twenty-first century about $700 million had been collected out of the $25 billion estimated to be needed to protect the 25 hot spots.

Edward Wilson has joined those who now believe that while protecting "hot spots" is important, and should be expanded, it may no longer be enough. To really protect biodiversity, we must be concerned with protecting the whole biosphere. The variety of species on earth and their habitats provide essential services to life that the market cannot put a price tag on, services such as "nutrient cycling, the formation and enrichment of soils, the detoxification of pollutants and other forms of wastes, the provision of freshwater, the regulation of the atmosphere and climate, and the stability of ecosystems." Wilson believes that to preserve species "we must push back the deserts, replant the forests, preserve water supplies, reduce pollution, restore topsoil, and stabilize the climate."[118]

Two metaphors have been used to help people understand what the loss of biodiversity means. One is that of the loss of rivets and the other is of the loss of threads. In the first metaphor the extinction of a species is seen as being like taking a rivet out of an airplane. One probably doesn't matter, nor two, but if you keep pulling out the rivets eventually the airplane will crash. This metaphor conveys the idea of a collapsed ecosystem. The other metaphor says that the loss of a species is like pulling a thread out of a beautiful tapestry. You won't even notice a few pulled out but the more you pull the less rich in color is the tapestry. If you pull too many out in one location the tapestry may even tear.

Some scientists say that while these two metaphors can be useful to understand what the loss of biodiversity can mean, neither fully explains the complexity of real life. They point out that while it is true that the greater the species loss, the simpler and duller nature becomes, it is also true that the complexity of life forms can actually keep an ecosystem from collapsing, especially during times of stress such as a drought. Also, all species are not of equal importance in keeping an ecosystem healthy. There may be key ones that provide vital services to the others. If they are lost, the others depending on them are in danger.[119]

The Extinction of Cultures

There are about 15,000 nations on our planet and about 200 nation-states. The nation-states are the political entities, what are commonly referred to as countries. They are often made up of several or many individual nations, or different cultures. The nation is a group of people that share a common history, a common ancestry, and usually a common language and a common religion. They often have common traditions, common ways of doing certain things and of interacting with each other and toward outsiders. Because of these similar features that make them different from other peoples, each nation's people see the world and their place in it differently than others, approach problems differently, and have arrived at different solutions to situations humans face. The unique language of the culture is used to pass the common history and traditions down to the young. Linguists now estimate that of the approximately 6,000 languages in the world, one-half of them are endangered – that is, like species, they are in danger of extinction.[120] The group that speaks the endangered language – which is often unwritten – is becoming so small that there is a real possibility the group will die out or become absorbed by the larger dominant culture around it and will disappear forever.

Should we care? What will be lost if a culture dies out? The answer to that question is in some ways similar to the answer this book has given to the growing extinction of species. Species represent the amazing variety of life forms on this planet. Their interrelationships are still imperfectly known – to put it mildly – and that can affect the health, and even survival, of one of the species, our own. Cultures also represent the amazing variety of life on earth. But here it is not the form of life that is different, but the different ways members of one species – the human species – have created to live. The culture represents the accumulated knowledge of one group, knowledge that is available to others to pick and choose from, so they can improve their own lives. In addition, as with species, the multitude of cultures makes life on earth extremely rich and varied. The discovery of that variety often leaves an observer with a sense of awe and with a realization that the death of any species or culture leaves life a little less wonderful.

Development, especially since World War II, has often been equated with the culture of the United States. The United States is the largest producer of goods and services and its culture is closely associated with material wealth. Freedom from the burdens of excessive control by government and freedom from the restrictions common in more traditional societies are also characteristics of US society. These characteristics have contributed to an emphasis on innovation and change that has led to many new products and services. So it is not surprising that development and US culture have seemed to go together. Many other cultures have found that their youth are more attracted to the US culture than to their own. So also within the United States, ethnic groups have found that it is extremely difficult to keep from being absorbed by the dominant culture. The youth of the ethnic groups want to become accepted by the majority and they know that this will come only if they are like the majority, not different from it.

Because of the worldwide popularity of US movies, music, fast food, clothes, and the English language it is common to read that the American culture is replacing local cultures in many countries. But some recent studies indicate that only some rather superficial aspects of the American culture are being adopted, such as Coca-Cola and Big Macs, while more important values are not. For example, a study has shown that the cultures of Northern Europe, such as those of the Nordic countries, are actually better representatives of "modernity" than is the American culture with its more widespread traditional religious and social values.[121] European cultures place a higher value on leisure and government social services than does the American culture, which emphasizes earning higher income so people can acquire more material objects. Many Europeans seem to be happy to trade income for more leisure to enjoy life and, maybe not surprisingly, Europeans have a higher "satisfaction with their lives" than Americans do.[122] Americans are surprised to learn that even with the European desire for more leisure, some European countries, including France, have a higher level of economic productivity than the US has.[123]

Some people in the developed countries are giving a new respect to what had formerly been labeled as "primitive" cultures. There is a slowly growing recognition that these cultures may have knowledge that developed countries need if they are going to survive – such as an ability to live in harmony with nature, a concern for future generations, and a knowledge of how to foster a sense of community.[124] Tribal people in tropical forests have been finally recognized as possessors of important knowledge regarding natural drugs in plants and of skills that have enabled them to live in the forests without destroying them. There is also a growing recognition that if you want to preserve the tropical forests and the multitude of species that they harbor, you must make it possible for the indigenous people living in them to survive. If these people cannot survive, probably the forests cannot either. If these people do survive, they will help protect the forests that are their homes.

Let us now focus for a moment on two cultures under stress at present and in danger of extinction. One culture, the Yanomami, is found in two developing countries – Brazil and Venezuela – and the other, the Estonians, is found in Northern Europe.

The Yanomami

In the Amazon region of Latin America live the Yanomami. It is believed that these people have lived in this region for thousands of years. The approximately 25,000 Yanomami represent the largest group of indigenous people living in the Americas who still follow Stone Age methods. Although they had very limited contact with other peoples for many years, this changed in the late 1980s when gold was discovered in the Brazilian Amazon region. Thousands of miners flew into the area where the Yanomami lived. The miners brought with them diseases to which the Yanomami had no natural immunity. Amnesty International estimates that from 1988 to 1990 about 1,500 Yanomami died.[125] In addition to the malaria that killed many, some Yanomami died from mercury poisoning, which came from eating fish poisoned by the mercury the miners had used in the streams to sift for gold. Others were killed by armed attack. Amnesty International reported: "These attacks are often carried out by private agents, including gunmen hired by land claimants, timber merchants or mining interests. They have gone almost entirely unpunished – in fact, state-level authorities have even colluded with them."[126]

The Yanomamis' situation became known throughout Brazil and around the world. Responding to pressures within Brazil and from some foreign countries (the attention given to Brazil because of the upcoming United Nations environmental conference probably played a role), the Brazilian government in 1991 set aside for the Yanomami about 36,000 square miles of land. When added to that set aside by Venezuela, which was slightly smaller than the Brazilian grant, this was an amount of land equal to the size of Portugal and the amount anthropologists said the Yanomami needed in order to survive. In 1993 Brazil used its police and military force to forcibly remove 3,000 miners who were still in Yanomami lands.

What is the fate of the Yanomami? No one knows, of course, but if history is a guide, one would have to say that their prospects of surviving are not bright. While the actions by the Brazilian and Venezuelan governments to reserve a large amount of land for the use of these people is a hopeful step, disturbing signs exist. The presence of gold in their lands is unfortunate. In 1990 the agency in charge of Indian affairs in the Brazilian government announced that it was forcibly removing *all* miners from Yanomami lands.[127] But in 1993, as mentioned above, 3,000 miners were still there. Any attempt to keep the miners out permanently will probably fail. And in 2002 the Brazilian army began building more bases along its largely undefended northern border which crosses Yanomami lands. The soldiers are getting Yanomami women pregnant and bringing venereal diseases. The fear is that AIDS will soon follow.[128]

Another disturbing fact is that there is abundant research now showing that when indigenous peoples come in contact with the modern world, they often lose the special knowledge possessed by members of their culture – such as natural healing methods and drugs – and develop a dependency on modern goods, which destroys their self-sufficiency and pride.[129] Suicide rates and alcoholism often soar.

The Estonians

The Estonians live in a tiny country in Northern Europe. With less than 1 million people, there are fewer Estonians than the population of most major cities in the world.

Their culture is a very old one. Estonians have lived in this same spot next to the Baltic Sea for 5,000 years. Many foreigners have ruled the land during the past 700 years – among them Swedes, Germans, and Russians – but Estonian culture has survived. Now, after having survived the latest foreign rule – an especially hard 50 years under the Soviet Union – Estonians are struggling to keep their culture alive. The country has a slightly negative rate of population growth, a situation that could be dangerous if it continues too long.

Estonians have survived in spite of efforts by their latest rulers to destroy their culture. As they did to other peoples they conquered, the Soviets shipped many Estonian intellectual, political, and military leaders to Siberia. Others were killed within the country. Stalin introduced measures to Russify the country. One way he tried to do this was to introduce heavy industry into Estonia and to import Russian workers to run the plants. In the latter part of the twentieth century, Estonians were aware that if they did not regain their independence soon, their culture would be destroyed.

The way Estonians regained their independence marks them as a very unusual people. When I visited the country in 1990 they were still under Soviet rule. As a political scientist I knew that their chances of winning back the independence they had enjoyed before Hitler and Stalin made a deal in 1939 to give the country to the Soviet Union were very slim. US political scientists knew that the Soviets could not agree to Estonian independence without threatening the very foundations of the Soviet Union itself. Yet within one year after my visit, Estonia had become independent. It did this through nonviolent opposition and by waiting for the right moment to declare its freedom. When a coup d'état was attempted in the Soviet Union by conservative forces in 1991, the Estonians moved. That move was followed by similar declarations of independence by the two other small Baltic nations of Lithuania and Latvia. Not long after that the Soviet Union itself broke up.

The Estonian fight to regain their independence has been called the "Singing Revolution."[130] Instead of using guns to push out the Soviets – an effort the Soviets probably would have welcomed to give them an excuse to crush the independence movement – the Estonians used songs. They had persuaded the Soviets to build a huge outdoor stadium in Tallinn, the Estonian capital city, where their song festivals – an important part of the Estonian culture – could be held. What the Russians didn't realize was that the song festivals were helping to keep alive the Estonians' love for their land and freedom.

The 1990 song festival in Estonia was the last one under Soviet rule. I was fortunate enough to attend it and found it to be an extremely moving experience. The two-day festival, the largest song festival in the world, was attended by about 500,000 people – about one-half of the nation. The 28,000 singers from all parts of the country, dressed in traditional clothes, sang of their love for their

Song festival in Tallinn, Estonia (*Aldo Bender*)

homeland and their desire to be free. Nine thousand dancers performed in colorful costumes, each unique to a different section of the country.

Two years after regaining its independence Estonia was being cited in the Western press as a model for the rest of the former republics of the Soviet Union to follow as they tried to pass from stagnant, centrally planned economies to those based on individual initiative and freedom. Today the country has made relatively good economic progress, but huge problems still exist. After many years of Soviet control that were designed to replace love for and loyalty to their own culture with loyalty to the Soviet state (with its supposedly new type of person), the Estonians have an uphill battle. Many Russians who have never learned the Estonian language or participated in the culture still live in Estonia. And in the early 1990s the Russian army was still there. (It finally left in 1994.)

Even if the threat from Russia recedes, the Estonians face another threat to their culture. As they develop and become part of the Western world, they fear that the dominant US-influenced culture will come to replace their own. They face the same challenge as do other nations with their distinctive cultures. Can they develop but yet retain their distinctiveness, their own culture? Or will the influences brought by the new opportunities to travel, increased contact with Western goods and tourists, and messages they receive from the Western media overwhelm their own ways? Cultures are always changing and this can be healthy. But can Estonia change in some ways but not change in others? Will they continue to be Estonians? Only time will tell.

Environmental Politics

In this final section we shall try to understand what makes environmental politics so controversial. Politics is a passionate business, but why are environmental issues often emotional? Obviously, conflicting interests and values must be involved. Politics involves the making of laws and decisions that everyone must obey in a society. These laws and decisions are directed at settling conflicts that arise among people living together in a community, and at achieving commonly desired goals. As we shall see, environmental politics does deal with very strongly held opposing values and interests. It also represents an effort by a community to achieve some goals – such as clean air and clean water – which cannot be reached individually, only by the community as a whole.

The political scientists Harold and Margaret Sprout believe that most participants in environmental politics show a tendency toward having one of two very different philosophies or worldviews and that these are at the root of most environmental conflicts. One they call "exploitive," and the other "mutualistic." Here is how they define them:

> A[n] . . . exploitive attitude would be one that envisages inert matter, nonhuman species, and even humans as objects to be possessed or manipulated to suit the purposes of the exploiter. In contrast, a . . . mutualistic posture would be one that emphasizes the interrelatedness of things and manifests a preference for cooperation and accommodation rather than conflict and domination.[131]

While conflicting worldviews are a part of environmental politics, so also is a conflict of basic interests. Economist Lester Thurow believes that environmental politics often involves a conflict between different classes having very different interests. He sees the environmental movement as being supported mainly by upper middle-class people who have gained economic security and now want to improve the quality of their lives further by reducing environmental pollutants. On the opposite side, he sees both lower income groups and the rich – lower income people because they see environmental laws making it more difficult for them to find jobs and obtain a better income, and the rich because they can often buy their way out of environmental problems and see pollution laws as making it more difficult for them to increase their wealth even further.[132]

Other conflicting interests are also involved in environmental politics. Antipollution laws often make it more difficult and costly to increase energy supplies, extract minerals, and increase jobs by industrial growth. Barry Commoner's Fourth Law of Ecology – There Is No Such Thing as a Free Lunch – means that for every gain there is some cost.[133] There are tradeoffs involved in making the air and water cleaner as there are in making more cars and television sets. Also, the costs of pollution control often increase substantially as you try to make the environment cleaner and cleaner. The cost required to make a 50 percent reduction in a pollutant is often quite modest, whereas if you try to reduce the pollutant by 95 percent, the cost usually increases dramatically.[134]

Much environmental destruction is extremely difficult for the political system to deal with, since the damage often shows up many years after the polluting action takes place. It is now clear that prevention is much cheaper than trying to clean up the damage after it has occurred, but the nature of politics does not lend itself to long-range planning. Generally, politicians have a rather short-term outlook, as do many business people. Both are judged on their performance in handling immediate problems; this promotes a tendency to take actions showing some immediate result. Such actions further the politician's chances for reelection and the business person's profits or chances for promotion. Yet environmental problems often call for actions before the danger becomes clear. A further complication is the fact that, even after action is taken to reduce a pollutant, because of the inherent delays in the system the harmful effects of the pollutant do not decrease until a number of years later. Thus the inclination of the public official – and the business person – is to do nothing and hope that something turns up showing that the problem was not as bad as feared or that there is a cheaper way to deal with it.

An additional factor in environmental politics is unique to the United States. The American dream has been one of continuing abundance. For much of the country's history, there has seemed to be an unlimited abundance of many things needed for the good life, such as land, forests, minerals, energy, clean air, and natural beauty. It is a country that seemed to offer unlimited opportunities for many to make a better life for themselves, and "better" has been usually defined as including more material goods. The setting of limits on consumption and production that environmentalists often promote is certain to cause dismay to many.

If the above were not enough to make environmental politics very difficult, there is also the fact that the costs in environmental matters are often very difficult to measure. One can calculate the cost of a scrubber on a coal-burning power plant, but how do you measure the cost of a shortened life that occurs if the scrubber is not used? How do you place a dollar figure on the suffering a person with emphysema experiences, or a miner with brown lung disease, or an asbestos worker with cancer? How do you measure the costs the yet unborn will have to pay if nothing is done now about climate change? And how do you put a dollar figure on the loss of natural beauty? Because it is so difficult to weigh the costs in conventional terms of measurement, the costs often were not weighed in the past.[135]

There is, of course, also the matter of values – the value individuals place on more material goods, the convenience of throwaway products, open spaces, and clean air. The resolution of conflicts over values can often be handled only by politics, in a democracy by the community as a whole making decisions through its representatives and then requiring all members of the community to obey them. That such stuff causes controversy and stirs passions should not be surprising. It is hard work.

Conclusions

Development is more than economic growth: it also includes the social changes that are caused by or accompany economic growth. As this chapter has shown, the increase in the production of goods and services that came with industrialization had, and still has, frightening costs. Poverty was basically wiped out in a number of countries by industrialization – obviously an impressive benefit of the new economic activity. But that activity harmed both people and the environment. Slowly and painfully, people in the developed countries have come to realize that economic growth is not enough. Attention has to be paid to its effect on the earth and on people. (If one gets cancer, for example, what good is material wealth?) And an awareness has grown in the industrialized nations, and continues to grow, that the question of how economic growth is affecting the environment needs to be asked and answered. The rich countries are slowly learning that it is cheaper and causes much less suffering to try to reduce the harmful effects of an economic activity at the beginning, when it is planned, than after the damage appears. To do this is not easy and is always imperfect. But an awareness of the need for such effort indicates a greater understanding and moral concern than did the previous widespread attitude that focused only on creating new products and services.

The less developed nations are also slowly realizing that the effects of economic activity on the environment should not be ignored. But here the new awareness is less widespread than in the rich countries. This is understandable because, except for some of the rulers and elite groups, the reduction of poverty is the first concern people have. It explains why some developing countries have welcomed polluting industries, such as factories that manufacture asbestos, since jobs today are more important than a vague worry that workers may contract cancer in 20 to 30 years. But also in developing countries, a slowly growing number of people realize that if the economic activity that gives jobs to people harms the environment at the same time, the benefits from that economic activity will be short-lived.

Poverty harms the environment, as we saw for example in the case of deforestation, where poor people searching for land to farm and for fuel are one cause of the extensive destruction of the remaining tropical rainforests. Economic growth that benefits the majority of people is needed to protect the environment. And a control on the rapidly expanding populations of many of the poorest countries is also needed to protect the environment, since increasing numbers of poor people hurt the land on which they live as they struggle to survive.

For both rich and poor nations, the environment is important. Economic growth is also important, especially for the poorer countries. The challenge remains for both poor and rich to achieve a balance between economic activity and protection of the land, air, and water upon which life depends.

Notes

1 Bill Clinton, who defeated President Bush, signed the biodiversity treaty, but the US Congress never ratified it.
2 James Dao, "Protesters Interrupt Powell Speech as UN Talks End," *New York Times*, national edn (September 5, 2002), p. A8.
3 Felicity Barringer, "United States Ranks 28th on Environment, a New Study Says," *New York Times*, national edn (January 23, 2006), p. A3.
4 "Chinese Official Sees Private Role on Environment," *New York Times*, national edn (June 6, 2006), p. A6.
5 Lester Brown, *The Twenty-Ninth Day* (New York: W. W. Norton, 1978), p. 44. One of the ways Britain reduced its air pollution was to build tall smokestacks, which has probably led to worse air in Scandinavia.
6 Stephen Klaidman, "Muddling Through," *Wilson Quarterly* (Spring 1991), p. 76.
7 Jon Luoma, "Sharp Decline Found in Arctic Air Pollution," *New York Times*, national edn (June 1, 1993), p. B7.
8 World Resources Institute, UN Environment Programme, UN Development Programme, and World Bank, *World Resources 1998–1999* (New York: Oxford University Press, 1998), p. 63.
9 Reed McManus, "Out Front in the Air Wars," *Sierra*, 89 (January/February 2004), p. 12.
10 Jocelyn Kaiser, "Mounting Evidence Indicts Fine-Particle Pollution," *Science*, 307 (March 25, 2005), p. 1858.
11 Gary Gardner, "Air Pollution Still a Problem," in *Vital Sign 2005* (New York:

Worldwatch Institute, 2005), p. 94, and Genevieve Wong, "Study: Asthma Rates Rise Dramatically," *Spartanburg Herald-Journal* (May 1, 2004), p. B5.

12 Roger-Mark De Souza et al., "Critical Links: Population, Health, and the Environment", *Population Bulletin*, 58 (September 2003), p. 18.

13 Worldwatch Institute, *Vital Signs 2005*, p. 94.

14 Ibid., p. 95.

15 Jim Yardley, "Bad Air and Water, and a Bully Pulpit in China," *New York Times*, national edn (September 25, 2004), p. A4.

16 Keith Bradsher, "China Pays a Price for Cheaper Oil," *New York Times*, national edn (June 26, 2004), pp. B1–2.

17 Nancy Riley, "China's Population: New Trends and Challenges," *Population Bulletin*, 59 (June 2004), p. 25.

18 World Resources Institute et al., *World Resources 1998–1999*, pp. 63–6.

19 Marlise Simons, "Rising Iron Curtain Exposes Haunting Veil of Polluted Air," *New York Times*, national edn (April 8, 1990), p. 1.

20 *New York Times*, late city edn (May 13, 1980), p. C3.

21 "Peddling Lead", *Science*, 299 (February 7, 2003), p. 795.

22 Ibid.

23 Jane Brody, "Even Low Lead Levels Pose Perils for Children," *New York Times*, national edn (August 5, 2003), p. D7.

24 *New York Times*, national edn (July 27, 1994), p. C20.

25 Brody, "Even Low Lead Levels Pose Perils for Children," p. D7.

26 "Lead Concentrations Down in Greenland Ice," *New York Times*, national edn (October 15, 1991), p. B8.

27 Marc Lacey, "Belatedly, Africa Is Converting to Lead-Free Gasoline," *New York Times*, national edn (October 31, 2004).

28 Brody, "Even Low Lead Levels Pose Perils for Children," p. D7.

29 John Heilprin, "Pollution Eases, but Lead Worries EPA," *Spartanburg Herald-Journal* (July 1, 2003), p. A3.

30 Barry Commoner, *The Closing Circle* (New York: Alfred A. Knopf, 1971), p. 39.

31 W. Wayt Gibbs, "How Should We Set Priorities?" *Scientific American*, 293 (September 2005), p. 110.

32 Ibid., pp. 110, 112; and Kevin Krajick, "Long-Term Data Show Lingering Effects from Acid Rain," *Science*, 292 (April 13, 2001), pp. 195–6.

33 World Resources Institute et al., *World Resources 1998–1999*, p. 182.

34 Yardley, "Bad Air and Water, and a Bully Pulpit in China," p. A4.

35 William Stevens, "Ozone Loss over US Is Found to Be Twice as Bad as Predicted," *New York Times*, national edn (April 5, 1991), p. A1.

36 Kenneth Chang, "Scientists Say Recovery of the Ozone Layer May Take Longer than Expected," *New York Times*, national edn (December 7, 2005), p. A25.

37 William Stevens, "New Survey Shows Growing Loss of Arctic Atmosphere's Ozone," *New York Times*, national edn (April 6, 2000), p. A19, and Richard Kerr, "Ozone Loss, Greenhouse Gases Linked," *Science*, 280 (April 10, 1998), p. 202.

38 David Karoly, "Ozone and Climate Change," *Science*, 302 (October 10, 2003), pp. 236–7.

39 World Resources Institute, *The 1993 Information Please Environmental Almanac* (Boston: Houghton Mifflin, 1993), pp. 38–40.

40 Katharine Seelye and Jennifer Lee, "EPA Calls US Cleaner and Greener than 30 Years Ago," *New York Times*, national edn (June 24, 2003), p. A28.

41 Lisa Foderaro, "Caution Urged in Eating of Fish from Mountains of New York," *New York Times*, national edn (April 16, 2005), p. A11.

42 John Cushman, "Michigan Balks at Tainted-Salmon Warning," *New York Times*, national edn (February 8, 1997), p. 7.

43 Michael Janofsky, "Mercury Taints Fish across US," *New York Times*, national edn (August 25, 2004), p. A19.

44 Richard Bernstein, "No Longer Europe's Sewer, but Not the Rhine of Yore," *New York Times*, national edn (April 21, 2006), p. A4.

45 Andrew Revkin, "Stream Tests Show Traces of Array of Contaminants," *New York Times*, national edn (March 13, 2002), p. A14, and Andrew Revkin, "FDA Considers New Tests for Environmental Effects," *New York Times*, national edn (March 14, 2002), p. A20.

46 Nels Johnson et al., "Managing Water for People and Nature," *Science*, 292 (May 11, 2001), p. 1071.

47 Ibid.
48 Riley, "China's Population: New Trends and Challenges," p. 25.
49 Mark Stevenson, "Sewage-Laced Mexico City to Host Water Conference," *Spartanburg Herald-Journal* (March 13, 2006), p. A9.
50 David Rohde, "Bangladesh Wells Pumping Poison as Cleanup Lags," *New York Times*, national edn (July 17, 2005), p. 6. For more on this problem see A. Mushtaque Chowdhury, "Arsenic Crisis in Bangladesh," *Scientific American*, 291 (August 2004), pp. 86–91, and Kirk Nordstrom, "Worldwide Occurrences of Arsenic in Ground Water," *Science*, 296 (June 21, 2002), pp. 2143–6.
51 William Stevens, "Expectation Aside, Water Use in US Is Showing Decline," *New York Times*, national edn (October 10, 1998), p. A1.
52 Johnson et al., "Managing Water for People and Nature," p. 1071.
53 "US Waste and Recycling," *New York Times*, national edn (August 20, 2002), p. D4.
54 De Souza et al., "Critical Links: Population, Health, and the Environment," p. 18.
55 Charles Moore, "Trashed," *Natural History*, 112 (November 2003), p. 51.
56 Ibid., pp. 46, 50.
57 Ibid., p. 50. For a report on the increase of plastic fibers in the beaches and seabed of the British Isles see Andrew Revkin, "Plastics Permeate Even the Seabed," *New York Times*, national edn (May 11, 2004), p. D2.
58 Moore, "Trashed," p. 50.
59 Rebecca Renner, "Tracking the Dirty Byproducts of a World Trying to Stay Clean," *Science*, 306 (December 10, 2004), p. 1887.
60 Paul Webster, "Study Finds Heavy Contamination across Vast Russian Arctic," *Science*, 306 (December 10, 2004), p. 1875.
61 Ibid.
62 Barbara Ward, *Progress for a Small Planet* (New York: W. W. Norton, 1979), pp. 65–6.
63 Michael Royston, "Making Pollution Prevention Pay," *Harvard Business Review* (November/December 1980), p. 12.
64 Heinrich von Lersner, "Commentary: Outline for an Ecological Economy," *Scientific American*, 273 (September 1995), p. 188.
65 Andrew Revkin, "China Is Bright Spot in Dark Report on the World's Diminishing Forests," *New York Times*, national edn (November 15, 2005), p. D3.
66 World Resources Institute et al., *World Resources 1998–1999*, p. 188.
67 Ibid., p. 186.
68 Howard French, "Billions of Trees Planted, and Nary a Dent in the Desert," *New York Times*, national edn (April 11, 2004), p. 3.
69 "Resurgent Forests Can Be Greenhouse Gas Sponges," *Science*, 277 (July 18, 1997), p. 316.
70 William Laurance, "Gaia's Lungs," *Natural History*, 108 (March 1999), p. 96.
71 "Resurgent Forests Can Be Greenhouse Gas Sponges," p. 315.
72 Larry Rohter, "Relentless Foe of the Amazon Jungle: Soybeans," *New York Times*, national edn (September 17, 2003), p. A3.
73 Jane Perlez, "Forests in Southeast Asia Fall to Prosperity's Axe," *New York Times*, national edn (April 29, 2006), p. A7.
74 Anne LaBastille, "Heaven, Not Hell," *Audubon*, 81 (November 1979), p. 91.
75 Christopher Uhl, "You Can Keep a Good Forest Down," *Natural History*, 92 (April 1983), p. 78.
76 "Cancer: Causes and Prevention," *Scientific American*, 275 (September 1996), pp. 79–101.
77 "Study Hailed as Convincing in Tying Fat to Cancers," *New York Times*, national edn (April 24, 2003), p. A23, and Jane Brody, "Another Study Finds a Link between Excess Weight and Cancer," *New York Times*, national edn (May 6, 2003), p. D7.
78 Solana Pyne, "Small Particles Add up to Big Disease Risk," *Science*, 295 (March 15, 2002), p. 1994.
79 "Study Links DDT and Cancer," *New York Times*, national edn (April 22, 1993), p. A10.
80 Marian Burros, "High Pesticide Levels Seen in US Food," *New York Times*, national edn (February 19, 1999), p. A12.
81 *New York Times*, late edn (April 21, 1986), p. A14.
82 Philip Hilts, "Results of Study on Pesticide Encourage Effort to Cut Use," *New York Times*, national edn (July 5, 1993), p. 8.
83 Matthew Wald, "Citing Children, EPA Is Banning Common Pesticide," *New York Times*, national edn (August 3, 1999), p. A1.

84 Michael Janofsky, "Environmental Groups Are Praising the EPA for Updating Cancer-Risk Guidelines," *New York Times*, national edn (April 4, 2005), p. A18.
85 *New York Times*, late edn (March 3, 1984), p. 10.
86 Commoner, *The Closing Circle*, p. 261.
87 Otto Pohl, "European Environmental Rules Propel Change in US," *New York Times*, national edn (July 6, 2004), p. D4.
88 Claudia Deutsch, "DuPont Looking to Displace Fossil Fuels as Building Blocks of Chemicals," *New York Times*, national edn (February 28, 2006), p. C1.
89 Rebecca Renner, "Scotchgard Scotched," *Scientific American*, 284 (March 2001), p. 18, and Jennifer Lee, "EPA Orders Companies to Examine Effects of Chemicals," *New York Times*, national edn (April 15, 2003), p. D2.
90 Hans Landsberg et al., "Nonfuel Minerals," in Paul Portney (ed.), *Current Issues in Natural Resource Policy* (Washington, DC: Resources for the Future, 1982), p. 83.
91 World Resources Institute et al., *World Resources 1998–1999*, p. 167.
92 Ibid.
93 Ibid.
94 Ibid., p. 168.
95 Samuel Loewenberg, "Old Europe's New Ideas," *Sierra*, 89 (January/February 2004), p. 43.
96 "Reduce, Reuse, Rejoice," *Sierra*, 90 (November/December 2005), p. 43.
97 "US Waste and Recycling," *New York Times*, national edn (August 20, 2002), p. D4.
98 Norimitsu Onishi, "How Do Japanese Dump Trash? Let Us Count the Myriad Ways," *New York Times*, national edn (May 12, 2005), p. A1.
99 John Markoff, "Technology's Toxic Trash Is Sent to Poor Nations," *New York Times*, national edn (February 25, 2002), p. C1.
100 Jon R. Luoma, "Trash Can Realities," *Audubon*, 92 (March 1990), p. 95.
101 Brown, *The Twenty-Ninth Day*, p. 284.
102 Elizabeth Brown, "Bottle Bills Proliferate in States and in Congress," *Christian Science Monitor* (March 5, 1991), p. 7.
103 Ted Williams, "The Metamorphosis of Keep America Beautiful," *Audubon*, 92 (March 1990), p. 132.
104 Robin Pogrebin, "Now the Working Class, Too, Is Foraging for Empty Cans," *New York Times*, national edn (April 29, 1996), p. B12.
105 Williams, "The Metamorphosis of Keep America Beautiful," pp. 128–9.
106 Charles Birch, *Confronting the Future* (New York: Penguin, 1976), p. 35.
107 Edward O. Wilson, *The Diversity of Life* (Cambridge, MA: Harvard University Press, 1992).
108 W. Wayt Gibbs, "On the Termination of Species," *Scientific American*, 285 (November 2001), p. 42.
109 Stuart Pimm and Clinton Jenkins, "Sustaining the Variety of Life," *Scientific American*, 293 (September 2005), p. 68.
110 Loren McIntyre, "Jari: A Billion Dollar Gamble," *National Geographic* (May 1980), p. 701.
111 Ibid., p. 711.
112 Larry Rohter, "A Mirage of Amazonian Size," *New York Times*, national edn (November 9, 1999), pp. C1–2.
113 Peter Raven, "Tropical Rain Forests: A Global Responsibility," *Natural History*, 90 (February 1981), p. 29.
114 Norman Myers, "The Exhausted Earth," *Foreign Policy*, 42 (Spring 1981), p. 143.
115 Norman Myers, "Room in the Ark?" *Bulletin of the Atomic Scientists*, 38 (November 1982), p. 48.
116 Paul Ehrlich and Anne Ehrlich, *Extinction* (New York: Random House, 1981), p. 94.
117 William Stevens, "The 'Hot Spot' Approach to Saving Species," *New York Times*, national edn (March 14, 2000), p. D3.
118 Norman Myers, "A Convincing Call for Conservation," *Science*, 295 (January 18, 2002), pp. 447–8.
119 A fuller discussion of the strengths and weaknesses of the two metaphors is contained in William Stevens, "Lost Rivets and Threads, and Ecosystems Pulled Apart," *New York Times*, national edn (July 4, 2000), p. D4.
120 W. Wayt Gibbs, "Saving Dying Languages," *Scientific American*, 287 (August 2002), p. 80.
121 Rodger Doyle, "Measuring Modernity," *Scientific American*, 289 (December 2003), p. 40. For a contrary view that the American culture is dominant in Europe see Alan Riding, "A Common Culture (from the USA) Binds Europeans Ever Closer," *New York Times*, national edn (April 26, 2004), p. B1.

122 Katrin Bennhold, "Love of Leisure, and Europe's Reasons," *New York Times*, national edn (July 29, 2004), p. A8.
123 Ibid.
124 As one author has written: "[indigenous peoples] may offer living examples of cultural patterns that can help revive ancient values for everyone: devotion to future generations, ethical regard for nature, and commitment to community among people." Alan Durning, "Supporting Indigenous Peoples," in Lester R. Brown et al., *State of the World – 1993* (New York: W. W. Norton, 1993), p. 100.
125 James Brooke, "Brazil Evicting Miners in Amazon to Reclaim Land for the Indians," *New York Times*, national edn (March 8, 1993), p. A4.
126 Ibid.
127 James Brooke, "In an Almost Untouched Jungle Gold Miners Threaten Indian Ways," *New York Times*, national edn (September 18, 1990), p. B6.
128 Larry Rohter, "A New Intrusion Threatens a Tribe in Amazon: Soldiers," *New York Times*, national edn (October 1, 2002), p. A1.
129 One encouraging sign is that when Western scientists seek information from the medicine men of indigenous peoples about natural drugs and health cures, the medicine men are given new respect. This new respect might help encourage some of their youth to study under them. But, all too often today, when the medicine men die, the knowledge they have acquired dies with them. See Daniel Goleman, "Shammans and Their Longtime Lore May Vanish with the Forests," *New York Times*, national edn (June 11, 1991), p. B5. As an example of a study showing the harmful effects Western contact can have on the culture of indigenous peoples, see Katharine Milton, "Civilization and Its Discontents," *Natural History* (March 1992), pp. 37–43.
130 See, for example, Walter C. Clemens Jr, "Baltics Sang Their Way to Independence," *Christian Science Monitor* (September 5, 1991), p. 19.
131 Harold Sprout and Margaret Sprout, *The Context of Environmental Politics* (Lexington: University Press of Kentucky, 1978), pp. 47–8.
132 Lester Thurow, *The Zero-Sum Society* (New York: Basic Books, 1980), pp. 104–5.
133 Commoner, *The Closing Circle*, pp. 45–6.
134 William Ophuls, *Ecology and the Politics of Scarcity* (San Francisco: W. H. Freeman, 1977), p. 75.
135 For an attempt to measure the hidden costs of today's energy, see Harold M. Hubbard, "The Real Cost of Energy," *Scientific American*, 264 (April 1991), pp. 36–42. For an explanation of how the accounting system presently used by economists allows policymakers to ignore the deterioration of the environment caused by economic activity, see Robert Repetto, "Accounting for Environmental Assets," *Scientific American*, 266 (June 1992), pp. 94–100.

Further Reading

Arthus-Bertrand, Yaun, *Earth from Above*, revised and expanded edition (New York: Henry Abrams, 2002). From a helicopter flying over 60 countries, the author records the beauty of the earth as well as its spoiled places. These stunning photographs give a rarely seen view of the earth.

Brown, Phil, and Edwin J. Mikkelsen, *No Safe Place: Toxic Waste, Leukemia, and Community Action* (Berkeley: University of California Press, 1990). This book documents the efforts of a small group of citizens – led by a mother whose son had acute leukemia – to uncover the source of illness in their community. It describes their battle to overcome resistance by their neighbors and the indifference of local and state government officials.

Daily, Gretchen, and Katherine Ellison, *The New Economy of Nature: The Quest to Make Conservation Profitable* (Washington, DC: Island Press, 2002). The authors show that the environment provides services that have an economic value. That recognition can lead to arrangements (like carbon trading) based on mutual self-interest and an understanding by all that conservation pays. Highly readable.

Easterbrook, Greg, *A Moment on the Earth: The Coming Age of Environmental Optimism* (New York: Viking, 1995). The author attacks the environmentalists for ignoring their own triumphs and instead preaching that we face constant crises. He does not advocate people continuing their insults on nature, but believes that nature takes the long view and is fairly good at defeating the insults on it by human beings.

Freese, Barbara, *Coal: A Human History* (Cambridge, MA: Perseus, 2003). Freese explains how vital coal has been in the development of modern civilization, but most of her book deals with what coal has done to people rather than what it has done for them. (See the further reading in chapter 4, "Energy," for another book on coal.)

Lerner, Steve, *Eco-pioneers: Practical Visionaries Solving Today's Environmental Problems* (Cambridge, MA: MIT Press, 1997). A reporter spent four years searching out what he calls "eco-pioneers," people working to reduce the pace of environmental destruction. Lerner presents case studies of people who are exploring practical and sustainable ways to log forests, grow food, save plant species, raise cattle, build houses, clean up cities, redesign rural communities, generate power, conserve water, protect rivers and wildlife, treat hazardous waste, reuse materials, and reduce waste and consumption.

McKibben, Bill, "A Special Moment in History," *Atlantic Monthly* (May 1998), pp. 55–78. In simple but powerful prose, McKibben argues that we live in a very important time in the history of the planet. He believes the fate of the planet for many years to come will be determined by the lifestyle, population, and technological choices we make in the next few decades.

McNeill, J. R., *Something New under the Sun: An Environmental History of the Twentieth-Century World* (New York: W. W. Norton, 2000). Liberals probably won't like this book because McNeill does not show moral outrage over what he reports, and conservatives probably won't like it because of what he reports. McNeill, a historian at Georgetown University, focuses on the tremendous economic growth occurring in modern times and the devastating impact it has had on the air, water, plants, animals, and human beings themselves.

Simon, Noel, *Nature in Danger: Threatened Habitats and Species* (New York: Oxford University Press, 1995). Simon, in association with the World Conservation Monitoring Centre in Cambridge, England, provides authoritative information on threatened habitats and species around the world. The book is beautifully illustrated.

Somerville, Richard, *The Forgiving Air: Understanding Environmental Change* (Berkeley: University of California Press, 1996). Using simple prose, the author presents a clear primer on the chemistry and physics of the earth's atmosphere, covering such items as global climate change, acid rain, and ozone depletion.

Speth, James Gustave, *Red Sky at Morning: America and the Crisis of the Global Environment* (New Haven, CT: Yale University Press, 2004). Speth explores some of the fundamental reasons for the environmental deterioration occurring in the world. He seeks to explain why a number of economic indicators continue to show growth at the same time that serious environmental deteriorating is occurring.

Steingraber, Sandra, *Having Faith: An Ecologist's Journey to Motherhood* (Cambridge, MA: Perseus, 2001). What's it like for a woman who knows a lot about chemicals and their effects on human beings to have a baby today? Streingraber explains but also analyzes how we got to where we accept that residues of many toxic chemicals in our world are OK.

Williams, Michael, *Deforesting the Earth: From Prehistory to Global Crisis* (Chicago: University of Chicago Press, 2003). A reviewer in *Science* has called this book "the most comprehensive account ever written of when, where, and how humans have wrought what is surely the most dramatic change in Earth's surface since the end of the Pleistocene 10,000 years ago . . ."

Wilson, Edward O., *The Diversity of Life* (Cambridge, MA: Harvard University Press, 1992). Wilson shows how the evolution of life has progressed on earth, with five great "spasms" of death occurring along the way. Wilson believes that, for the first time, humans are causing an extinction spasm – the sixth – but he leaves the reader with hope that humans may come to realize what they are doing before it is too late.

CHAPTER 6

Technology

Will mankind murder Mother Earth or will he redeem her? He could murder her by misusing his increasing technological potency.

Arnold J. Toynbee (1889–1975), *Mankind and Mother Earth* (1976)

To many people, technology and development are synonymous. Technology is what makes economic growth and social change happen. The limited use of high technology by the less developed nations is sometimes given as one of the reasons why they are less developed and less prosperous than the industrialized nations. But the relationship between technology and development is a complicated one. At times the negative features of technology seem to outweigh the positive features. Technology can cause a society to change in some very undesirable ways. In this chapter, after a short section on the benefits of technology, we will look closely at some of the negative relationships between technology and development.

Benefits of Technology

A book such as this one, whose readers will probably be mostly from the developed nations, does not need to dwell on the benefits of technology. Advertising and the mass media herald the expected joys that will come with a new product, technique, or discovery. In the United States people are socialized to like new things; they are also pragmatic, which means that science and its application, technology, are commonly used to solve problems, to make things work "better." They would have to be foolish not to recognize the benefits that technology has brought.

One of the main reasons much of the world envies the United States is that its technology has in many real ways made life more comfortable, stimulating, and free of drudgery. People in the United States know this and need to remember it. But they and others also need to learn several other lessons: (1) short-term benefits from using a technology can make it impossible to achieve some long-term goals; (2) there can be unanticipated consequences of using a technology; (3) the use of some types of technology in certain situations can be inappropriate; and (4) there are many problems that technology cannot solve. The inability to learn these lessons could lead to our destruction, as the case study in this chapter on the threat of nuclear weapons will show.

Without modern technology to help, necessary tasks can be difficult. A woman in Nepal breaks up clumps of soil to prepare the land for planting (*Ab Abercrombie*)

Benefits of technology

In personal terms, technology has allowed me to visit about 40 countries; to see a photograph of the earth taken from space; to write this book on a personal computer that greatly facilitated its composition; to wear shirts that don't need ironing; and to keep my glaucoma under control so that I do not go blind. What items would your list include?

Short-Term versus Long-Term Benefits -- the Tragedy of the Commons

Garrett Hardin, the late biologist, coined the phrase "the tragedy of the commons" to describe what can happen when the short-term and long-term interests of people are in conflict.[1] Hardin shows how it is rational and in the best short-term interest of each herdsman in a village to increase the number of cattle he has grazing on the "commons," the open-access, commonly owned lands in the village. The apparent short-term benefit to an individual herdsman of increasing the number of cattle he has there seems to him greater than the long-term harm resulting from the overgrazing that the additional cattle create; the cost of the overgrazing will be shared by all the herdsmen using the commons, while the individual herdsman will reap the profit that comes from selling additional cattle. Also, if the individual herdsman does not increase his cattle but others increase theirs, he loses out since the overgrazing harms his cattle. Thus the tragedy occurs. Each herdsman, acting rationally and in his own best short-term interest, increases his stock on the commons. Soon there is so much overgrazing that the grass dies and then the cattle die.

The global commons today are those parts of the planet that are used by many or all nations: the oceans, international river systems, the seabed, the atmosphere, and outer space. Technology can give some nations an advantage over others in exploiting these commons and it is clearly in their short-term interest to do so.

So it is with commercial fishing in the world's oceans. Technology has made possible bigger and more powerful fishing boats, equipped with sonar to locate schools of fish. It has also led to the creation of huge drift nets – some up to 40 miles long – which critics claim were used to "strip mine" the seas. These nets allowed a relatively small number of fishermen to catch large quantities of fish. (The United Nations in 1992 banned drift nets over 1.5 miles long but six years later nets much longer than this were still being used in the Mediterranean Sea and parts of the Atlantic Ocean.)[2] New technology also allows trawlers to drag dredges the size of football fields over the bottom, scraping it clean. Ninety percent of all large predator fish such as sharks, swordfish and tuna have been caught and there are fewer different kinds of fish in the oceans than before, putting ecosystems at more risk when they are confronted with disruptions such as climate change. There is every indication that many fisheries worldwide are being overfished, or to put it another way, "overgrazed," and are threatened with collapse.[3] If this is not controlled, all nations using the oceans for fishing will be hurt. Not only will their fishing industries be hurt, but unique forms of life on earth will probably become extinct. This could well be the fate of many species of the fish-like mammal, the whale, unless international efforts to reduce drastically the numbers of whales killed succeed in allowing whale populations to increase.

Another example of a tragedy of the commons situation is pollution. Individuals gain a short-term advantage by polluting – for example, by disconnecting the pollution control device on their cars to decrease gasoline consumption (some of my students confess to doing this) – but the long-term interests of the whole community are hurt by the polluting of the air. In fact, the lungs of the individual doing the polluting may be hurt in the long run by his or her auto's pollution. While this is true, the attractiveness of the short-term benefit over the long-term interests for any one individual can be overwhelming. Such was the case when I bought a car in 1979. I had the choice of buying a 1979 model, which used leaded gasoline, or a 1980 model, which used unleaded gasoline. Although I knew in a general way at that time that using unleaded gasoline was better for the environment, I bought the 1979 model because leaded gasoline was cheaper.

A way to deal with situations in which individuals gain benefits from polluting is to use political solutions – solutions designed by the community or its representatives, and which all members of the community will have to obey. In the example involving my students, a possible solution would be more effective auto inspections and the use of steep fines for removing the pollution control device. In the example involving me, a simple solution would have been for the government to have placed a higher tax on leaded gasoline than on nonleaded gasoline to equalize their prices so that there would not have been a monetary advantage for me to pick the 1979 model over the 1980 one.

Finally, possibly one of the most important examples in human history of human beings immersed in a tragedy of the commons situation is occurring today. The world is confronted with a change of its climate caused by human activity, as discussed in the energy chapter, with possibly disastrous consequences for life on the planet. One of the richest nations on earth and the largest producer of the main gas causing the change of the climate – CO_2 – refuses so far to cut back on this pollution. The Bush administration sees a short-term advantage for the US economy by not taking effective steps to deal with this problem in the face of widespread warning by scientists of likely long-term dangers. China, a rapidly developing nation and large polluter, uses similar reasoning, while some other nations, and even some US states and cities, recognize the danger and are beginning to take action to try to prevent this tragedy occurring.

Unanticipated Consequences of the Use of Technology

Ecology is the study of the relationships between organisms and their environments. Without a knowledge of ecology, we are tempted to use technology to solve

The case of the parachuting cats

The following situation, which occurred in Borneo, illustrates the unanticipated consequences of the use of technology. There, the efforts of health officials to destroy malaria-carrying mosquitoes by spraying houses with DDT led to the collapsing of the roofs of village houses and to the need to parachute cats into the villages:

> [Shortly after the spraying] the roofs of the natives' houses began to fall because they were being eaten by caterpillars, which, because of their particular habits, had not absorbed very much of the DDT themselves. A certain predatory wasp, however, which had been keeping the caterpillars under control, had been killed off in large numbers by the DDT. But the story doesn't end here, because they brought the spraying indoors to control houseflies. Up to that time, the control of houseflies was largely the job of a little lizard, the gecko, that inhabits houses. Well, the geckos continued their job of eating flies, now heavily dosed with DDT, and the geckos began to die. Then the geckos were eaten by house cats. The poor house cats at the end of this food chain had concentrated this material, and they began to die. And they died in such numbers that rats began to invade the houses and consume the food. But, more important, the rats were potential plague carriers. This situation became so alarming that they finally resorted to parachuting fresh cats into Borneo to try to restore the balance of populations that the people, trigger happy with the spray guns, had destroyed.

Source: "Ecology: The New Great Chain of Being," *Natural History*, 77 (December 1968), p. 8

a single problem. But there are many examples to illustrate the truth that we cannot change one part of the human environment without in some way affecting other parts. Often these other effects are harmful, and often they are completely unanticipated,[4] as the box about cats nicely illustrates.

DDT

The use of DDT in the United States has also had major unanticipated effects since it is persistent (it does not easily break down into harmless substances) and poisonous to many forms of life.[5] According to one study, many of the effects of the use of DDT could not have been predicted before its use.[6] The author of this study believes that only through the close monitoring of the effects of new chemicals and through an open debate on those effects can chemicals such as DDT be controlled.

Factory farms

Let's look at factory farms and the unanticipated consequences that have come with the adoption of factory techniques to produce animals for human consumption. Such techniques have been adopted to raise poultry, pigs, veal calves, and cattle. The techniques allow large numbers of animals to be raised in a relatively small space. (Many of these animals never see the light of day until they are removed for slaughter.) The crowding of many animals in a small space and the confinement of individual animals in small stalls creates stress, frustration, and boredom in the animals. Stress can lower the natural defenses of the animals to diseases, and the crowded conditions facilitate the rapid spreading of diseases among the animals. It is common in the US for factory-raised animals to receive large doses of antibiotics in their feed to prevent the outbreak of diseases and to promote growth.

There is now evidence that the abundant use of antibiotics in animal food is creating bacteria that are resistant to treatment by modern drugs and that these bacteria can cause illness in humans.[7] In 2001 researchers in the US reported in the *New England Journal of Medicine*, one of the country's main medical journals, that antibiotic-resistant bacteria were widespread in meats and poultry sold in the country and could be found in consumers' intestines. This means that many food-borne illnesses will not respond to the usual treatments and that some may be resistant to all current drugs.[8]

Of the 50 million pounds of antibiotics produced in the United States each year about 40 percent is given to animals, especially poultry and livestock, mostly to promote growth. Six of the 17 classes of antibiotics given to animals in their feed in the United States to promote growth are also used to treat illnesses in human beings. Drug-resistant infections, some of them fatal, have been increasing in the country and many scientists believe this is because of the wide use of antibiotics in animals as well as humans. A study in the late 1990s showed that the percentage of salmonella food poisoning cases with bacteria that were resistant to five commonly used antibiotics rose from about 1 percent in 1980 to about 35 percent in 1996.[9]

Following the recommendations of the UN's World Health Organization, the European Union in 1998 banned the placing of antibiotics that are used to treat illnesses in human beings in animal food to promote growth of the animal and to prevent diseases. Denmark stopped their use in 1999 and found that by improving the sanitary conditions of its animals there were generally no negative consequences of the ban. The only exception was with pigs, where there were more intestinal infections and thus a small increase in production costs.[10]

McDonald's made an important decision in 2003 that could help reduce the use of antibiotics in foods. (Because McDonald's buys such a large amount of food, its decisions influence the practices of the food industry.) It announced that it was going to require its poultry suppliers to eliminate the use of medically important antibiotics for growth promotion. It did not require this of its beef suppliers, but encouraged them to follow this policy also.

Controversy regarding factory farms in the United States is growing. In the late 1990s special attention was being given to large hog farms that were tending to dominate the industry. The number of hogs grown on an average swine farm rose from about 90 in the late 1970s to about 400 in the late 1990s.[11] Some of the largest farms held about 2,500 hogs. The complaints came from the small farmers who could not compete with the large factory farms and also from nearby residents. The foul smell from the farms was at times very powerful and the large amount of animal waste presented a real danger to underground water supplies. Some of the large hog farms produce as much raw sewage as a middle-sized city but without sewage treatment plants. Farm waste has always been treated more leniently in the United States than urban waste. The EPA in 2002 announced new regulations to reduce pollution from large farms by 25 percent, but the farms were allowed to write their own plans to meet the goals of the regulations and to keep their plans secret from the public.[12]

In the late 1990s, according to the EPA, there were about 450,000 factory-style farms in the United States, including about 7,000 that can raise 1,000 cattle, 2,500 hogs, or at least 30,000 chickens.[13] The defenders of the large hog farms cite the demand by US consumers for low-cost, lean pork. Their farms can produce this because the pigs are artificially inseminated and genetically designed to produce an identical cut of meat. The defenders also cite the need for jobs in rural areas and tax revenue which their farms provide.

The manure from large chicken factory farms is often spread as a fertilizer on crop farms. The manure has a high content of the nutrients nitrogen and phosphorus. Some of these nutrients wash into rivers and bays during heavy rains and are suspected of helping to cause blooms of toxic algae in Chesapeake Bay and to cause the lack of dissolved oxygen in the Gulf of Mexico off Louisiana.[14]

Foreign aid

The unanticipated consequences of the use of technology can be seen in a situation of which I have some personal knowledge. When I was in Iran in the late 1950s with the US foreign aid program, one of our projects was to modernize the police force of the monarch, the Shah of Iran. We gave the national police new communications equipment so that police messages could be sent throughout the country quickly and efficiently. The United States gave this kind of assistance to the Shah to bolster his regime and help him to maintain public order in Iran while development programs were being initiated. All fine and good, except for the fact that the Shah used his now efficient police – and especially his secret police, which the US CIA helped train – not just to catch criminals and those who were trying to violently overthrow his government, but to suppress all opponents of his regime. His secret police, SAVAK, soon earned a worldwide reputation for being very efficient – and ruthless. Such ruthlessness, which often involved torturing suspected opponents of the Shah, was one of the reasons why the Shah became very unpopular in Iran and was eventually overthrown in 1979 by the Ayatollah Khomeini, a person who had deep anti-American feelings.[15]

Inappropriate Uses of Technology

In 1973, E. F. Schumacher published his book *Small Is Beautiful: Economics as if People Mattered*.[16] This book became the foundation for a movement that seeks to use technology in ways that are not harmful to people. Schumacher argued that the developing nations need intermediate (or "appropriate") technology, not the high (or "hard") technology of the Western industrialized nations. Intermediate technology lies in between the ineffective and primitive technology common in the rural areas of the less developed countries – where most of the world's people live – and the technology of the industrialized world, which tends to use vast amounts of energy, pollutes the environment, requires imported resources, and often alienates the workers from their own work. The intermediate technology movement seeks to identify those areas of life in the South, and also in the industrialized West, where a relatively simple technology can make people's work easier while remaining meaningful, that is, giving them a feeling of satisfaction when they do it.

It is this sense of satisfaction, or contentment, which is often absent in workers in developed nations. A good example of this can be seen in the "workers' revolt" which took place in the ultramodern automobile plant in Lordstown, Ohio, which was to produce Vegas for General Motors and which incorporated the latest in automated technology. The revolt led to a vote by 97 percent of the workers to strike over working conditions. The workers' discontent with the new plant and its mass-production techniques can be summed up by the suggestion of one of the strikers that the workers ought to take a sign that was attached to some of the machines, "Treat Me with Respect and I will give you Top Quality Work with Less Effort," and print it on *their* T-shirts.[17]

The high technology of the West is often very expensive, and thus large amounts of capital are needed to acquire it, capital that most developing nations do not have. This technology is referred to as being capital intensive instead of labor intensive. This means that money – but not many people – is needed to obtain it and maintain it. In other words, high technology does not give many workers jobs. (This is the essence of the mass-production line: lots of products by a relatively small number of workers.) But the main problem in nations that are trying to develop – and, in fact, in the United States also when its economy is in a recession – is that there are not enough jobs for people in the first place. It is the absence of jobs in the rural areas that is causing large numbers of the rural poor in the South to migrate to the cities looking for work, work that is often not there.

While it is fairly obvious, and widely recognized, that developing nations should select technologies that are appropriate to their needs, why don't they always do this? Why has this seemingly simple "lesson" not been learned? The authors of a study of World Bank experiences over nearly four decades explain why they believe inappropriate technology is frequently chosen:

> Why does this happen? Foreign consultants or advisers may advocate the technology with which they are most familiar. Local engineers, if educated abroad or the heirs of a colonial legacy, may have acquired a similar bias in favor of advanced technology, or they may simply presume, as do their superiors, that what is modern

> is best. Special interest groups may favor a particular technical approach. . . . Deep-seated customs and traditions may favor certain solutions and make others unacceptable. Economic policies that overprice labor (through minimum wage or other legislation) or underprice capital (through subsidized interest rates or overpriced currency) may send distorted signals to decision makers. A simple lack of knowledge or reluctance to experiment may limit the range of choice. . . . When aid is tied to the supply of equipment from the donor country . . . freedom to choose an appropriate technology may be compromised. With so many factors at work, it is not surprising that a "simple" lesson – such as selecting an appropriate technology – may prove far from simple to apply.[18]

I have witnessed the inappropriate use of high technology in both Liberia and the United States. As part of US economic assistance to Liberia, we gave the Liberians road-building equipment. That equipment included power saws. As I proceeded to turn some of this equipment over to Liberians in a small town in a rural area, I realized that the power saws we were giving them were very inappropriate. To people who had little or no experience with power tools – which applied to nearly all the Liberians in that town – the power saw was a deadly instrument. Also, they would not be able to maintain them or repair them when they broke down. Their noise would ruin the peacefulness of the area. A much more appropriate form of assistance would have been crates of axes and hand saws, tools that they could easily learn to use safely, that they would be able to maintain and repair themselves, and that would have provided work for many people.

In the United States I became aware of the inappropriate use of high technology as my wife and I began to prepare for the birth of our children. Most children in the world are born at home, but in the United States and in many other developed countries nearly all births take place in hospitals. An impressive number of studies now show that moving births into hospitals has resulted in unnecessary interventions in the birth process by doctors and hospital staff, which upset the natural stages of labor and can jeopardize the health of both the mother and the baby.[19] As many as 85 to 90 percent of women can give birth naturally, without the use of technology being required.[20] Prenatal care can usually identify the 10 to 15 percent that cannot deliver normally, and for them the use of technology can help protect the lives of the mother and baby. But the major error that has been made is that procedures that are appropriate for these few are now routinely used for most births.

The intermediate technology movement is not against high technology as such (it recognizes areas where high technology is desirable – there is no other way to produce vaccines against deadly diseases, for example), but only against the use of such technology where simpler technology would be appropriate.

Limits to the "Technological Fix"

In US society, which makes wide use of technology, there is a common belief that technology can solve the most urgent problems. It is even believed that the problems that science and technology have created can be solved by more science and technology. What is lacking, according to this way of thinking, is an adequate use of science and technology to solve the problem at hand. In other words, we must find a "technological fix."

While the ability of technology to solve certain problems is impressive, there are a number of serious problems confronting humans – in fact, probably the most serious problems which humans have ever faced – which seem to have no technological solution. Technology itself has often played a major role in causing these problems. Let's look at a few of them.

The population explosion appears to have no acceptable technological solution. Birth control devices can certainly help in controlling population growth; without such devices a solution to the problem would be even more difficult than it is. But as we have seen in chapter 2, the reasons for the population explosion are much more complicated than the lack of birth control devices. Economic, social, and political factors play a significant role in this situation and must be taken into consideration in any effort to control the explosion. A technological advancement was one of the causes of the population explosion – the wiping out of major diseases, such as smallpox, which used to kill millions. Some people, such as Garrett Hardin, have argued that many of those people who are advocating technological solutions to the population problem such as farming the seas, developing new strains of wheat, or creating space colonies, "are trying to find a way to avoid the evils of overpopulation without relinquishing any of the privileges they now enjoy."[21]

Huge municipal sanitation plants were once considered the solution to our polluted streams, rivers, and lakes, but the rising costs of these plants and the fact that they treat only part of the polluted water are bringing this solution into question.[22] As much water pollution is caused by agricultural and urban runoffs, both of which are not treated by the plants, as by sewage. To talk about a technological fix for this problem is to talk about spending astronomical sums of money to treat all polluted water, and even then the solution would still be in doubt.

A final example will be given to illustrate the limits to the technological fix. As we shall see in the case study below, the nuclear arms race between the Soviet Union and the United States after World War II threatened the world with a holocaust beyond comprehension. Many believed that technology would solve this problem; all that was needed to gain security was better weapons and more weapons than the other side. But the history of the arms race, which lasted nearly half a century until the disintegration of the Soviet Union in the early 1990s, clearly shows that one side's advantage was soon matched or surpassed by new weapons on the other side. Momentary feelings of security by one nation were soon replaced by deepening insecurity felt by both nations as the weapons became more lethal. "Security dilemma" is the phrase that has been coined to describe a situation where one nation's efforts to gain security led to its opponent's feeling of insecurity. This insecurity causes the nation that believes it is behind in the arms race to build up its arms, but it also causes the other nation to feel insecure. So the race goes on. The temptation to believe that a new weapon will solve the problem is immense.

A brief history of the arms race shows how both superpowers were caught in a security dilemma.

The United States exploded its first atomic bomb in 1945 and felt fairly secure until the Soviets exploded one in 1949. In 1954 the United States tested the first operational thermonuclear weapon (a hydrogen or H-bomb), which uses the A-bomb as a trigger, and a year later the Soviets followed suit. In 1957 the Soviets successfully tested the first intercontinental ballistic missile (ICBM) and launched the earth's first artificial satellite, Sputnik. The United States felt very insecure, but within three years had more operational ICBMs than the Soviet Union. (This "missile gap," in which the Soviets trailed, could have been the reason they put missiles in Cuba in 1962, which led to the Cuban missile crisis, the world's first approach to the brink of nuclear war. The humiliation the Soviet Union suffered when it had to take its missiles out of Cuba may have led to its buildup of nuclear arms in the 1970s and 1980s, which caused great concern in the United States.)

The Soviet Union put up the first antiballistic missile system around a city – around Moscow – in the 1960s, and in 1968 the United States countered by developing MIRVs (multiple, independently targetable reentry vehicles), which could easily overwhelm the Soviet antiballistic missiles. The Soviets started deploying their first MIRVs in 1975, and these highly accurate missiles with as many as ten warheads on a single missile, each one able to hit a different target, led President Reagan in 1981 to declare that a "window of vulnerability" existed, since the land-based US ICBMs could now be attacked by the Soviet MIRVs. Reagan began a massive military buildup.

The technological race was poised to move into space when President Reagan in 1983 announced plans to develop a defensive system, some of which would probably be based in space, which could attack any Soviet missiles fired at the United States. This system (formally known as the Strategic Defense Initiative, and informally called "Star Wars") was criticized by many US scientists as being not feasible and by the early 1990s it had been greatly reduced in scope.

An unexpected end to the nuclear arms race between the Soviet Union and the United States came in the late 1980s with the collapse of the Soviet empire in Europe and with the breakup of the Soviet Union itself in the early 1990s. The huge financial strain on its economy caused by the arms race undoubtedly contributed to its collapse. But the nuclear arms race also placed serious strains on the US economy. The end of the Cold War brought a nearly miraculous release to the world from the danger of a third world war, which likely would have been the world's last one.

As the twenty-first century began, a nuclear arms race of a sort threatened to start up again. The United States was considering building a limited defense against nuclear missiles, supposedly from so-called "rogue states," or from an accidental launch. Countries such as North Korea and Iran were unfriendly to the United States and had some ability already, or would have in the future, to build nuclear missiles. China and Russia believed the US plan would undermine their deterrence abilities so they strongly opposed it and threatened to respond in some way if the US proceeded to carry it out.

War

Why do human beings make war? Some of the people who have studied the causes of war believe that war is caused by the negative aspects of human nature, such as selfishness, possessiveness, irrationality, and aggressiveness. Other students of war have come to the conclusion that certain types of government – or, more formally, how political power is distributed within the state – make some countries more warlike than others. And other analysts have concluded that international anarchy, or the absence of a world government where disputes can be settled peacefully and authoritatively, is the main cause of war. Kenneth Waltz, a respected US student of war, has concluded that human nature and/or the type of government are often the immediate causes of war, but that international anarchy explains why war has recurred throughout human history.[23]

War reflects the relatively primitive state of human political development. When Albert Einstein, the German-born American theoretical physicist who is considered to have been one of the most brilliant people of the twentieth century, was reportedly asked why it is that we are able to create nuclear weapons but not abolish war, he responded that the answer was easy: politics is more difficult than physics.

In 2003 the nations of the world spent about $1 trillion a year on military expenditures. The United States spent about 50 percent of that amount. Following the US was Japan with about 5 percent, while Britain, France and China came next with about 4 percent each. In 2002 the average cost for each US citizen for the country's military budget was about $1,200, while the average cost per US citizen for international peacekeeping was about $2.[24]

Since World War II there have been more than 150 wars, with 90 percent of those occurring in the less developed nations. Wars have been frequent in the South since 1945 for a number of reasons. During the Cold War the United States and the Soviet Union supported with arms various political groups in the less developed nations that favored their side in the East/West conflict. Although the Cold War has now ended, the huge amounts of weapons supplied by the superpowers are now circulating within the South (and even in conflicts in Europe such as in the former Yugoslavia). Conflicts have been frequent in the developing world also because many of these nations received political independence relatively recently and territorial disputes, power struggles, ethnic and religious rivalries, and rebellions caused by unjust conditions are common.

With the exception of the Persian Gulf wars – which could be called "resource wars" – wars since the end of the Cold War have been mainly civil wars involving three categories of participants: first, ethnic groups that are fighting for more autonomy or for a state of their own, such as the Kurds in Turkey and the Chechens in Russia; second, groups trying to get control of a state, such as in Afghanistan; and third, so-called "failed states" where the central government has collapsed or is extremely weak and fighting is occurring over political and/or economic "spoils," such as the recent war in Liberia.

A characteristic of modern war is that often more civilians are killed than soldiers. In many wars in the past the military combatants were the main casualties but this has now changed so that civilians often bear the greatest burden. In the first half of the twentieth century about 50 percent of the war-related dead were civilians. In the 1960s the proportion of the war dead who were civilians rose to about 65 percent, and in the 1980s it reached about 75 percent. It probably went even higher in the 1990s.[25] If one adds to the number of civilians killed and wounded during the fighting the vast number of civilians who flee the fighting and become refugees – sometimes finding no place which will accept them – civilians indeed bear the largest burden of modern war. Also the destruction of the land by the fighting is often immense so that when the fighting finally ends, civilians return to an ecologically damaged land.

Another characteristic of modern wars is that technology has been used to greatly increase the destructive capacity of the weapons. The case study on nuclear weapons that follows this section will illustrate that point well, but even so-called conventional weapons are now much more destructive than they used to be. (See the box on

Modern high-tech weapons

The following is a description by Paul Walker, a military specialist, of some of the new weapons used by the United States during the 1991 war to push Iraq out of Kuwait:

> The BLU-97/B is a three-quarter pound bomblet which carries a triple punch: a pre-fragmented anti-personnel casing to spray deadly shrapnel; a hollow-charge anti-tank warhead; and a disc of incendiary zirconium. Whatever is left after the shrapnel and bomb is lapped up by fire.
>
> The laser-guided bombs which destroyed the air-raid shelter in Iraq, a refuge for over 500 Iraqi civilians, had high penetration noses with delayed fuses mounted on the tails so as to explode only after entering hardened targets like the bunker.
>
> The CBU-87B is a 950-pound bomb which carries 202 small bomblets. One B-52 plane loaded with the CBU-87B's can carpet-bomb over 176 million square yards, equal to 27,500 football fields.
>
> The MADFAE (mass air delivery fuel-air explosive) mimics small nuclear explosions. It consists of 12 containers of ethylene oxide or propylene oxide. Trailed behind utility helicopters, the containers release a cloud of highly volatile vapors which, when mixed with air and detonated, can cover an area over 1,000 feet long with blast pressures five times that of TNT.

Source: Ruth L. Sivard, *World Military and Social Expenditures 1993*, 15th edn (Washington: World Priorities, 1993), p. 18

modern, high-tech weapons.) In addition to the increase in destructive capacity, technology has been used to increase the weapons' accuracy, penetration ability, rates of fire, range, automation, and armor.

At the beginning of the twenty-first century there are both positive and negative signs regarding war. On the positive side, according to the respected Stockholm International Peace Research Institute, there was a decrease in the number of wars in the late twentieth century and early twenty-first century, dropping from a high of about 30 in 1991 to about 15 in 2005. For the first time NATO was used several times to end fighting and killing in the former Yugoslavia. Other peacekeeping military forces under the United Nations were active in many locations throughout the world – 14 in 2006. Under nuclear arms reduction agreements between the United States and Russia, by the mid-1990s nuclear arsenals had been reduced from about 18,000 megatons of explosive power to about 8,000 megatons.[26]

On the negative side one must recognize that although war among the great powers has increasingly become unlikely because of the threat it would become a nuclear war, war is still a political instrument in the world. In the Democratic Republic of the Congo nearly 4 million people have died since a conflict began there in 1998. Sudan's civil war since 1983 has led to about 2 million deaths.[27] Since the 2003 US–British invasion of Iraq an estimated 50,000 people have died. The reduced stockpile of nuclear weapons still represents over 700 times the explosive power used in the twentieth century's three major wars which killed about 44 million people. As we will see in the next section, nuclear weapons represent the darkest part of the "dark side" of our species.

The Threat of Nuclear Weapons: A Case Study

The threat of nuclear weapons is a subject that touches on many of the themes we have examined in this chapter. It is the "ultimate" development subject since it is the achievements of weapons technology by the developed nations that have brought the survival of human life into question. It is a problem that cries out for a political solution. Karl von Clausewitz, the famous Prussian author of books on military strategy, described war as a continuation of politics by other means. But, given the probable consequences of a nuclear war as presented below, one must ask whether war between nations with nuclear weapons can remain a way of settling their disputes. Let us look at the nature of the threat created by nuclear weapons and then at four contemporary problems related to these weapons.

The threat

It has taken 4.5 billion years for life to reach its present state of development on this planet. The year 1945 represents a milestone in that evolution since it was then that the United States exploded its first atomic bombs on Hiroshima and Nagasaki, Japan, and demonstrated that humans had learned how to harness for war the essential forces of the universe. After 1945, when the United States had no more than two or three atomic bombs, the arms race continued until the two superpowers, the United States and the Soviet Union, had a total of about 50,000 nuclear weapons, the equivalent of 1 million Hiroshima bombs – or, to put it another way, about 3 tons of TNT for every man, woman, and child in the world. The Hiroshima bomb was a 15 kiloton device (a kiloton having the explosive force of 1,000 tons of TNT); some of the weapons today fall in the megaton range (a megaton being the equivalent of 1 million tons of TNT).[28]

Today, in addition to the United States and Russia, Britain, France, China, India, and Pakistan have nuclear weapons that could be used in a nuclear war. In 2004 the United States had an estimated 7,000 operational nuclear warheads and Russia had about 7,800.[29] Both countries tentatively agreed to reduce these to 3,500 warheads each by 2007 but the implementation of this agreement was in doubt in 2000 because of the controversy over a possible US antimissile defense system. Russia said it would not carry out the new reductions if the United States proceeded to build the defense system. In 2002 President George Bush ordered a partial missile defense system to be built even though tests had failed to show it would reliably work.[30]

What would happen if these weapons were ever used? We cannot be sure of all the effects, of course, since, as the author Jonathan Schell has stated, we have only one earth and cannot experiment with it.[31] But we do know from the Hiroshima and Nagasaki bombings, and from the numerous testings of nuclear weapons both above and below ground, that there are five immediate destructive effects from a nuclear explosion: (1) the initial radiation, mainly gamma rays; (2) an electromagnetic pulse, which in a high altitude explosion can knock out electrical equipment over a very large area; (3) a thermal pulse, which consists of bright light (you would be blinded by glancing at the fireball even if you were many miles away) and intense heat (equal to that at the center of the sun); (4) a blast wave that can flatten buildings; and (5) radioactive fallout, mainly in dirt and debris that is sucked up into the mushroom cloud and then falls to earth.

The longer-term effects from a nuclear explosion are three: (1) delayed or worldwide radioactive fallout, which gradually over months and even years falls to the ground, often in rain; (2) a change in the climate (possibly a lowering of the earth's temperature over the whole Northern Hemisphere, which could ruin agricultural crops and cause widespread famine); and (3) a partial destruction of the ozone layer, which protects the earth from the sun's harmful ultraviolet rays. If the ozone layer is depleted, unprotected Caucasians could stay outdoors for only about ten minutes before getting an incapacitating sunburn (black people, because of the color of their skin, could go somewhat longer), and people would suffer a type of snow blindness from the rays which, if repeated, would lead to permanent blindness. Many animals would suffer the same fate.[32]

Civil defense measures might save some people in a limited nuclear war but would not help much if there were a full-scale nuclear war.[33] Underground shelters in cities hit by nuclear weapons would be turned into ovens since they would

Underground nuclear weapons testing site in the USA (*Los Alamos National Laboratory*)

tend to concentrate the heat released from the blast and the firestorms. Nor does evacuation of the cities look like a hopeful remedy in a full-scale nuclear war, since people would not be protected from fallout, or from retargeted missiles, and could not survive well in an economy that had collapsed.

Since most of our hospitals and many doctors are in central-city areas and would be hit by the first missiles in an all-out nuclear war, medical care would not be available for the millions of people suffering from burns, puncture wounds, shock, and radiation sickness. Many corpses would remain unburied and would create a serious health hazard, which would contribute to the danger of epidemics spreading among a population whose resistance to disease had been lowered by radiation exposure, malnutrition, and shock.

What could be the final result of all of this? Here is how Jonathan Schell answers that question in probably the longest sentence you have ever read, but in one with no wasted words:

> Bearing in mind that the possible consequences of the detonations of thousands of megatons of nuclear explosives include the blinding of insects, birds, and beasts all over the world; the extinction of many ocean species, among them some at the base of the food chain; the temporary or permanent alteration of the climate of the globe, with the outside chance of "dramatic" and "major" alterations in the structure of the atmosphere; the pollution of the whole ecosphere with oxides of nitrogen; the incapacitation in ten minutes of unprotected people who go out into the sunlight; the blinding of people who go out into the sunlight; a significant decrease in photosynthesis in plants around the world; the scalding and killing of many crops; the increase in rates of cancer and mutation around the world, but especially in the targeted zones, and the attendant risk of global epidemics; the possible poisoning of all vertebrates by sharply increased levels of vitamin D in their skin as a result of increased ultraviolet light; and the outright slaughter on all targeted continents of most human beings and other living things by the initial nuclear radiation, the fireballs, the thermal pulses, the blast waves, the mass fires, and the fallout from the explosions; and considering that these consequences will all interact with one another in unguessable ways and, furthermore, are in all likelihood an incomplete list, which will be added to as our knowledge of the earth increases, one must conclude that a full-scale nuclear holocaust could lead to the extinction of mankind.[34]

New dangers

Despite the end of the Cold War and of the threat of a cataclysmic war between two superpowers, nuclear weapons still remain a danger for the world. Four problems exist with which the world will have to deal: (1) the proliferation of nuclear powers; (2) the control of the nuclear weapons remaining in Russia; (3) the cleanup of the huge amount of toxic wastes produced in both the United States and the former Soviet Union when they built their large numbers of nuclear weapons; and (4) the threat of nuclear terrorism.

Nuclear proliferation

The spread of nuclear weapons to new countries represents a growing danger because the larger the number of countries that have these weapons the greater the likelihood that they will be used. Many of these new nuclear powers – either actual or potential – are authoritarian regimes that have serious conflicts with their neighbors in the less developed world. For example, the Middle East is a region plagued by conflict. It is widely believed that Israel has already acquired nuclear weapons and has them ready for use or could have them ready in a very short time. After the defeat of Iraq in the Persian Gulf War in 1991, UN inspectors discovered that Iraq had been making major efforts to build both atomic weapons and the much more powerful hydrogen weapons. This was in spite of the fact that Iraq had signed the Nuclear Nonproliferation Treaty, in which it had agreed not to acquire nuclear weapons, and in spite of the fact that officials from the International Atomic Energy Agency had inspected nuclear facilities in Iraq just prior to the war and had found no evidence that Iraq was building nuclear weapons.

Another example of proliferation is in South Asia. In this region two countries – India and Pakistan – have already fought each other three times in the past 40 years and both tested nuclear weapons in 1998. A dispute over the territory of Kashmir, which was the central issue in two of their previous wars, flared up again in the late 1990s and the fear was raised that if the two countries fight again it could be with nuclear weapons.

Regional conflicts in which these weapons could be used are not the only concern; also disturbing is the possibility of accidental or unauthorized use of nuclear weapons by these countries.[35] In 2006 the list of actual and potential nuclear powers was as follows:

Acknowledged nuclear powers: United States, Russia, United Kingdom, France, China, India and Pakistan.
Suspected nuclear power: Israel.
Past suspected aspiring nuclear powers: Algeria, Argentina, Brazil, Iraq, Libya, South Korea, Taiwan, South Africa.
Present suspected aspiring nuclear powers: North Korea, Iran.

Control of nuclear weapons in Russia

With the collapse of communism has come frightening political, economic, and social instability in Russia and the other former Soviet republics. Until these states obtain a new stability, it is disquieting, to say the least, to remember that thousands of nuclear weapons and much nuclear material remain in this area. (More will be said about this subject under "The threat of nuclear terrorism.") Also disturbing are indications that because the Russian conventional military forces have seriously deteriorated, Russia is now relying more on the possible use of its nuclear forces in any future conflict.[36]

The cleanup

The production of vast quantities of nuclear weapons in both the Soviet Union and the United States led to huge environmental contamination with highly toxic chemical and nuclear wastes. In both countries wastes from the plants producing components for the nuclear weapons were released into the air and dumped onto the ground, and they have leaked from temporary storage facilities. The extent of this contamination did not become public until the late 1980s in the US case, when the US government released a number of reports outlining the huge extent of the problem, and in the Soviet case in the late 1980s and the early 1990s, in the last years of the Soviet communist state.[37]

It is painful to read about the deliberate inflicting of harm by a government on its own citizens. Although the Soviet contamination is probably greater than the American,[38] both governments used "national security" to justify their actions and to keep them secret. In the United States the plants were exempt from state and federal environmental laws, and actions were carried out that had long before been declared illegal for private industry and individuals. An estimated 70,000 nuclear weapons were made in the United States over a 45-year period, in 15 major plants covering an area equal in size to the state of Connecticut. They cost about $300 billion (in 1991 dollars). Estimates by various government agencies of the cost of cleaning up the environmental damage at the plants, which will take decades to accomplish, range from $100 billion to $300 billion.[39]

At the beginning of the twenty-first century the National Academy of Science, the most prestigious scientific group in the United States, declared that most of the sites related to the production of nuclear weapons are so contaminated that they can never be cleaned up. Of the 144 sites, the Academy believes that only 35 can be cleaned up enough so there is no potential harm to human beings. The remaining 109 sites will stay dangerous for tens and even hundreds of thousands of years. The Academy found that the government's plans for guarding permanently contaminated sites are inadequate and that the government does not have the money or technology to keep the contamination from "migrating" off the sites.[40]

The threat of nuclear terrorism

When Mohamed ElBaradei, the head of the UN's International Atomic Energy Agency (IAEA) accepted the Noble Peace Prize in 2005 for his agency's work in preventing the spread of nuclear weapons, he warned that terrorists are actively trying to obtain nuclear weapons. Russia has reported that its customs service has detected 500 illegal shipments of nuclear and radioactive material across its borders, and the IAEA has reported that there have been 18 confirmed seizures of stolen plutonium or highly enriched uranium, both of which can be used to construct nuclear weapons.[41]

One fear is that terrorists might make a so-called "dirty" nuclear bomb that would spread radioactive material over a large area, making that area uninhabitable for decades. Especially vulnerable to air attack are nuclear power plants, where in the United States nuclear waste that is still highly radioactive is stored above ground on the site.

Terrorism is a major problem today. The modern industrial state is relatively open to anyone who wants to harm the public. Globalization with its greatly increased trade and contacts among people has dramatically increased the possible targets and ways to deliver explosive devices.

The civil defense priority of the US government before the terrorist attack of September 11, 2001 on the World Trade Center was on antimissile defense. In 2000, the year before the attack, 500 million people crossed the US borders, along with 125 million passenger vehicles, 800,000 airplanes, 12 million trucks, 2 million railroad cars, 200,000 ships, and 12 million containers on ships.[42] This huge movement of people and goods, and the difficulty of checking it for dangerous materials, may explain partly why the government focused on missiles rather than its land borders. But also history shows that nations often prepare for the last war instead of a possibly new kind of war. (The attack on the US naval base at Pearl Harbor in 1941 by Japan greatly influenced US defense policy for many years. The defense priority in the US became the preventing of a surprise air attack by enemy military forces.)

Terrorism is not new. What is new is that modern technology gives one or a few terrorists a way to kill large numbers of people and destroy large amounts of property, sometimes with a relatively small and easily made device. Throughout history terrorism has often been the weapon of weaker groups attacking a stronger enemy by unconventional means and often directed at noncombatants for a psychological purpose, such as to create fear and publicity for their cause. Suicide terrorism is also not new. The Jewish sect of Zealots used it against the Romans when the Romans occupied Judea. The Islamic Order of Assassins fought against the Christians in the early Crusades. Japan used suicide pilots against American ships near the end of World War II to try to prevent defeat. Two thousand kamikaze hit 300 US ships during one sea battle late in the war, killing 5,000 Americans, an event that was probably a factor in the American decision to drop the atomic bombs on Japan in order that a costly invasion would not be necessary to end the war.[43]

According to some analysts the attack on the World Trade Center was designed to get the US into a guerrilla war in a Muslim country (Afghanistan) which would help Islamic extremists recruit more terrorists. A report attributed to the Egyptian who became the chief military advisor to Osama bin Laden, the founder of the terrorist organization al-Qaeda and the architect of the attack on the World Trade Center, states that the purpose of the attack was "to draw the United States out of its hole" and get it directly involved in a military conflict in an Islamic country.[44] If this analysis is correct, the US invasion of Afghanistan and especially Iraq served bin Laden's goal very well, adding new grievances to the many that Muslims already feel against the US.[45]

After the World Trade Center attack there was a revulsion against terrorism in many nations of the world. That feeling changed with the US-led invasion of Iraq, which was condemned by most nations – Muslim and non-Muslim. Pictures of the destruction of the World Trade Center were now used to show the effectiveness of terrorism. Since the destruction of the Center's two towers, terrorist attacks have killed thousands in Tunisia, Bali, Mombasa, Riyadh, Istanbul, Casablanca, Jakarta, Madrid, Sharm El Sheikh, and London. In 2004 there were 650 significant terrorist attacks, according to the US State Department, triple the number the year before. Nearly 200 of these were in Iraq, not counting the many attacks against US troops there. In 2005, in one month alone – May – there were 90 suicide bombings in Iraq.[46]

The challenge for the modern world is to decide how to defend itself against those who hold extremist beliefs, both religious and political, which call for the elimination of all those who do not believe as they do. There is a great need to strengthen defense against terrorists, which includes better intelligence and a greater ability to prevent attacks before they occur. At the same time, I believe, there is a need to resist the desire for revenge, which can easily create more hatred and more terrorists as innocent people are killed during the revenge action. And to try to guard every vulnerable place and spy on every potential terrorist could lead to the creation of a police state, the use of uncivilized means the Western world has rejected, such as torture, and the loss of freedoms. A long-term effort is also needed to remove legitimate grievances of oppressed groups from which terrorists recruit their members.

Conclusions

This chapter has focused on the negative aspects of technology. It has done so because many of the readers of this book will probably be citizens of developed countries who already have a strong belief in the advantages of technology. It is not my intent to undermine that belief, because technology has benefited human beings in countless ways, and its use is largely responsible for the high living standards in the industrialized nations. Rather, my intent is to bring a healthy caution to the use of technology. An ignoring of the negative potential of technology has brought harm to people in the past and could cause unprecedented harm in the future. Much technology is neither good nor bad. It is the use that human beings make of this technology that determines whether it is mainly beneficial or harmful. But other technology is basically harmful or excessively dangerous and should be rejected. It is of course not always easy to place technologies in these categories, but an effort should be made.

The less developed nations need technology to help them solve many of their awesome problems. But often intermediate technology should be used by them rather than the high technology of the industrialized nations. The temptation to imitate the West is strong, but ample evidence exists to show that this could be a serious mistake for developing nations. The South needs to remember that its conditions and needs are different from those of the West, and that it should take from Western science only what is appropriate.

The industrial nations face another task. They must become more discriminating in their use of technology and lose some of their fascination with and childlike faith in it. The fate of the earth is now literally in their hands. The wisdom or lack of wisdom these nations show in using military and industrial technology affects all – the present inhabitants of earth, both human and nonhuman, and future generations, who depend on our good judgment for their chance to experience life on this planet.

Notes

1 Garrett Hardin, "The Tragedy of the Commons," *Science*, 162 (December 13, 1968), pp. 1243–8.
2 Marlise Simons, "Boats Plunder Mediterranean with Outlawed Nets," *New York Times*, national edn (June 4, 1998), p. A3.
3 Daniel Pauly and Reg Watson, "Counting the Last Fish," *Scientific American*, 289 (July 2003), pp. 42–7, and Cornelia Dean, "Scientists Warn Fewer Kinds of Fish Are Swimming the Oceans," *New York Times*, national edn (July 29, 2005), p. A6.
4 A description of 50 case studies of development projects in the developing world that had harmful and unanticipated effects on the environment is contained in the following conference report: M. Taghi Farvar and John P. Milton (eds), *The Careless Technology: Ecology and International Development* (Garden City, NY: Natural History Press, 1972).
5 Although its use was banned in the United States in 1972, residues of DDT could still be found in most Americans 20 years later. DDT is stored in the human body for decades.
6 Thomas R. Dunlap, *DDT: Scientists, Citizens, and Public Policy* (Princeton, NJ: Princeton University Press, 1981), p. 8.
7 Denise Grady, "A Move to Limit Antibiotic Use in Animal Feed," *New York Times*, national edn (March 8, 1999), p. A1, and Dan Ferber, "Livestock Feed Ban Preserves Drugs' Power," *Science*, 295 (January 4, 2002), pp. 27–8.
8 Jane Brody, "Studies Suggest Meats Carry Resistant Bacteria," *New York Times*, national edn (October 18, 2001), p. A12.
9 Grady, "A Move to Limit Antibiotic Use in Animal Food," p. A1.
10 Denise Grady, "WHO Finds Use of Antibiotics in Animal Feed Can Be Reduced,"

New York Times, national edn (August 14, 2003), p. A5.
11 Dirk Johnson, "Growth of Factory-Like Hog Farms Divide Rural Areas of the Midwest," *New York Times*, national edn (June 24, 1998), p. A12.
12 Elizabeth Becker, "US Sets New Farm-Animal Pollution Curbs," *New York Times*, national edn (December 17, 2002), p. A28.
13 John Cushman, "Pollution Control Plan Views Factory Farms as Factories," *New York Times*, national edn (March 6, 1998), p. A1.
14 Ibid., p. A10, and "Poultry Growers Unite to Address Waste Issue," *New York Times*, national edn (August 25, 1998), p. A14.
15 For a fuller discussion of the unanticipated consequences of American aid to the Shah, see John L. Seitz, "The Failure of US Technical Assistance in Public Administration: The Iranian Case," in Eric Otenyo and Nancy Lind (eds), *Comparative Public Administration: The Essential Readings* (Oxford, UK: Elsevier, 2006), pp. 321–34.
16 E. F. Schumacher, *Small Is Beautiful: Economics as if People Mattered* (New York: Harper & Row, 1973).
17 Emma Rothschild, *Paradise Lost: The Decline of the Auto-Industrial Age* (New York: Random House, 1973), p. 119.
18 Warren C. Baum and Stokes M. Tolbert, *Investing in Development: Lessons of World Bank Experience* (Oxford, UK: Oxford University Press, 1985), p. 574.
19 See, for example, Suzanne Arms, *Immaculate Deception: A New Look at Women and Childbirth in America* (Westport, CT: Bergin & Garvey, 1984); Dr Robert A. Bradley, *Husband-Coached Childbirth* (New York: Harper & Row, 1974); Robbie E. Davis-Floyd, *Birth as an American Rite of Passage* (Berkeley: University of California Press, 1992).
20 Dr John S. Miller, "Foreword," in Lester D. Hazell, *Commonsense Child-birth* (New York: Berkley Books, 1976), p. x.
21 Hardin, "The Tragedy of the Commons," p. 1243.
22 Jon R. Luoma, "The $33 Billion Misunderstanding," *Audubon*, 83 (November 1981), pp. 111–27.
23 Kenneth A. Waltz, *Man, the State and War: A Theoretical Analysis* (New York: Columbia University Press, 1959).
24 "Defense When Money Is No Object," *Bulletin of the Atomic Scientists*, 57 (September/October 2001), pp. 36–7.
25 Ruth L. Sivard, *World Military and Social Expenditures, 1996*, 16th edn (Washington, DC: World Priorities, 1996), p. 7.
26 Ibid., p. 20.
27 Michael Renner, "Violent Conflicts Unchanged," in Worldwatch Institute, *Vital Signs 2005* (New York: W. W. Norton, 2005), pp. 74–5.
28 A train transporting 1 million tons of TNT would be about 250 miles long.
29 "US Nuclear Forces, 2004," *Bulletin of the Atomic Scientists*, 60 (May/June 2004), p. 68, and "Russian Nuclear Forces, 2004," *Bulletin of the Atomic Scientists*, 60 (July/August 2004), p. 72.
30 Eric Schmitt, "Bush Ordering Missile Shields at Sites in West," *New York Times*, national edn (December 18, 2002), p. A1.
31 Jonathan Schell, *The Fate of the Earth* (New York: Avon Books, 1982).
32 For a fuller description of the effects of a nuclear war see Schell, *The Fate of the Earth*, ch. 1; Ruth Adams and Susan Cullen (eds), *The Final Epidemic: Physicians and Scientists on Nuclear War* (Chicago: Educational Foundation for Nuclear Science, 1981); and "Nuclear War: The Aftermath," *Ambio: A Journal of the Human Environment* (Royal Swedish Academy of Sciences and Pergamon Press), 11/2–3 (1982).
33 For an interesting discussion of the negative American attitude toward civil defense, see Freeman Dyson, *Weapons and Hope* (New York: Harper & Row, 1984), especially ch. 8.
34 Schell, *The Fate of the Earth*, p. 93.
35 Good discussions of these problems are contained in Bruce Blair, Harold Feiveson, and Frank von Hippel, "Taking Nuclear Weapons off Hair-Trigger Alert," *Scientific American*, 277 (November 1997), pp. 74–81, and in William Broad, "Guarding the Bomb: A Perfect Record, but Can It Last?" *New York Times*, national edn (January 29, 1991), pp. B5, B8.
36 Michael Gordon, "Maneuvers Show Russian Reliance on Nuclear Arms," *New York Times*, national edn (July 10, 1999), p. A1.
37 For details of the contamination at US plants as described by US government agencies, see the following articles in *New York Times*, national edn: Keith Schneider, "Candor on Nuclear Peril" (October 14, 1988), p. 1; Kenneth Noble, "US, for Decades, Let Uranium Leak at Weapons Plant" (October 15, 1988), p. 1; Keith Schneider, "Wide Threat Seen in Contamination at Nuclear Units"

(December 7, 1988), p. 1; Matthew Wald, "Waste Dumping that US Banned Went on at Its Own Atom Plants" (December 8, 1988), p. 1; Matthew Wald, "Wider Peril Seen in Nuclear Waste from Bomb Making" (March 28, 1991), p. A1.

38 Matthew Wald, "High Radiation Doses Seen for Soviet Arms Workers," *New York Times*, national edn (August 16, 1990), p. A3.

39 Jim Bencivenga, "Scale-Down of US Nuclear Arsenal Will Affect Widespread Industry," *Christian Science Monitor* (November 29, 1991), p. 4.

40 "Nuclear Sites May Be Toxic in Perpetuity, Report Finds," *New York Times*, national edn (August 8, 2000), p. A12.

41 Jennifer Lee, "Report Says Plan to Safeguard Nuclear Material Is Lacking," *New York Times*, national edn (March 13, 2003), p. A13.

42 Stephen Flynn, "America the Vulnerable," *Foreign Affairs* (January/February 2002), p. 64.

43 Scott Atran, "Genesis of Suicide Terrorism," *Science*, 299 (March 7, 2003), p. 1535.

44 Mark Danner, "Taking Stock of the Forever War," *New York Times Magazine* (September 11, 2005), p. 50.

45 For an example of such an analysis and a discussion of the many grievances Muslims feel against the US and the West in general see ibid., pp. 45–86. This article also contains a description of the extreme religious goals of some of the terrorist organizations, such as a desire to set up Islamic states following the practices of the Prophet Mohammed and his early followers in the seventh century.

46 Ibid., p. 46.

Further Reading

Allison, Graham, *Nuclear Terrorism: The Ultimate Preventable Catastrophe* (New York: Times Books/Henry Holt, 2004). This Harvard scholar believes that without bold measures to prevent terrorists from getting a nuclear weapon, they will get one and will use it. He outlines the aggressive steps the world will need to take to prevent this disaster.

Chiles, James, *Inviting Disaster: Lessons from the Edge of Technology* (New York: Harper Business, 2001). Human beings are prone to error. Chiles takes us inside many famous technological disasters, showing us what led up to the events. With our increasing technological power comes the potential for greater disasters. We are unable to eliminate all risks – no system is perfect – so there are some technologies that we should not accept because they are very dangerous.

Drengson, Alan, *The Practice of Technology: Exploring Technology, Ecophilosophy, and Spiritual Disciplines for Vital Links* (Albany: State University of New York Press, 1995). Drengson explores the connections between technology, ecologically centered philosophy, and the wisdom of spiritual disciplines. He shows what happens when technology is created that ignores these connections.

Hall, Brian, "Overkill Is Not Dead," *New York Times Magazine* (March 15, 1998), pp. 42–9, 64, 76, 78, 84–5. US nuclear forces continue to improve while Russia's continue to deteriorate. This creates a dangerous instability that is not being addressed by arms control agreements.

Hamburg, David, *No More Killing Fields: Preventing Deadly Conflict* (Lanham, MD: Rowman & Littlefield, 2002). As preventive medicine is better than treating an illness, so is preventing conflict better than fighting a war. By analyzing the rise of war makers such as Hitler and Milosevic, Hamburg believes the international community could have prevented their wars. He shows how an early warning and response system could be created today.

Klare, Michael, *Resource Wars: The New Landscape of Global Conflict* (New York: Henry Holt, 2001). Klare focuses on oil, water, timber, and minerals, such as diamonds. In this readable book the author shows that many wars that seem to be ethnic or sectarian are really over resources.

Lewis, H. W., *Technological Risk* (New York: W. W. Norton, 1990). Lewis examines the perceived risks associated with our technological society and argues that bad policy, misused resources, and lack of education rather than technology itself are the culprits. He argues that many things people fear the most actually pose no real risk to them, whereas some things that do pose a real risk to people are not perceived as being dangerous.

McKibben, Bill, *Enough: Staying Human in an Engineered Age* (New York: Times Books/Henry Holt, 2003). McKibben is not against genetic engineering to repair defective genes but he warns against genetic engineering that tampers with fundamental behavioral traits. He also looks at the darker side of nanotechnology and robotics.

Pacey, Arnold, *Technology in World Civilization* (Cambridge, MA: MIT Press, 1990). Pacey explores the history of technology and its diffusion around the world, arguing that a lack of understanding of how this diffusion works has led to misguided programs in developing nations.

Rochlin, Gene, *Trapped in the Net: The Unanticipated Consequences of Computerization* (Princeton, NJ: Princeton University Press, 1997). Rochlin explores the changes that computerization and networking have wrought. He presents many examples of unanticipated consequences such as the role of computer-programmed trading in the stock market crash of 1987 and the shooting down of an Iranian civilian airliner by a US naval warship. He found that even though the equipment worked perfectly and the people involved acted as expected, the novel circumstances caused the systems to fail.

Schell, Jonathan, *The Gift of Time: The Case for Abolishing Nuclear Weapons Now* (New York: Henry Holt, 1998). The nuclear powers promised the nonnuclear states in the Nonproliferation Treaty that they would make a good-faith effort to get rid of their nuclear weapons if the nonnuclear states agreed not to obtain them. It is time that this promise be carried out, says Schell, because even though the Cold War has ended, the nuclear danger remains.

Shambroom, Paul, *Face to Face with the Bomb: Nuclear Reality after the Cold War* (Baltimore, MD: Johns Hopkins University Press, 2003). The author visited US military installations from 1992 to 2001 to photograph components of the country's nuclear arsenal. Included are photos of missiles, warheads, planes, submarines, command centers, and the people who operate them.

Stern, Jessica, *Terror in the Name of God: Why Religious Militants Kill* (New York: Ecco/HarperCollins, 2003). For four years Stern interviewed terrorists to try to find out what motivates them to kill.

Tenner, Edward, *Why Things Bite Back: Technology and the Revenge of Unintended Consequences* (New York: Alfred Knopf, 1996). This book is based on a huge amount of research. Tenner says this of his book: "I am arguing not against change, but for a modest, tentative and skeptical acceptance of it." To deal with what Tenner calls "revenge effects," he calls for more use of our brains, not more "stuff." By examining a large number of cases, from low-tar cigarettes, black-lung disease, Chernobyl, and Windows 95, Tenner recommends we use technology cautiously.

Winner, Langdon, *The Whale and the Reactor: A Search for Limits in an Age of High Technology* (Chicago: University of Chicago Press, 1986). The sight of a whale surfacing near a nuclear reactor causes the author to contemplate the connections between nature and technology and to call for a more conscious effort by people to think about how technology can affect human life.

CHAPTER 7

Alternative Futures

In human affairs, the logical future, determined by past and present conditions, is less important than the willed future, which is largely brought about by deliberate choices – made by the human free will.

René Dubos (1901–82)

Where is development leading us? What can we say about the future? Probably the wisest thing we can say is that the future is essentially unknowable; it cannot be predicted. If this is so – and the dismal record of past predictions leads us to believe it is – then we might ask, "Does it make any sense to think about the future at all?" I would answer, "Yes, it does." Although the complexity of life and natural and spiritual forces make the future unknowable, human actions can make one future more likely than another. Human beings and societies are given free will. Their options are not unlimited because of the times in which they live and their individual circumstances. But they do have some freedom to make choices. It is this ability to influence the future that concerns us in this final chapter.

If we can accurately recognize some of the major trends and currents in the past and the present, we can make an educated guess about where we are heading. And if we do not like the direction in which the world is heading, we can examine our individual behaviors and governmental policies to determine if they should be changed so that they contribute to a more desired future. As an old Chinese proverb states, if you do not change the direction in which you are headed, you will end up where you are headed.

René Dubos, the late well-known bacteriologist, coined the phrase, "Think globally, act locally." This is the way, according to Dubos, that an individual can help bring about a desirable future. Dubos no doubt realized that the great benefit of local action is that not only does it help solve problems, but it helps the individual's spirit grow. It also effectively combats the sense of powerlessness and depression that can come by "thinking globally," by becoming aware of the immense and serious problems the world faces. Wendell Berry, the US author who writes about the need for a connection to the land, gives the following tribute to "local action": "The real work of planet-saving will be small, humble, and humbling, and (insofar as it involves love) pleasing and rewarding. Its jobs will be too many to count, too many to report, too many to be publicly noticed or rewarded, too small to make anyone rich or famous."[1] It is my belief that politics – the process a society uses to achieve commonly desired goals and to settle conflicts among groups with different interests – will also play a central role in determining what the future will be like.

Of the many possible futures the human race faces on earth, three look most likely at present: doom, growth, and sustainable development. Some catastrophe might lead to the death of hundreds of millions of people; economic growth might continue into the future; or the world might achieve development that can continue indefinitely because it does not undermine the environment and resource base upon which it rests. There could, of course, be a combination of two of these or of all three. For example, one part of the world might experience a harsher life in the future while another part continues to expand economically. Or part of the world could reach sustainable development while another part continues to grow. Or even within one country, two or all three scenarios might exist at different times in the future: a period of growth could be followed by a catastrophe that is followed by sustainable development. In the rest of the chapter we will examine what the main proponents of these three views say, and then end with my assessment of the future.

Doom

There are a number of writings today warning that if humankind does not change its ways some kind of disaster will occur in the future. The implicit or explicit purpose of most of these authors is to help prevent the expected disaster by suggesting changes in human behavior or policies. Thus these works are not really predictions of what the future will be like, but rather what it could be like if present trends continue. The most frequently discussed disasters are those caused by nuclear weapons, climate change, overpopulation and overconsumption, the depletion of nonrenewable resources, and cosmic collisions.

Nuclear weapons

Jonathan Schell, author of *The Fate of the Earth*,[2] is probably the best-known writer who argues that the nation-state system, with its competition among big nations armed with nuclear weapons and with the proliferation of nuclear weapons to developing nations, is leading the world to a nuclear war. Schell argues, as was pointed out more fully in chapter 6, that such a war could bring about the extinction of human life on our planet. With the end of the Cold War and the collapse of the former Soviet Union, the danger of a full-scale nuclear war has been significantly reduced, but the spread of nuclear arms to new nations allows the nuclear danger to continue. And the efforts of terrorists to obtain nuclear weapons represent a real danger today. Any use of a nuclear weapon would result in huge casualties and extensive environmental damage.

Global climate change, overpopulation, resource depletion

Since the early 1960s, several widely publicized books have been written that forecast some sort of disaster coming because of industrial pollution, scarcity of food, overpopulation, or depletion of nonrenewable resources. One of the first was Rachel Carson's *Silent Spring*, which predicted premature death to humans and other animals because of the growing use of pesticides and other chemicals.[3] Another book, which received about as much publicity as Carson's book, was *The Population Bomb* by Paul Ehrlich. Ehrlich spelled out the threat of overpopulation in the following terms:

> The battle to feed all of humanity is over. In the 1970s and 1980s hundreds of millions of people will starve to death in spite of any crash programs embarked upon now.... No changes in behavior or technology can save us unless we can achieve control over the size of the human population. The birth rate must be brought into balance with the death rate or mankind will breed itself into oblivion.[4]

Paul Ehrlich, along with his wife, Anne Ehrlich, continued with this theme in their 1990 book *The Population Explosion*.[5]

Another well known book in the 1970s was *The Limits to Growth*, which was a report by the Club of Rome, a private group concerned with world problems. The report was based on a computer analysis of the world's condition by a research team at the Massachusetts Institute of Technology. The book emphasized that the earth is finite and that there are definite limits to its arable land, nonrenewable resources, and ability to absorb pollution. The study's main conclusion was as follows:

> If the present growth trends in world population, industrialization, pollution, food production, and resource depletion continue unchanged, the limits to growth on this planet will be reached sometime within the next 100 years. The most probable result will be a rather sudden and uncontrollable decline in both population and industrial capacity.
>
> It is possible to alter these growth trends and to establish a condition of ecological and economic stability that is sustainable far into the future.[6]

In 1992 a sequel to *The Limits to Growth* was published by three of the original authors. This book, *Beyond the Limits: Confronting Global Collapse, Envisioning a Sustainable Future*, still presents the possibility of "overshoot and collapse" if present trends continue, but unlike their earlier book, *Beyond the Limits* focuses more on how collapse can be averted.[7]

In the late 1970s the US government conducted a three-year study of what the world would be like in the year 2000 if present trends continued. The conclusions of the *Global 2000 Report to the President* were consistent with those of the earlier Club of Rome studies:

> If present trends continue, the world in 2000 will be more crowded, more polluted, less stable ecologically, and more vulnerable to disruption than the world we live in now. Serious stresses involving population, resources, and environment are clearly visible ahead. Despite greater material output, the world's people will be poorer in many ways than they are today.
>
> For hundreds of millions of the desperately poor, the outlook for food and other necessities of life will be no better. For many it will be worse. Barring revolutionary advances in technology, life for most people on earth will be more precarious in 2000 than it is now – unless the nations of the world act decisively to alter current trends.[8]

The doomsday scenario is clearly seen in a major book of the early 1990s by the Yale University historian Paul Kennedy. *Preparing for the Twenty-First Century* outlines the major challenges the world will have to deal with in the next century: the population explosion, the globalizing of the world's economy and the increasing power of multinational corporations, technological advances in agriculture and in industry (mainly biotechnology and robotics), and environmental damage. Kennedy sees these challenges as having a potentially catastrophic effect on many poor nations, but even the rich ones will be harmed by them if they do not recognize them and come up with imaginative ways to deal with them.[9]

Jared Diamond, a respected US geographer, published a major book in 2005 titled *Collapse: How Societies Choose to Fail or Succeed.*[10] Diamond investigated many societies in the past which failed, such as the well-known Easter Island, and concluded that, among other things, inappropriate values and destructive environmental practices led to their doom. None of these societies wanted to fail, but important choices they made led to their collapse.

In the late twentieth century and early twenty-first century many articles were written by scientists warning of the dangers of the climate changes they saw occurring around the earth and the changes they forecast would occur as the effects of global warming became clearer. The American journals *Science* and *Scientific American* and the British journal *Nature* contain such articles. No single book has yet been written that has effectively captured the world's attention to this potential disaster.

Triage and lifeboat ethics

A forecast of doom for the future of humanity has led some authors to recommend policies designed to deal with such situations. One called "triage" was discussed in *Famine-1975!* by William and Paul Paddock.[11] Triage is a procedure that was used in World War I when doctors in battlefield hospitals had to decide which of the many wounded would receive the limited medical care available. The wounded were divided into three categories. The first were those soldiers who were only slightly wounded and, although in pain, would probably survive even if untreated. The second category consisted of soldiers who were so seriously wounded that even if they received medical attention, they would probably die. In the third category were soldiers who were seriously wounded but who could probably be saved if the doctors treated them. It was to the last category that the military doctors first turned their attention. The Paddock brothers recommended that the United States place the countries of the world who were requesting food aid into three categories similar to those in the triage procedure and give aid only to those countries in the third category, that is, to those which would have a good chance of progressing to a state of being able to survive by their own efforts if they received some aid.

"Lifeboat ethics" is a policy suggested by the late biologist Garrett Hardin in a world of desperately poor and overcrowded countries.[12] Hardin used the metaphor of lifeboats at sea, some of which are threatened to be swamped by people in the water trying to get in. According to Hardin, the people in the lifeboats that are not completely filled have three choices. The first is to take in everyone who wants to get on board; but that would lead to the lifeboats being swamped and everyone drowning. The second choice is to take on only a few to fill the empty seats; but that would lead to the loss of the small margin of safety and make for a very difficult decision as to which few will be selected. The third choice is to take no further people on board and to protect against boarding parties. Hardin saw the rich nations of the world as being in partially filled lifeboats and the poor nations as being in overcrowded boats with people spilling into the water because of their inability to control their population growth. Hardin recommends the third choice for the United States. He admits this is probably unjust, but recommends that those who feel guilty about it can trade places with those in the water. Good-willed but basically misguided efforts by the United States to aid poor countries, such as by giving them food during famines, can lead to more suffering in the long run. The emergency food aid contributes to a larger population eventually and thus a deeper crisis in the future.

Growth

Julian Simon, the late professor of business administration at the University of Maryland, was one of the main spokespersons for the second position on the future, that economic growth will and should continue indefinitely into the future. The rest of this section on "Growth" is my summary of the argument Simon presented in his book *The Ultimate Resource*.[13]

Natural resources are not finite in any real economic sense. When there is a temporary scarcity of a mineral, prices rise and the increased price stimulates new

Cosmic collisions

In 1989 an asteroid, a half-mile in diameter, missed hitting earth by six hours. The asteroid was spotted only after it had already passed earth. After that near miss the US Congress instructed the National Aeronautics and Space Administration (NASA) to study the possibility that an object from space – an asteroid or a comet – could collide with earth, and to offer recommendations about how to deal with the danger. NASA's reports were received with a fair degree of skepticism, but that skepticism all but disappeared in the summer of 1994 when astronomers on earth recorded the spectacular fireballs (some as large as earth) on Jupiter as it was hit by the fragments of a comet. Although there is still debate about how likely it is such a collision could take place with earth (the NASA report said the risk of a major collision soon was slim but not negligible), the danger is now being taken seriously. In 1998 Congress requested NASA to locate by 2008 90 percent of the objects 1 kilometer and larger that could hit earth. By 2002 NASA had discovered about 600 such objects, but only one had a slim chance of hitting earth in 200 years. NASA is now halfway to accomplishing Congress's goal.

Some scientists now believe it is possible that a collision with an object from space 65 million years ago led to the extinction of the dinosaurs and many other forms of life as a huge amount of dust rose to block sunlight and drastically lowered the temperature on earth. Since that time, numerous objects have hit earth, some with the force of many nuclear bombs.

Sources: William Broad, "When Worlds Collide: A Threat to the Earth Is a Joke No Longer," *New York Times*, national edn (August 1, 1994), p. A1; Richard Kerr, "A Little Respect for the Asteroid Threat," *Science*, 297 (September 13, 2002), pp. 1785–7

efforts to find more ore and more efficient methods to process it. The higher price also leads to the search for substitutes that are able to provide the same service as the temporarily scarce mineral. In fact, the cost of most minerals has actually been decreasing, so in a real sense minerals are becoming less scarce rather than more scarce. There are large deposits of minerals in the sea and even on the moon that have not yet been tapped. There is no need to conserve natural resources because of the needs of future generations or of poor nations. The present consumption of natural resources stimulates the production of them and improves the efficiency with which they are produced. Both of these developments will aid future generations. Poor nations are not helped by the rich nations using fewer resources; what the poor nations need is economic growth and that growth depends on their increased use of resources.

As with natural resources, the long-run future of energy looks very promising. Aside from temporary price increases caused by the political maneuvering of some countries, the long-run trend of the cost of energy has been downward. Over time, an hour's work has bought more rather than less electricity. This means that energy has become less scarce rather than more scarce. It is likely that an expanding population will speed the development of cheap energy supplies that are almost inexhaustible. In the past, increased demand for energy led to the discovery of new sources, new types of energy, and improved extraction processes. There is no reason why this trend should not continue into the future. Much of the world has not even been systematically explored for oil.

It is true that the more developed an economy becomes the more pollution it produces, but overall we live in a healthier environment than ever before. The best indicator of the level of pollution is length of life, indicated by the average life expectancy of the population. Life expectancy is rising, not falling, around the world. In the United States and other developed nations it has been rising for the past several centuries, and in the less developed nations for the past several decades. Although a rising income in a country often means more pollution, it also means a greater desire to clean up the pollution and an increased capacity to pay for cleaning it up. If one doesn't believe this, one can compare the cleanliness of streets in rich countries with those in poor countries.

Since World War II, the per capita supply of food in the world has been improving. Famines have become fewer during the past century. The price of wheat has fallen over the long run. The trend toward cheaper grains should continue into the future. Overall, nutrition has been improving and there is no reason why it should not continue to improve into the indefinite future. The amount of agricultural land on the planet has been increasing, especially as irrigation spreads. The amount of arable land is likely to continue to increase and it is not unrealistic to think about land becoming available on other planets. The colonizing of space is not an impossible dream any more.

Additional children mean costs to the society in the short run, but in the long run these children become producers, producing much more than they consume. For both less developed and more developed countries, a moderately growing population is likely to lead to a higher standard of living in the future than is a stationary or rapidly increasing population. When additional children are born, both the mothers and the fathers work harder and spend less time in leisure. Also, a larger population means a bigger market that makes economies of scale possible; that is, industries can adopt more efficient procedures since they are producing more products at a time.

Past studies of animal behavior have often been cited as evidence that crowding is unhealthy for human beings, both psychologically and socially. This is probably true for animals, but not for human beings. Isolation is what harms human beings, not crowding. In fact, a dense population makes necessary and economical an efficient transportation system. Such a system is essential for economic growth. A dense population also improves communications, something anyone can see by comparing a newspaper in a large city with one produced in a small city. A growing population spurs the adoption of existing technology and the search for new technology, as well as the search for and production of new natural resources and energy.

The main question we should ask ourselves when considering population is, "What value do we place on human life?" Who is to say that the life of a poor person is not of value? Who is to say that a country of 50 million people with a yearly per capita income of $4,000 is better than a country with 100 million people and a per capita income of $3,000? The most important resource we have on earth is the human mind. The human mind is the source of knowledge. The more human minds we have, the more knowledge we will have to solve the problems we face.

Sustainable Development

A widely accepted definition of sustainable development was given by the World Commission on Environment and Development in its 1987 report *Our Common Future*: sustainable development enables current generations to "meet their needs without compromising the ability of future generations to meet their own needs."[14]

The term "sustainable development" probably originated with Lester Brown, head of the Worldwatch Institute, a research group established to analyze global problems. In his book *Building a Sustainable Society*,[15] which was published in 1981 – the same year Julian Simon's book was published – Brown claimed that present economic growth in the world is undermining the carrying capacity of the earth to support life. The Worldwatch Institute believes that if the world does not achieve sustainable development in about 35 years, environmental deterioration and economic decline will probably be feeding on each other, leading to a downward spiral in the human condition. The rest of this section is a summary of what sustainable development could look like in 2030, according to the Worldwatch Institute.[16]

Today, in 2030, birth rates around the world have fallen dramatically, so that the world's population is now about 8 billion instead of 9 billion, which the UN in the mid-1990s had projected for this year. The world's population is either stable or slowly declining to a number that the earth can sustain indefinitely.

Renewable and clean solar energy and geothermal energy have replaced fossil fuels as the main energy sources for the world. The type of solar energy used varies according to the climate and natural resources of the region. Northern Europe is relying heavily on wind power and hydropower. Northern Africa and the Middle East are using direct sunlight as their main energy source. Japan, Indonesia, Iceland, and the Philippines are tapping their ample geothermal energy reserves, while Southeast Asia is using a combination of wood, agricultural wastes, and direct sunlight. Solar thermal plants stretch across the deserts in the United States, North Africa, and Central Asia. Most villages in the less developed world now receive electricity from photovoltaic solar cells, which are much less expensive than they were in the mid-1990s. Wind power is common not only in Northern Europe but also in Central Europe, and vast wind farms exist in the Great Plains of the United States.

The efficient use of energy is widespread. Many of the technologies to achieve high energy efficiency already existed in the mid-1990s. The fuel efficiency of cars has been doubled from that in the mid-1990s, the energy efficiency of lighting systems has been tripled, and typical heating requirements have been cut by 75 percent. Refrigerators, air conditioners, water heaters, and clothes dryers are all highly energy efficient. Improvements in the design of electric motors have made them highly efficient and easy to maintain. Cogeneration, the combined production of heat and power in the same plant, is widespread, thus greatly reducing the use of energy by industry.

The transportation of people has undergone major changes from what it was 35 years ago. The typical European and Japanese cities in the mid-1990s had already developed the first stage of the new urban transport system. Fast and efficient rail and bus systems are used to move the urban population rather than the automobile. When automobiles are used in the cities, they are nonpolluting and highly efficient electric or hydrogen-powered "city cars." Larger cars are available for rent by families for use on long trips. As in some European cities and in many Asian cities in the mid-1990s, bicycles are widely used, for example by commuters to reach rail lines that are connected to the city center.

People live closer to their work in 2030 than they did 35 years ago, thus contributing to the reduced use of energy. The widespread use of computers to shop from one's home has greatly reduced the use of energy. Telecommunications are permitting many people to work from their homes or in satellite offices, thus greatly reducing the need for travel. Employees and supervisors are connected by computers and other telecommunication devices instead of by crowded highways.

The throwaway mentality that was widespread in the mid-1990s has been replaced by the recycling mentality. Many countries have a comprehensive system of recycling metal, glass, paper, and other materials, with the consumers separating the material for easy collection. The principal source of materials for industry is recycled goods. Waste reduction is greatly aided by pervasive recycling, but even more by the elimination of waste by industry in the production stage. Waste produced by industry has been cut by at least 30 percent from that produced 35 years ago, as many industries have redesigned their industrial processes. Food packaging has been simplified, thus also reducing much waste. And human wastes, after health procedures have been followed to remove the threat of disease, are used in fish farms and in greenbelts surrounding cities for vegetable growing. This practice was already being followed in the mid-1990s in a number of Asian countries. Many households have also helped reduce wastes by composting their yard wastes.

Because the world's population is larger in 2030 than it was in the mid-1990s, the land is used more intensively than it was in the earlier period. Land that was degraded by overuse and neglect is being restored so that it can be brought into use for agriculture. Improved varieties of crops and planting methods are helping conserve fertile soil and water. Major efforts are being taken to halt desertification: one of the best methods has been the eliminating of overgrazing by cutting back on the unnecessarily large herds of animals. Because of the warmer climate produced by global warming, a larger variety of crops is being grown, many of which are specifically salt-tolerant and drought-resistant.

The forest cover is stable or expanding in most parts of the earth. The cutting down of tropical rainforests has been largely halted and many "extractive reserves" have been established where local people are harvesting nuts, fruits, rubber, and medicines on a sustainable basis. A widely dispersed network of reserves has been established to protect unique and rare animals and plants. The replanting of trees in large areas deforested back in the twentieth century has begun, and efforts are in progress to learn how to harvest the forest for timber without decreasing its productivity, its diversity of species, or its overall health.

Restoring the land has been aided by establishing a pattern of land ownership much more equitable than it was in the mid-1990s. The large land holdings of the few in countries with a majority of poor landless people have been broken up, so that more people now have a chance to earn a livelihood. Also, many government-owned parcels of forests and pastures have been turned over to villages for their management on a sustainable basis.

A shift of employment has taken place so that losses of jobs in coal mining, auto production, road construction, and metals prospecting have been offset by gains in the manufacturing and sale of solar cells, wind turbines, bicycles, mass transport equipment, and recycling technologies.

The trend toward ever larger cities has been reversed. Decentralized, low-cost, renewable energy sources have fostered greater local self-reliance and made smaller human settlements more attractive. GNP – the amount of goods and services produced – is no longer accepted as the measure of human progress because it is now seen to undervalue what is needed for a sustainable society. A sustainable society needs qualities such as the durability of products and resource preservation; it does not need waste and planned obsolescence.

As the recognition has grown that environmental threats to people's security are greater than military threats, the amount of money spent on the military in the world – which totaled about $1 trillion a year in the mid-1990s – has been drastically reduced. This has freed needed funds for energy efficiency, soil conservation, tree planting, family planning, and other sustainable development activities. More countries are now relying on the UN for protection against aggression and have supported its greatly strengthened peacekeeping capabilities. Some countries have even followed the example of Costa Rica and have abolished their army. At the same time that nations have moved to decentralize power and decision making within their borders, they have also expanded their efforts to cooperate internationally in order to monitor the environment and coordinate efforts to attack global problems.

A slow change in people's values is taking place, so that concern for future generations now has a higher priority than it did earlier. Materialism and consumerism are gradually being replaced by voluntary simplicity, a recognition that personal self-worth cannot be measured by the amount of goods one owns. More efforts are being made to form richer human relationships, and stronger communities, and more emphasis is now placed on music and the arts. As the desire to amass more personal and national wealth has subsided, the gap between the rich and the poor in the world has gradually been reduced. This has led to a reduction in social tensions. There is an expanding recognition in the world of the value of democracy, human rights, diversity, and the freedom to innovate, as these are seen as helping nations achieve sustainability.

Conclusions

Could the world's future contain parts of all three alternative futures presented in this chapter? I believe it could. The dangers that the "doom" alternative conveys are real. Some of them have already taken place in parts of the earth, such as overpopulation in India and China, famine in Africa, and toxic poisoning in the United States, Europe, and Japan. The early effects of climate change can already be seen. The threat of nuclear weapons is widely recognized. If actions are not taken to reduce greenhouse gasses that are changing our climate, reduce the nuclear threat, reduce population growth in the poorer nations, and to end many forms of environmental deterioration, it is possible that huge loss of life could occur in the future. The positive feature of the doom scenario is that it causes us to recognize real dangers and to try to prevent them from occurring. The negative aspect of this imagined future is that for some people it can weaken their will to act. It frightens them so much that they literally give up on the future, becoming either depressed, numb, or inclined to live for the present. Too much preaching of doom can be a self-fulfilling prophecy if its effect is to discourage action.

Some aspects of the "growth" future appear to be good ones for the developing nations to strive for. They need more economic growth in order to raise their living standards. To attempt to achieve a more equal distribution of income in many of these countries today without further growth would create tremendous political turmoil and would probably bring little benefit to their societies, since in many of these countries there is not that much economic wealth to redistribute. To argue that these countries should not grow economically would be to condemn them to living forever with their present poverty. But to advocate more economic growth for the less developed nations does not deny the need for many of these nations to achieve a more equal distribution of income. Nor does it deny their vital need to slow down their population growth and to move toward a stable population. The less developed nations need to learn from the mistakes of the industrial countries and to make their growth as nonpolluting as possible. The industrial countries have demonstrated well that it is much less costly in the long run to make efforts to prevent environmental destruction from taking place than to try to clean up the damage after it has occurred.

But does more economic growth make sense in the developed nations? It is unpopular today to suggest that it does not, but this may indeed be the case. The desire to acquire more and more material possessions in the wealthy nations has placed a tremendous strain on the planet. This book has been concerned with documenting that strain. The developed nations have achieved for most of their citizens the goal that only kings and queens could achieve in the past – material comfort, an abundance of food, a relatively long life, and leisure. But these countries are finding it very difficult to learn when enough is enough. Obsession with materialism has been condemned by many of the great religions of the world, but it is still an obsession experienced by many. This obsession is not healthy – for the individual or for the planet – and human beings need a different goal for their lives.

The sustainable development future appears to be the one the developed nations should strive for, since the economic growth they are pursuing is destroying the resource base and environment upon which life rests. In the language of economics, it is using up the earth's natural capital – the clean air, clean water, fertile soil, favorable climate, etc. – for a short-term profit. And a basic principle of economics is that if you expend your capital – the financial and physical resources that allow you to produce goods and services – you will soon go bankrupt. It is also clear that this type of development shows little or no concern for future generations. One of the basic rules the Native Americans of the Iroquois confederacy followed in North America before their demise was the rule of the seventh generation: "consider how your decisions will affect the lives of the seventh generation to come." Sustainable development is a powerful concept because it is hard to argue against it. How can one publicly defend unconcern for future generations? Also it is a useful concept because it sets up a standard against which one's present actions can be judged.

A sustainable world would not mean the absence of growth, but the growth that would be emphasized would be intellectual, moral, and spiritual growth rather than the growth of material objects.

The human race does seem to be at a critical juncture. Will it realize the destructive things it is doing to life on earth and pursue a new course before it is too late? Is it a species that is developing intellectually, morally, and spiritually? The uncertainty of those answers is what makes the present day an exciting and challenging day in which to be alive. The stakes are high. We are changing our world at an unprecedented rate. I struggle to comprehend the fact that the photograph of earth on the cover of this book is no longer accurate. Since that photo was taken just a few decades ago, the northern ice field and parts of Antarctica have shrunk, some deserts have expanded, and parts of the coast of the Gulf of Mexico have sunk beneath the water.

For the first time in human history, human beings have the technology to enable them to monitor the planet, to see how their actions are changing the forests, the air, and the water. And they are learning to think of the earth as a single system, a system in which they are just one of the parts. British scientist James Lovelock's Gaia hypothesis, which holds that the earth appears to behave like a single living organism, is an example of this new thinking.[17] Out of Lovelock's thinking has sprung a new, but still primitive, science called Earth System Science, that focuses on the interrelationships among the key disciplines that study earth, such as ecology, oceanography, and atmospheric physics.[18] Because human beings are the only species to have high intellect and to have knowledge that their time on earth is limited by death, they are learning that they have a special responsibility to all of life on the planet. Whether they are learning this fast enough to prevent irreversible destruction is uncertain.

Although there is little prospect at present of the developed nations adopting sustainable development as their real goal – the one they are putting most of their efforts into achieving – it may come. It became the goal that most nations, at least publicly, subscribed to at the Earth Summit in Rio de Janeiro in 1992. The United States is already moving toward an economy in which occupations that utilize new knowledge will soon be more common than blue-collar manufacturing jobs. And nearly all industrialized nations have already achieved a birth rate that is at or below replacement level, thus leading to a relatively stable population for them in the not too distant future.

And there is a slowly growing awareness in the developed nations – as well as in parts of the developing world – that human beings need to live in harmony with nature, to move beyond their compulsion to dominate it. For people who do not yet have this awareness, one of the best ways they can learn this is by personal experience. When they learn, for example, that the water they have been drinking contains cancer-producing chemicals, they learn a lesson about ecology better than any textbook could teach them. Human beings can also learn by using their reason; their use of this capacity can make personal experience less needed as a teaching tool. But either way, human beings can and do learn and sometimes change their ways when their own survival depends on it.

We do not know if life exists anywhere else in the universe. If we are the only intelligent life in the universe, as some respected scientists believe,[19] our actions are obviously of extreme importance. On the other hand, life may exist on some planets around the billions of stars in our galaxy or in the billions of other galaxies. But even if our present efforts to monitor radio signals from space do reveal other life, life as it has developed on earth is probably unique. And preserving this life on our planet in its wonderful diversity and beauty and improving the human condition must be the goal of development.

What central idea can guide us during our individual journeys through life? Each person must search for her or his own guiding vision, of course, but one that may be satisfying for those who are concerned about the issues that our world faces today is the Buddhist vision, as given by the Dalai Lama:

> I believe that cultivating compassion is one of the principal things that make our lives worthwhile. . . . it is the foundation of a good heart, the heart of one who acts out of a desire to help others. . . . We cannot escape the necessity of love and compassion. As long as we have compassion for others and conduct ourselves with restraint – out of a sense of responsibility – there is no doubt we will be happy.[20]

Issues or problems have two sides. One side is the task that must be solved. Dealing with this side can be painful since some of these tasks present us with difficult choices. But the other side can light us up, for the issues also present us with opportunities. They give us an opportunity to grow – intellectually, morally, and spiritually. They give us a chance to become more loving, both to our fellow human beings and to the planet itself. And as we grow, so can our society. Not a bad deal.

Notes

1 Wendell Berry, "Out of Your Car, Off Your Horse," *Atlantic Monthly*, 267 (February 1991), p. 63.
2 Jonathan Schell, *The Fate of the Earth* (New York: Avon Books, 1982).
3 Rachel Carson, *Silent Spring* (Greenwich, CT: Fawcett Books, 1962).
4 Paul R. Ehrlich, *The Population Bomb*, revised edn (New York: Ballantine Books, 1971), pp. xi–xii.
5 Paul R. Ehrlich and Anne H. Ehrlich, *The Population Explosion* (New York: Simon & Schuster, 1990).
6 Donella Meadows et al., *The Limits to Growth*, 2nd edn (New York: Universe Books, 1974), p. 24.
7 Donella H. Meadows, Dennis L. Meadows, and Jorgen Randers, *Beyond the Limits: Confronting Global Collapse, Envisioning a Sustainable Future* (Post Mills, VT: Chelsea Green, 1992).
8 Council on Environmental Quality, and the Department of State, *The Global 2000 Report to the President: Entering the Twenty-First Century*, vol. 1 (New York: Penguin Books, 1982), p. 1.
9 Paul Kennedy, *Preparing for the Twenty-First Century* (New York: Random House, 1993).
10 Jared Diamond, *Collapse: How Societies Choose to Fail or Succeed* (New York: Viking, 2005).
11 William Paddock and Paul Paddock, *Famine-1975!* (Boston: Little, Brown, 1967), p. 9.
12 Garrett Hardin, "Living on a Lifeboat," *Bioscience*, 24 (October 1974), pp. 561–8.
13 Julian Simon, *The Ultimate Resource* (Princeton, NJ: Princeton University Press, 1981).
14 World Commission on Environment and Development, *Our Common Future* (New York: Oxford University Press, 1987).
15 Lester R. Brown, *Building a Sustainable Society* (New York: W. W. Norton, 1981).
16 Lester R. Brown, Christopher Flavin, and Sandra Postel, "Picturing a Sustainable Society," in Lester R. Brown et al., *State of the World – 1990* (New York: W. W. Norton, 1990), pp. 173–90.
17 For an explanation of how the Gaia hypothesis has inspired interdisciplinary research see Jill Neimark, "Using Flows and Fluxes to Demythologize the Unity of Life," *New York Times*, national edn (August 11, 1998), p. B8.
18 John Lawton, "Earth System Science," *Science*, 292 (June 15, 2001), p. 1965.
19 For articles about scientists who believe intelligent life may be unique to our planet, see an interview with Britain's top

astronomer in Claudia Dreifus, "Tracing Evolution of Cosmos from Its Simplest Elements," *New York Times*, national edn (April 28, 1998), p. B15, and an article about two prominent US scientists in William Broad, "Maybe We Are Alone in the Universe, After All," *New York Times*, national edn (February 8, 2000), p. D1. For an article on the search for life on other planets see William Broad, "Astronomers Revive Scan of the Heavens for Signs of Life," *New York Times*, national edn (September 29, 1998), pp. B11, B13.

20 Dalai Lama, "Foreword," in Rimpoche Gehlek, *Good Life, Good Death* (New York: Riverhead Books, 2001).

Further Reading

Adamson, David, *Defending the World* (London: I. B. Tauris, 1990). Adamson expresses doubts about the possibilities for sustainable development and argues that the developed world will be too pressed to save itself to worry about the developing world.

Bossel, Hartmut, *Earth at a Crossroads: Paths to a Sustainable Future* (Cambridge, UK: Cambridge University Press, 1998). Bossel argues that humanity faces a choice between continuing on its present unsustainable path or shifting to a sustainable path that will call for significant changes. He offers his views of what ethical and organizational changes must be made to achieve a sustainable path and gives recommendations of what individuals and organizations can do to aid the transformation.

Buzan, Barry, and Gerald Segal, *Anticipating the Future: Twenty Millennia of Human Progress* (London: Simon & Schuster, 1998). Even though there have been many setbacks, the authors believe there has been human progress over the past 20,000 years. The human species has spread over the globe, abolished slavery, and reduced its appetite for large-scale wars. That progress should continue, although there will be ups and downs. They see the world gradually solving its environmental problems, and eventually moving into space.

Daly, Herman, *Beyond Growth: The Economics of Sustainable Development* (Boston: Beacon Press, 1996). An unconventional economist, Daly questions the value of economic growth. He believes that traditional economists have a false faith in growth because they do not accurately calculate the cost of depleting national resources. While still far from being accepted by the academic mainstream, Daly's ideas have inspired a new academic subdivision called "ecological economics."

Diamond, Jared, *Collapse: How Societies Choose to Fail or Succeed* (New York: Viking, 2005). Diamond investigates why some human societies in the past failed. Among other causes he found inappropriate values and environmental destruction. None of these societies wanted to fail but they lacked the ability to make the proper choices needed for their continued prosperity. It's a big book; rich in detail. Diamond hopes our own society can learn from the mistakes of the past to save itself.

Hammond, Allen, *Which World? Scenarios for the 21st Century* (Washington, DC: Island Press, 1998). This book does not try to predict the future. Rather it presents three possible futures. One is the "Market World: A New Golden Age of Prosperity"; the second is the "Fortress World of Instability and Violence"; and the third is the "Transformed World: Changing the Human Endeavor." The main theme of the book is that the future is not already determined. Trends can be changed quite rapidly if people so wish and can bring forth the political will to act.

Kates, Robert, "Sustaining Life on the Earth," *Scientific American*, 271 (October 1994), pp. 114–22. The author argues that evolving institutions and technology and a widespread global concern for the environment are indications that we may already be moving toward an environmentally sustainable future.

Lewis, John S., *Rain of Iron and Ice: The Very Real Threat of Comet and Asteroid Bombardment* (Reading, MA: Addison-Wesley, 1996). Lewis examines the evidence of past impacts on earth of comets and asteroids. He assesses the risks of such collisions in the future and argues that sudden, dramatic extinctions on earth are possible. The book also describes what scientists are doing today to search for possible threats to our planet from space.

McKibben, Bill, *Hope, Human and Wild: True Stories of Living Lightly on the Earth* (New York: Little, Brown, 1995). McKibben admits that after he finished his widely discussed book *The End of Nature*, he felt depressed. To help counter that depression he searched for examples of where people are living lightly on the land and not destroying it. He found three examples of such places – Curitiba, Brazil; Kerala, India; and his own Adirondack Mountains in the United States – and presents their stories here.

Musser, George, "The Climax of Humanity," 293, *Scientific American* (September 2005), pp. 44–7. How we manage the next few decades could lead to sustainability or collapse. This article is part of an issue titled "Crossroads for Planet Earth." It contains nine articles (on population, poverty, biodiversity, energy, food and water, disease, economics, and policy) focused on how we can successfully pass beyond 2050.

Myers, Norman, and Julian Simon, *Scarcity or Abundance: A Debate on the Environment* (New York: W. W. Norton, 1995). Before Simon died in 1998 he had a debate with Norman Myers at Columbia University which is printed in this book along with pre- and post-debate statements by both men. Simon presents a defense of his pro-growth position while Myers presents the contrary environmental position.

Raven, Peter, "Science, Sustainability, and the Human Prospect," *Science*, 297 (August 9, 2002), pp. 954–8. Raven, in his Presidential Address to the American Association for the Advancement of Science, presents his analysis of the condition of the world at present and his views of what we can do to help us pass through these difficult times.

Rees, Martin, *Our Final Hour: A Scientist's Warning: How Terror, Error, and Environmental Disaster Threaten Humankind's Future in This Century – on Earth and Beyond* (New York: Basic Books, 2003). Rees is Britain's Astronomer Royal and a professor at Cambridge University. He gives our civilization a 50/50 chance of surviving the twenty-first century. Rees believes the choices we make in the next several decades could decide our fate.

Shi, David, *The Simple Life: Plain Living and High Thinking in American Culture* (Oxford: Oxford University Press, 2001). Voluntary simplicity is a component of sustainable development. Shi explores the roots of the idea in the lives and writings of Socrates, Plato, Aristotle, Jesus, St Francis, Buddha, Leo Tolstoy, Marcus Aurelius, Gandhi, Confucius, and Thoreau, among others. Its advocates hold that the simple life frees a person for real intellectual, moral, and spiritual growth.

Wilson, Edward, *The Future of Life* (New York: Alfred A. Knopf, 2002). We still have time and choices we can make to save life on our planet. Here is how Wilson states it: "The race is now on between the technoscientific forces that are destroying the living environment and those that can be harnessed to save it. We are inside a bottleneck of overpopulation and wasteful consumption. If the race is won, humanity can emerge in far better condition than when it entered, and with most of the diversity of life still intact."

APPENDIX 1

Studying and Teaching Global Issues

For the Student

You may find it useful to learn how the concept "development" can be used to study global issues and to have an overview of the topics covered in this textbook. One way to do this is to examine the structure of the course I teach in which *Global Issues: An Introduction* is the principal textbook.

Introduction

The first two or three days of the course are spent explaining what "development" means and how development and global issues are related. I define development as economic growth plus the social changes caused by or accompanying that economic growth. In this short introduction I try to help my students understand some of the main differences in the social and economic conditions of the rich and poor countries.

Wealth and poverty

The first full week of the course is spent on getting students to consider the extremely difficult question "Why are some nations rich and some poor?" Students examine three of the most widely accepted approaches or views of economic development: the market approach (also called the neoclassical or capitalist approach), the state approach (also called the command economy or socialist approach), and the civil society approach (decentralized development by community organizations, and grassroots movements). Globalization, which has greatly expanded international trade, is explained and the uneven effects it is having on poor and rich nations are discussed.

Population

For two weeks we look at the relationship between population and development. The changing population of the world is described, and the causes of the population explosion in the less developed nations are given. Students learn how population growth affects development (rapid population growth hinders development by putting a large stress on resources, health and education facilities, the environment, etc.), and how development affects population growth (development at first makes it greater as it lowers death rates, but later it reduces birth rates as the education level of women increases and children become less desirable economically and socially). The demographic transition is explained and students become familiar with the factors that lower birth rates. Some attention during this period is paid to the population policies of major countries, such as China. This segment of the course ends with a consideration of the future – whether a stabilization of the world's population will occur and whether the carrying capacity of the earth will be exceeded.

Food

For two weeks food holds our attention. World food production trends are examined, and a tentative answer is given to the question of how many are hungry in the world today. We investigate the causes of hunger in parts of the South (poverty – the lack of development – is one cause). Students learn how the availability and quality of food affect development (malnourished people are not good producers), and how development affects both the production of food and the type of food consumed (industrialized agriculture produces a large amount of food, but wealthy people often do not have a healthy diet). A short history of the Green Revolution is given. The food policies of the United States and a few other countries are examined. Finally, we think about how changes in the climate, biotechnology, the amount of arable land, and the cost of energy could affect future food supplies.

Energy

Two weeks is not enough time to investigate thoroughly the relationship between energy and development, but it is enough time to introduce students to this vital subject. A description of the energy crisis caused by the developed world's dependency on a polluting and highly insecure energy source – oil – is followed by a summary of the responses to that crisis by the United States, Western Europe, and Japan. The effect of the energy crisis on the South's development plans is explained. As we explore the relationship between energy use and development, students learn about the shift in the types of energy sources that took place as the Industrial Revolution progressed and how there has been a partial decoupling of energy consumption and economic growth – a new ability to produce economic growth with less energy. The subject of climate change (global warming) could be examined in the next section of the course, but I include it in the energy section because it serves as a good bridge to my discussion of nonrenewable and renewable energy sources. The role of conservation during the present period of energy transition is also explored. I end this section of the course with a presentation of the main arguments for and against nuclear power, which allows me to demonstrate how difficult and complicated are the choices the political system must make when dealing with energy.

The environment

As is the case with energy, two weeks is not much time to explore the effects development has on the environment – and the reverse, the effects the environment has on development – but significant information on the subject can be passed to students. (Poor people are hard on the environment as they struggle to survive, but the rich may or may not treat the environment well.) A brief history of the awakening in the United States to threats to the environment caused by industrialization introduces the subject, and provides the setting for an examination of the threats to the air, water, and land that have come with development. Airborne lead, acid rain, and the depletion of the ozone layer illustrate some of the main concerns we have at present with air pollution. The current concern with threats to our groundwater by migrating chemicals presents an example of water pollution caused by development, and the problem of how to handle huge amounts of solid and toxic wastes demonstrates well to the students the extremely difficult tasks the political system faces as it tries to preserve the land.

The problem of deforestation in the developing countries is briefly examined so that students become aware of the harm deforestation can bring to the land, its connection to the extinction of species, as well as the changes it can make in the climate. The connection between development and the extinction of cultures is also examined, as, for example, in the name of development the forest homes of numerous indigenous peoples are being destroyed. Chemicals, cancer, and pesticides are considered under a section in which we focus on the workplace and the home. Finally, the effect development has on the use of natural resources gains our attention, and students learn – often with some surprise – that development has often, at least so far, made many natural resources more available and cheaper.

Recycling, substitution, and the mining of low-grade ores are subjects presented at the end of this section of the course as we consider future supplies. The concept "overdevelopment" (consuming and polluting at a rate that cannot be maintained indefinitely) is also presented, as students consider reducing needs as a possible response to scarcities.

Technology

To many people, technology and development are synonymous. Technology is what makes economic growth and social change happen. Students are reminded of the many benefits that technology has brought to our lives. But because they are more aware of the benefits than the harm technology can produce, the course focuses on the dangers. Students learn that the decision of whether or not to use a certain technology can be a difficult one, especially in "tragedy of the commons" situations where short-term interests and long-term interests conflict. Illustrations of the unanticipated consequences of the use of technology are given, as are examples of the inappropriate uses of technology. Limits to the "technological fix" are illustrated. The issue of war is introduced, with technology making the destructive capacity of weapons greater. The threat of nuclear weapons is presented as a case study under the technology section.

Alternative futures

I end the course by focusing on different possible futures. During the last week we examine the main arguments that advocates make for the possibilities that our present type of development is leading us to "doom," or to continued "growth," or to "sustainable development" in the future.

For the Teacher

The problem

Improving and increasing international studies has become a priority on many campuses,[1] but as a report for the American Council on Education concludes, "the internationalizing of undergraduate education still has a long way to go."[2] How far it has to go can be easily shown. Reports of the shocking ignorance of people in the United States about other countries are well known, but less well known, and of some embarrassment to the college teaching profession, is that college-age people in the country are the most ignorant of all adults. Adding to the insult is the fact that attending college for four years reduces that ignorance only slightly.[3] Young people aged 18–24 in the United States in the late 1980s possessed *less* information about the world than the same age group had 40 years earlier.[4]

This information is especially surprising given the new emphasis many colleges are placing on international studies. Also surprising is the fact that the average student in a four-year college or university course takes several international studies courses, outside of foreign language instruction, before he or she graduates.[5] But a close look at these international studies courses reveals that most of them still focus on only one country or one region (often Western Europe), and only a few focus on a problem or issue that is found throughout the world. Also, few are interdisciplinary, and *only a minority deal with the world as it is today.*[6]

We indeed seem to be far from achieving what one report called an important characteristic of the truly internationalized university: it is a school where "no student graduates who has never been asked to think about the rights and responsibilities of this country in the world community, or who has never been brought to empathize with people of a different culture."[7]

Preparing students so that they will be able to function in an increasingly complex and interdependent world is a huge task, one which will require a better trained and more committed faculty and college administration. No easy answers, solutions, or quick fixes are possible, but many different methods and approaches are being tried, with varied degrees of success. As the American Council on Education study found, what we do not have now in the United States is a way to know what works, and what does not, and why it does or does not.[8] What we need are reports of successes and failures in the attempts to achieve the important characteristic of the truly internationalized university that the above quotation appropriately identifies.

A solution

While attending a conference on the developing world, I heard college teachers complain that they could not get their students interested in studying the South, where most of the world's people live. As I thought about this complaint, I realized that I had discovered an answer to the question "How do you get American students to want to study the non-Western world?" I know that you don't do it by reminding them that their bananas come from that world. The student's reaction to that statement is "So what? Who cares?" The way you get them interested is by introducing real global problems and exploring their possible solutions. You demonstrate that global problems are American problems, that our actions help create or solve the problems, and that the problems affect our lives, in the present as well as in the future.

Over the past 20 years I have taught a course for undergraduates called "Global Issues." The course, outlined in the section addressed to the student, focuses on many of the most important global issues today, issues that both the more developed and the less developed nations can no longer ignore.

I believe that one reason many social science teachers do not teach a course on global issues is that they do not know how to deal with these issues in a respectable, scholarly way – in a manner that will prevent the class from becoming just a forum for the discussion of current events. But I have found that there is a concept – "development" – which can serve as the tool we need for treating these issues in a responsible manner.

Social scientists commonly use this concept only with reference to the poorer nations, but "development" can also be a powerful tool for analyzing conditions in and actions of the richer nations.

Teaching techniques

How does one teach the above material? I have used a combination of techniques. I have adopted as the basic textbook this book, *Global Issues: An Introduction*. I have also used the latest edition of the Worldwatch Institute's *State of the World*.[9] This book is an excellent annual updating of many of the topics covered in my course, although the large amount of detailed, factual information it contains overwhelms some undergraduates. At times, in place of *State of the World*, I have used the United Nations Development Programme's *Human Development Report*, which covers many development-related subjects.[10] Students read selections from the latest edition of *Annual Editions: Global Issues*, which is a collection of articles from many different sources – some with opposing viewpoints – on many of the issues presented in the course.[11] Students are also required to subscribe to the *New York Times*, which allows them to follow current developments in all of the subjects covered in the course. Weekly quizzes are given on the *Times*.

All possible examination questions are given to the students (see appendix 3) and we use these questions to guide our discussion of the textbook. I do not give lectures. The questions on the examinations are randomly selected from these questions. I find that students learn the material better when they know what they will be tested on.

Videotapes play an important role in the course. Many excellent programs related to topics in our course appear on public television (see appendix 2). The experience of seeing an interesting, current portrayal of a topic we are studying is a powerful teaching technique. The tapes reinforce what the students are learning and broaden their knowledge. Also, the tapes serve another important role. Studying global issues can be depressing. The problems are numerous and serious, and at first glance appear to be unsolvable. The tapes help counter that depression by often showing what some individuals are doing to attack these problems. I try to show at least two tapes related to each of the five main subtopics in the course.

Students write a five- to eight-page typewritten research paper. In the paper they focus on an issue in greater depth than we have been able to in the course. The students are required to use at least one, but not more than two, sources from the internet. Appendix 4 gives some relevant internet sources.

A course of instruction following the above outline utilizes three levels of analysis, which contribute to its effectiveness: the individual, the nation, and the international system. To understand the issues one must look at the behavior of individuals, the actions and policies of nations, and the condition of the world's environment as well as of its economic and political systems. Solutions to the global problems require individual efforts, new national policies, and international agreements.

Such a course of instruction has three main goals. The first is to increase student knowledge of some of the most important problems facing the world today, a knowledge that the student learns comes from many different disciplines. The second goal is to help students learn of the complex interrelationships among the issues. The third is to evaluate possible solutions to the problems studied. As the students consider possible solutions, they learn the vital fact that human actions (including their own) can change the world in very different ways.

Can these goals be achieved? Certainly they cannot for every student, nor will every student who achieves one, achieve all three. But many can achieve one or more of these goals. Students appreciate an effort that helps them understand the complicated and rapidly changing world in which they live. When we help them acquire this information, we are giving them both the knowledge they will need to live in today's world, but more importantly, the knowledge that will enable them, if they so desire, to add their talents to the efforts being made to solve many of these global problems.

Student comments

For the past 20 years I have at the end of the course asked students to write, in a short unsigned essay, what they felt was the most important thing they learned in the course. These three responses give some common conclusions:

> The most important thing I learned is that problems concerning population, food, energy, etc. are *real*. I feel that most people don't realize the magnitude of these problems. However, by taking this course, I now see that all these problems are greater than I originally thought . . . This course taught me the first step in combating these problems, and that is to recognize that they are REAL!

> I had . . . known about the environmental movement and even considered myself an environmentalist. Sure I wanted to take care of my environment; new energy sources sounded cool; pollution was bad and needed to be stopped, etc. However, I never really knew how *interconnected* all of this was until I took this course. . . . I learned how changes in one area can drastically affect what I previously thought were unrelated things. . . . I learned that all of these problems are interconnected and must be studied as such if any real (long-term) solution is ever to be found for them.

> The most important thing I learned was to stop thinking like an American and only think about self-interest. Rather now I think about my neighbor be it in Converse Heights or my neighbor in South America. Professor Seitz, you focused my mind to look at the big picture instead of the small one. When I . . . [threw away an empty] can of Coke previously I would say, "What can I do about recycling?" Now I see that even a little effort to make a difference does just that, it makes a difference. Now when I get in my car to go to the store, I think twice and now I usually will walk. Before when I said [what's wrong] with one more light on, it's just 20 cents a day lost. Now I think about how [the production of] electricity pollutes the atmosphere, so now I conserve electricity and other fossil fuels as well. To sum it all up, I have learned to be more responsible to this precious world we call earth. For that, whatever grade I receive, I thank you for opening not just my eyes but my mind.

Notes

1 Ann Kelleher, "One World, Many Voices," *Liberal Education*, 77 (November/December 1991), pp. 2–7.
2 Richard D. Lambert, *International Studies and the Undergraduate* (Washington, DC: American Council on Education, 1989), p. 153.
3 Ibid., p. 107.
4 Ibid., p. 106.
5 Ibid., p. 126.
6 Ibid., pp. 115–27.
7 Humphrey Tonkin and Jane Edwards, "Internationalizing the University: The Arduous Road to Euphoria," *Educational Record*, 71 (Spring 1990), p. 15.
8 Lambert, *International Studies and the Undergraduate*, p. 157.
9 Lester Brown et al., *State of the World* (New York: W. W. Norton, annual).
10 United Nations Development Programme, *Human Development Report* (New York: Oxford University Press, annual).
11 Robert M. Jackson (ed.), *Annual Editions: Global Issues* (Guilford, CT: Dushkin, annual).

APPENDIX 2

Relevant Videos

All in the Genes [genetic research and biotechnology], produced by Christopher Martin, distributed by Filmakers Library, 1998, 52 minutes.

Alternative Power Sources and Renewable Energy, distributed by Films for the Humanities and Sciences, 1999, 22 minutes.

Amazon Journal [state of indigenous people in Brazilian Amazon and visit to Yanomami Territory], Geoffrey O'Connor, producer/director, a Production of Realis Pictures, distributed by Filmakers Library, 1996, 58 minutes.

Armed to the Teeth: The Worldwide Plague of Small Arms [award-winning film on what the UN considers to be a world crisis; case studies show what is being done to curb the small arms proliferation], a United Nations Production, distributed by Films for the Humanities and Sciences, 2000, 56 minutes.

Arming the Heavens [examines all sides of the space weapons debate], Glenn Baker, writer/producer, distributed by Azimuth Media, 2004, 25 minutes.

Baby Crash: Causes and Consequences of Declining Birthrates [why young people in Europe, Japan, and Canada are postponing or deciding not to have children at all], distributed by Films for the Humanities and Sciences, 2002, 46 minutes.

Batak: Ancient Spirits, Modern World [Philippine tribe struggles to preserve its eco-friendly, hunter/gatherer way of life], distributed by Films for the Humanities and Sciences, 2000, 50 minutes.

The Bells of Chernobyl: Ten Years After [coverup of the effects of the Chernobyl nuclear disaster], a coproduction for Tele Images International, distributed by Filmakers Library, 2000, 52 minutes.

Be Prepared for Global Warming [although global warming is irreversible, its progress can be slowed, its impacts managed], distributed by Films for the Humanities and Sciences, 2003, 51 minutes.

Bhopal: The Search For Justice [after 15,000 people were killed and hundreds of thousands more were permanently maimed by the leak of poisonous gas at a pesticide plant in India, the search for justice for the survivors is still going on], produced by The National Film Board of Canada, 2004, 53 minutes.

The Bottom Line: Privatizing the World [controversy over business's rush to commodify the world's common resources such as drinking water, and human and plant genes], distributed by Films for the Humanities and Sciences, 2002, 53 minutes.

Can We Learn to Live without Nuclear Weapons? Center for Defense Information, 1998.

Cappuccino Trail: The Global Economy in a Cup [follows the trail of two coffee beans, one ending up on the open market and one in a company dedicated to paying fair prices to farmers], distributed by Films for the Humanities and Sciences, 2001, 50 minutes.

Changing Nature: Population and Environment at a Crossroads [urbanization, industrialization, and agriculture are depleting the earth], distributed by Films for the Humanities and Sciences, 2001, 58 minutes.

Chernobyl: The Taste of Wormwood [onsite images and interviews with survivors], distributed by Films for the Humanities and Sciences, early 1990s, 52 minutes.

Children of Shadows [in Haiti parents are forced by poverty to give away their children to work as unpaid domestic servants or slaves], produced by Karen Kramer, distributed by Filmakers Library, 2003, 54 minutes.

Children without Childhood, distributed by Films for the Humanities and Sciences, 1999, 45 minutes each.

India: The Little Serfs [child labor].
Mexico: Back Door to the Promised Land [stories by children of desperately poor families].
Philippines: Angels of the Night [child prostitutes].
Uganda: The War of the Children [child soldiers].

China: From Poverty to Prosperity [part of the series Three Dynamic Economies], distributed by Films for the Humanities and Sciences, 1998, 30 minutes.

City Life [22-part series examines the effect of globalization on people and cities worldwide], produced by Television Trust for the Environment, distributed by Bullfrog Films, 2001, 27 minutes each.

Conservation and Energy Alternatives: Powering the Future [different views about the future are given, then a focus on wind power's pros and cons], distributed by Films for the Humanities and Sciences, 2001, 30 minutes.

Countdown to Hope: Opposing the Threat of Nuclear War [deals with the threat of nuclear war caused by the decline of nuclear security in Russia, the conflict between India and

Pakistan, and threats by rogue states and zealot factions], distributed by Films for the Humanities and Sciences, 2001, 57 minutes.

The Curse of Oil [a global history of the oil industry], distributed by Films for the Humanities and Sciences, 2003, 52 minutes.

Dodging Doomsday: Life beyond Malthus [two classic questions – Can the earth sustain its expanding population? Is population growth the cause of earth's social and environmental problems? – are discussed by Paul Ehrlich, Julian Simon, Lester Brown and other experts], BBC production, distributed by Films for the Humanities and Sciences, 1992, 50 minutes.

The Doomsday Asteroid [evidence of asteroids that could collide with earth and what, if anything, can be done about it], a NOVA production, distributed by PBS Video, 1997.

Effective Government in the Developing World, distributed by Films for the Humanities and Sciences, 1997, 30 minutes each.

Governments Caring for People [how governments can provide services for their people even when resources are scarce].
Regulation of Industry [examples from Korea, Chile, Uganda, and India].
Cleaning Up Corruption [examines anticorruption programs in Brazil, Uganda, and Singapore].

Endangered Biodiversity and Economic Development [a Fred Friendly Seminar in which a panel of experts and lay people try to determine, in a fictional case, what sacrifices should be made to protect biodiversity – and who should make them], distributed by Films for the Humanities and Sciences, 2001, 60 minutes.

The Energy Conspiracy [influential organizations have successfully lobbied for the coal, oil, and nuclear power industries against sustainable energy and have convinced the public, with inaccurate information, that global warming is not a problem], produced by Hans Bulow and Poul-Eric Heilburth, distributed by Filmakers Library, 1999, 59 minutes.

The Environment: A Historical Perspective [reacting to the human impact on the environment, Lester Brown and others give their views on ecological stewardship for the twenty-first century], distributed by Films for the Humanities and Sciences, 1999, 52 minutes.

The Environment: When Politics and Industry Intersect [two-part series investigates who might be profiting from manipulating environmental laws and who might be molding public opinion and the legislative process], distributed by Films for the Humanities and Sciences, 2000, 30 minutes each.

1 *Scientific Spin Doctors* [on pressing issues such as ozone depletion and global warming, some special interest groups are striving to bend science to their agendas].
2 *Green Pacts and Greenbacks* [to what extent are environmental standards achieving the goal of purifying the US's air, land, and water, and are new firms helping some industries meet the legal minimums in environmental protection laws simply to escape penalties?]

Epidemic! A Fred Friendly Seminar [experts discuss the causes, spread, and control of infectious diseases], distributed by Films for the Humanities and Sciences, 1999, 57 minutes.

Extreme Oil: The Wilderness [search for oil in fragile wilderness areas in Canada and Alaska leads to political controversy], distributed by Films for the Humanities and Sciences, 2004, 57 minutes.

Failed Nation Building: A Case Study of Haiti [one of the world's poorest nations where US intervention failed], an ABC News program, distributed by Films for the Humanities and Sciences, 2004, 22 minutes.

Fighting the Tide: Developing Nations and Globalization [what globalization means for the citizens of five developing nations], distributed by Films for the Humanities and Sciences, 2004, 26 minutes each.

1 *Malawi; A Nation Going Hungry.*
2 *Ecuador: Divided over Oil.*
3 *Nicaragua: Turning away from Violence.*
4 *India: Working to End Child Labor.*
5 *Guatemala: The Human Price of Coffee.*

Food or Famine? [is population going to outstrip food supply?], two-part series, produced by Canadian Broadcasting Corporation, distributed by Filmakers Library, 1997, 49 minutes each.

Free Market Economies: The Commanding Heights [Daniel Yergin, Milton Friedman, and John Kenneth Galbraith discuss the tension between free markets and managed economies with Ben Wattenberg], distributed by Films for the Humanities and Sciences, 1998, 26 minutes.

Frontline: Missile Wars [explores the US's missile defense program], produced by Azimuth Media with PBS's Frontline, distributed by PBS Video, 2002, 60 minutes.

The Gaia Hypothesis [its creator explains the development and evolution of the hypothesis and counterarguments], distributed by Films for the Humanities and Sciences, 1990, 25 minutes.

Gift of a Girl: Female Infanticide [India], produced by Mayyasa Al-Malazi, distributed by Filmakers Library, 1998, 24 minutes.

Global Capitalism and the Moral Imperative [globalization and the growing rich/poor gap], distributed by Films for the Humanities and Sciences, 1998, 29 minutes.

The Global Generation: The Human Face behind Globalization [six-part series], distributed by Films for the Humanities and Sciences, 2000, 26 minutes each.

1 *A World without Borders: What Is Happening with Globalization?* [middle-class luxury and abject poverty coexist side by side].
2 *The Global Marketplace: The Benefits of Globalization.*
3 *Global Partnerships: The Effects of Globalization* [volunteers and grassroots activity].
4 *The Global Neighborhood: What Can Happen with Globalization?* [humanitarian intervention is gaining acceptance; international work of nongovernmental organizations is growing.]
5 *Global Grassroots: The Ramifications of Globalization* [community-based initiatives tackle two downsides of globalization: loss of social safety nets and widening gap between haves and have-nots].

6 *The Global Dimension: The Risks of Globalization* [despite unprecedented growth in the world economy, 1.5 billion people still live in extreme poverty and 3 billion are very poor].

The Global Impact of AIDS, distributed by Films for the Humanities and Sciences, 1998, 50 minutes.

Globalization: Winners and Losers, distributed by Films for the Humanities and Sciences, 2000, 40 minutes.

Global Jihad [forces behind Islamic terrorism], an ABC News program, distributed by Films for the Humanities and Sciences, 2004, 20 minutes.

The Global Trade Debate [attempts to offer a balanced look at the realities of globalization and to examine the issues that divide those who support and criticize growing world trade], distributed by Films for the Humanities and Sciences, 2001, 42 minutes.

Global Warming: Global Policy? [focuses on the politics and agencies involved in formulating an international response to global warming], distributed by Films for the Humanities and Sciences, 1999, 30 minutes.

Global Warming: The Signs and the Science, produced by Public Broadcasting Service (PBS), distributed by PBS Video, 2005(?), approx. 60 minutes.

God's Earth: A Call for Environmental Stewardship [narrated by a Catholic priest and ecologist, presents a philosophical, ethical, and practical challenge to modern Western religions to accept their role as stewards of the environment], distributed by Films for the Humanities and Sciences, 1997, 57 minutes.

Greenhouse Earth: An Uncontrolled Experiment [global warming], a Deutsche Welle Production, award-winning film, distributed by Films for the Humanities and Sciences, 2001, 26 minutes.

Guns, Germs, and Steel [based on the Pulitzer Prize-winning book with this title by Jared Diamond, a three-part program presents Diamond's controversial theory that geography is the main reason the world is divided into haves and have-nots], produced by Lions Television, London, for National Geographic Television and Films, Washington, DC and distributed by PBS Video, 2005, three 60-minute programs.

Half Lives: History of the Nuclear Age [a balanced look at how and why the world entered the nuclear age and related issues today], distributed by Films for the Humanities and Sciences, 1996.

Journey to Planet Earth [programs explore the necessity of achieving a balance between the needs of people and the needs of the environment], produced by Emmy Award filmmakers Marilyn and Hal Weiner in association with South Carolina Educational Television (http://www.pbs.org/journeytoplanetearth/). Distributed by Screenscope (screenscope@screenscope.com), 25 minutes each (educational cut), or 60 minutes each.

1 *Rivers of Destiny* [environmental problems of four rivers – the Amazon, the Jordan, the Mekong, the Mississippi – and how they affect people who depend on the rivers for their livelihood], 1999.

2 *Land of Plenty, Land of Want* [can we feed the world's growing population without ruining the environment? Visits Zimbabwe, France, China, and US], 1999.
3 *Urban Explosion* [how to sustain the world's exploding urban population without destroying the environment. Visits Mexico City, Istanbul, Shanghai, and New York City], 1999.
4 *On the Brink* [severe environmental problems can produce political crises and more hostilities. Visits Haiti, Peru, South Africa, Mexico, and US], 2003.
5 *Seas of Grass* [some grasslands are in grave danger. Visits Kenya, South Africa, Argentina, China, US], 2003.
6 *Hot Zones* [changes in global and local ecosystems are connected to increased spread of infectious diseases. Visits Kenya, Peru, Bangladesh, US], 2003.
7 *Future Conditional* [spread of toxic pollution. Visits the Arctic, Mexico, Uzbekistan, US], 2005.
8 *The State of the Planet: Global Warming*, 2005.
9 *State of the Planet's Wildlife*, 2006.

The Jungle Pharmacy: Nature's Remedy [will the jungle pharmacy disappear before it is understood?], a Deutsche Welle production, distributed by Films for the Humanities and Sciences, 1995, 27 minutes.

The Last Warriors: Seven African Tribes on the Verge of Extinction [five part series], distributed by Films for the Humanities and Sciences, 2000, 54 minutes each.

Left Behind: Kenyan AIDS Orphans [award-winning film looks at the lives of children orphaned by AIDS], distributed by Films for the Humanities and Sciences, 2002, 36 minutes.

Legacies of War [focuses on efforts to repair physical, social, and personal damage in several postwar situations around the world]. A United Nations Production, Distributed by Films for the Humanities and Sciences, 2000, 32 minutes.

Life [30-part series about how globalization is affecting ordinary people. Series takes us to India, Africa, Asia, Brazil, Mexico, the Pacific Islands and the US], produced by Television Trust for the Environment for BBC Worldwide Television, distributed by Bullfrog Films, 2000, 24 minutes each.

Life III [12-part series about how globalization is affecting ordinary people. Series takes us to Russia, Guatemala, Bangladesh, India, Zambia, Nepal, South Africa, and Ghana. Programs focus on children, health and nutrition, the HIV/AIDS epidemic, poverty, agriculture, trade, sustainable development, women's issues, and human rights.] Produced by Television Trust for the Environment for BBC Worldwide Television, distributed by Bullfrog Films, 2002, 25 minutes each.

Life 4 [27-part series about global efforts to achieve the UN Millennium Development Goals], produced by Television Trust for the Environment, distributed by Bullfrog Films, 2004, 25 minutes each.

Mexico City: The Largest City [winners and losers], distributed by Films for the Humanities and Sciences, 2004, 26 minutes.

My Father's Garden [the case for sustainable agriculture], two parts, distributed by Bullfrog Films, 1995, 56 minutes each (34/23 minutes also available).

Nowhere Else to Live [Mexico's urban poor], Alan Handel Productions, distributed by Filmakers Library, 1998, 26 minutes.

Of Hopscotch and Little Girls: Stolen Childhood [stories of abuse and neglect of girls around the globe], distributed by Films for the Humanities and Sciences, 2000, 53 minutes.

Oil in Iraq: Curse or Blessing? [the politics of oil in the Middle East], produced by Robert Mugnerot in collaboration with Baudoing Koenig, distributed by Filmakers Library, 2003, 52 minutes.

One Day of War [follows combatants in 16 wars in the same 24-hour period], a BBCW Production, distributed by Films for the Humanities and Sciences, 2004, 47 minutes.

Our Hiroshima [eyewitness account, archival footage taken before and after the event, and the politics involved in developing and promoting the use of the bomb), distributed by Films for the Humanities and Sciences, 1995, 43 minutes.

Our Planet Earth [views from space and reflections by astronauts and cosmonauts from ten nations], distributed by Bullfrog Films, 1990, 23 minutes.

Paul Ehrlich and the Population Bomb [footage from around the world and interviews with Ehrlich and his critics], produced by Sam Hurst, presented by KQED/San Francisco, distributed by Films for the Humanities and Sciences, 1996, 60 minutes.

The Peacekeepers [UN peacekeeping force in the Democratic Republic of Congo to quell ethnic fighting], produced by The National Film Board of Canada, 2005, 83 minutes.

Population Six Billion [examines the grim realities of life in Third World nations where population continues to expand and includes population control initiatives in Vietnam, Uganda, and Mexico], distributed by Films for the Humanities and Sciences, 1999, 58 minutes.

Precious Earth: Mapping the Human Condition [eight-part series using global data-mapping to help analyze current issues], distributed by Films for the Humanities and Sciences, 2004, 31 minutes each.

1 *Life Expectancy: Geography as Destiny.*
2 *Infectious Diseases: More Mobility, Greater Danger.*
3 *Decaying Cities: Reclaiming the Rust Belt.*
4 *Empty Oceans: Global Competition for Scarce Resources.*
5 *Crime in the Cities: Public Safety at Risk.*
6 *Birthrate: New Options for Parenthood.*
7 *China's Prosperity: Behind the Scenes of Progress.*
8 *Extinct Species: Red Alert to Humanity.*

Race against Time: The AIDS Crisis in Africa, produced by Canadian Broadcasting Corporation from the Nature of Things Series, distributed by Filmakers Library, 2002, 48 minutes.

Rachel Carson's "Silent Spring" [*Silent Spring* was attacked by the chemical industry and medical establishment, but the book helped bring a revolution in environmental policy in the US and create a new ecological consciousness], produced by Neil Goodwin, distributed by PBS Video, 1993, about 60 minutes.

Refugees in Africa: Another Quiet Emergency [the plight of people – especially children – displaced and endangered by war], an ABC News program, distributed by Films for the Humanities and Sciences, 2004, 20 minutes.

Religion, War, and Violence: The Ethics of War and Peace [experts, scholars, and religious leaders from a variety of faiths discuss terrorism and its roots, fundamentalism, just war, holy war, pacifism, and the use of violence in the name of God], distributed by Films for the Humanities and Sciences, 2002, 90 minutes.

Sacrifice [poverty and other forces lead to children from Burma working in Thailand as prostitutes], produced by Ellen Bruno, distributed by Film Library, 1996, 50 minutes.

Scared Scared [people seeking positive ways to react to disasters such as those at the minefields of Cambodia, in post-9/11 New York City, at the toxic wasteland of Bhopal, in wartorn Afghanistan, at Hiroshima, in Bosnia, and in Palestine and Israel], produced and distributed by The National Film Board of Canada, 2004, 104 minutes.

Scientific American Frontiers XV: Hot Planet-Cold Comfort, narrated Alan Alda [global warming is already affecting Alaska and could affect the great Atlantic Conveyor], distributed by Public Broadcasting Service (PBS), 2004(?), 30 minutes.

Star Wars Dreams [missile defense system], produced by Leslie Woodhead, distributed by Filmakers Library, 2003, 50 minutes.

Stealing the Fire: The New Nuclear Weapons Underground [a story of international intrigue in the quest to obtain nuclear weapons], directed by John Friedman and Eric Nadler, distributed by Filmakers Library, 2003, 58 minutes.

Stolen Childhoods [child exploitation], an ABC News program, distributed by Films for the Humanities and Sciences, 2005, 20 minutes.

A Tale of Modern Slavery [caste systems and other archaic traditions perpetuate slavery in some poor countries], an ABC News program, distributed by Films for the Humanities and Sciences, 2005, 20 minutes.

Tapoori: Children of Bombay [the lives of two street boys], Alan Handel Productions, distributed by Filmakers Library, 1998, 26 minutes.

Thinking Globally, Acting Locally about Your Environment [major environmental threats in the US and positive steps individuals and communities can take to reduce them], distributed by Films for the Humanities and Sciences, 1998, 28 minutes.

Thirsting for War [conflict among Turkey, Syria, and Iraq over the water of the Euphrates River], written and directed by Christopher Mitchell, distributed by Filmakers Library, 2000, 50 minutes.

27 Dollars: Banking for the Poor [Grameen Bank in Bangladesh], produced by Andrea Beretta, distributed by Filmakers Library, 2003, 61 minutes.

Waging War against the New Terrorism [how Germany, Italy, and Egypt have combated terrorism in recent decades, and US strategies], an ABC News program, distributed by Films for the Humanities and Sciences, 2002, 23 minutes.

Was Malthus Right? Population and Resources in the 21st Century [contrasting views], distributed by Films for the Humanities and Sciences, 1998, 27 minutes.

Witness to Hate: Reporting on al Qaeda [report by a BBC correspondent who was shot by al-Qaeda], an ABC News program, distributed by Films for the Humanities and Sciences, 2005, 22 minutes.

A World without Borders: What Is Happening with Globalization [industrialized and developing nations are becoming somewhat similar with middle-class luxury and abject poverty coexisting side by side], distributed by Films for the Humanities and Sciences, 2000, 26 minutes.

You Can't Eat Potential: Breaking Africa's Cycle of Poverty, directed by Tony Freeth for Images First, South Africa, distributed by Filmakers Library, 1997, 56 minutes.

The following is a partial list of relevant video distributors in the USA:

1 Azimuth Media, 1779 Massachusetts Ave., NW, Washington, DC 20036-2109, telephone: (202) 232-8003; email: info@azimuthmedia.org; website: www.azimuthmedia.org.
2 Bullfrog Films, PO Box 149, Oley, PA 19547; telephone: 1-800-543-3764; email: video@bullfrogfilms.com; website: www.bullfrogfilms.com.
3 Film Library (Bruno Films), 22-D Hollywood Ave., Ho-Ho-Kus, New Jersey 07423; telephone: 1-800-343-5540; website: www.brunofilms.com.
4 Filmakers Library, 124 East 40th Street, New York, NY 10016; telephone: (212) 808-4980; fax: (212) 808-4983; email: info@filmakers.com; website: www.filmakers.com.
5 Films for the Humanities and Sciences, PO Box 2053, Princeton, NJ 08543-2053; telephone: 1-800-257-5126; fax: (609) 671-0266; email: custserv@films.com; website: www.films.com.
6 National Film Board of Canada, 1123 Broadway, Suite 307, New York, NY 10010; telephone: 1-800-542-2164; website: www.nfb.ca/store.
7 PBS (Public Broadcasting Service) Video, 1320 Braddock Place, Arlington, VA 22314; telephone 1-800-344-3337, website: http://teacher.shop.pbs.org.
8 Screenscope, Inc., 4330 Yuma St, NW, Washington, DC 20016; telephone: (202) 364-0055; email: screenscope@screenscope.com; website: www.screenscope.com.
9 Video Finders, telephone: 1-800-328-7271, a service of Public Broadcasting Service station KCET/Los Angeles. Provides availability and ordering information on more than 70,000 videocassette titles, including 3,000 programs broadcast by PBS.

Information on how to rent or purchase available videos is contained in a reference book found in many libraries: *Bowker's Complete Video Directory*, annual, which is published by R. R. Bowker, a Reed Reference Publishing Company, New Providence, New Jersey.

APPENDIX 3

Study and Discussion Questions for Students and Teachers

Chapter 1. Wealth and Poverty

1. Discuss the growing gap between the rich and poor on earth and the improvement of living standards for many, giving examples and explaining why these two trends are occurring.
2. Explain the meaning of the first Millennium Development Goal, the progress so far in achieving it, and the relative contribution of nations to development assistance and foreign aid.
3. Explain what the market approach to economic development says about why some nations are rich and some are poor.
4. Discuss some of the main arguments that advocates for the market approach make and some of the main criticisms of the approach.
5. Explain what the state approach to economic development says about why some nations are rich and some are poor.
6. Discuss some of the main arguments that advocates for the state approach make and some of the main criticisms of the approach.
7. Explain how the civil society approach would promote development in the poorest nations.
8. Discuss some of the main arguments that advocates for the civil society approach make and some of the main criticisms of the approach.
9. Explain how geography affects the wealth and poverty of nations.
10. Describe what globalization is and discuss its main positive and negative aspects.
11. List the main conclusions the author comes to about the relative strengths and weaknesses of the market, state, and civil society approaches, and give some of the main reasons he comes to these conclusions.

Chapter 2. Population

1. In a discussion of the changing population of the world, explain what is happening, where growth is taking place, and why there is movement of population within developing nations.
2. Discuss the causes of the population explosion, and explain why birth rates are high in less developed nations.
3. Discuss some of the main negative features of a rapidly growing population.
4. Discuss the main problems that can be created when a country's population growth is low and when the elderly begin to increase in number.
5. Discuss what UN conferences have concluded about the relationship between population growth and poverty, and how population growth can be reduced.
6. Explain the demographic transition and show the differences between the experiences of the more developed and the less developed nations with the transition.
7. Discuss the factors that cause birth rates to decline.
8. Discuss governmental policies that are designed to control the growth of population; include the experiences of Mexico, Japan, India, and China.
9. Explain why some governments have wanted to promote population growth in their countries, giving some specific examples of such countries and their reasons for having such a policy.
10. When and at what level will the world's population probably stabilize? In a discussion of whether the planet will be able to support this population, explain and use the concept "carrying capacity."
11. What are some of the factors that must be considered when one tries to answer the question "How many people can the earth support?" Explain why it is difficult to answer the question "What is the optimum size of the earth's population?"
12. Identify some of the most likely population-related problems that may occur in the future.

Chapter 3. Food

1. Discuss how much food is being produced at present and how many people are hungry. Who are the hungry and where do they live?
2. Discuss the causes of world hunger.
3. Discuss how the availability of food affects development in the less developed world.
4. Discuss how development has affected the amount of food produced in the developed nations and how food is produced.
5. Discuss some of the most important negative features of modern Western agriculture.
6. Explain how a nation's diet changes as the nation develops, and some of the more harmful characteristics of this diet.
7. Explain what the Green Revolution is, and identify some of its most important positive and negative features.
8. Explain why many less developed nations have not placed a high priority on rural development and the development of agriculture in their countries.
9. Discuss the governmental policies toward agriculture and rural development of the following: Northeast Asian countries (Japan, South Korea, Taiwan – treat these three as one), China, the former Soviet Union, United States.

10 Discuss how climate, amount of arable land, energy costs, and alternative (sustainable/organic) agriculture could affect future food supplies.
11 Discuss how new technology (such as biotechnology), fishing, and aquaculture could affect future food supplies. Explain why the prospects for food production in the future are both promising and troubling.

Chapter 4. Energy

1 In a discussion of the energy crisis, explain what it is, and describe the first, second, and third oil shocks.
2 Discuss the responses by the US government to the energy crisis.
3 Discuss the responses by the governments of Western Europe, Japan, and China to the energy crisis.
4 Explain how the energy crisis has affected the South's development plans.
5 Discuss the shift in sources of energy and the increase in the use of energy that have taken place as development occurred in the world.
6 In a discussion of the relationship between economic growth and energy growth, explain what the past one-to-one relationship was, why it has changed in the developed nations, and why energy efficiency has been less in the United States than in Europe and Japan.
7 Discuss climate change (global warming), explaining what it is, and the evidence it is occurring.
8 Discuss what climate change might do, and what can be done about it.
9 Discuss the main nonrenewable sources of energy, indicating their most important positive and negative features.
10 Discuss the main renewable sources of energy, indicating their most important positive and negative features.
11 In a discussion of conservation/energy efficiency as a form of energy, explain why it can be considered a source of energy, how it works, and its advantages.

Chapter 5. The Environment

1 Explain how the four UN conferences on the environment since 1972 show changing attitudes toward the environment in the world.
2 Discuss how development produces air pollution and give some evidence of the harmful effects of this pollution. Include in your answer the negative effects of airborne lead.
3 Explain acid rain and its causes. What effects does it have, and what actions are being taken to combat it?
4 Identify the problems caused by the depletion of the ozone layer and explain its causes. How extensive is this depletion? What actions are nations taking to combat it?
5 Discuss how development has affected the water on the planet, including the drinking water in the United States and in some European countries.
6 Discuss the relationship between development and the production of solid waste. Identify some of the ways governments can control solid wastes.
7 Discuss the relationship between development and the production of toxic waste. Identify some of the ways governments can control toxic waste.

8 What is causing deforestation? Where is it occurring? How extensive is it, and what harm is it doing?
9 What are the main causes of cancer and how are they related to development? What is the relationship between the production of chemicals and cancer?
10 What problems do pesticides cause? Explain how and where these problems arise.
11 How has development affected the availability and price of nonfuel natural resources? Why has this happened?
12 Discuss four of the five steps a country can take to counteract the shortages of a needed material if it cannot locate new rich deposits of the ore.
13 What is causing the extinction of species? Where is it mostly taking place? What harm does it do, and how can it be combated?
14 In a discussion of the extinction of cultures, explain why it is bad, and its relationship to development. What can be done to combat it? Show how the Yanomami or the Estonian culture is under siege.
15 Discuss what makes environmental politics so controversial.

Chapter 6. Technology

1 What is the "tragedy of the commons"? How do some present conflicts between short-term versus long-term benefits from using a technology (include global warming in your response) represent this type of situation?
2 Discuss the unanticipated consequences of the use of technology. Include in your discussion the use of DDT, and factory farms.
3 Discuss the inappropriate uses of technology, showing why intermediate or appropriate technology in the less developed nations could be more beneficial than high or hard technology. Discuss another situation that illustrates the inappropriate use of technology.
4 Explain the meaning of "technological fix" and give two illustrations of its limits.
5 Discuss why war can be considered a problem in the use of technology, giving special attention to the characteristics of modern war.
6 Explain how the threat of nuclear weapons can be considered a problem in the use of technology.
7 Discuss the new dangers that the world faces in the post–Cold War period because of nuclear weapons, including nuclear proliferation, control of nuclear weapons in Russia, and the cleanup of the production of nuclear weapons.
8 Show how technology is related to the problem of terrorism and the threat of nuclear terrorism.

Chapter 7. Alternative Futures

1 Discuss the most commonly mentioned disasters included in the "doom" future.
2 Discuss the policies of triage and lifeboat ethics that some have recommended as ways to deal with doom situations.
3 Discuss the main components of the "growth" future.
4 Discuss the main components of the "sustainable development" future.
5 Explain why the author believes the world's future may contain parts of all three alternative futures presented in the final chapter.

APPENDIX 4

Relevant Internet Websites

Acid Rain Data and Reports (by US Geological Survey): http://bqs.usgs.gov/acidrain
African Crop Improvement (an activity of the Rockefeller Foundation):
  http://www.africancrops.net
American Council for an Energy Efficient Economy: http://www.aceee.org/
American Water Works Association (international drinking water):
  http://www.awwa.org
Arms Control Association: http://www.armscontrol.org/
Atomic Archive (development, use and consequence of dropping atomic bomb):
  http://www.atomicarchive.com
Biodiversity – Hotspots (by Conservation International):
  http://www.biodiversityhotspots.org/xp/Hotspots
Canadian Institute for Health Information: http://www.cihi.com/
Canadian International Development Agency (CIDA): http://www.acdi-cida.gc.ca
CARE: http://www.care.org
Center for the Defense of Free Enterprise: http://www.cdfe.org
Center for Defense Information: http://www.cdi.org/
Center for International Earth Science Information Network (CIESIN) at Columbia
  University: http://www.ciesin.org/
Centers for Disease Control and Prevention: http://www.cdc.gov/
Climate Change (by United Nations Environment Programme):
  http://www.unep.org/themes/climatechange/
Conservation International: http://www.conservation.org/
CREST (Center for Renewable Energy and Sustainable Technology): http://crest.org/
Earth and Moon Viewer: http://www.fourmilab.ch/earthview/
Earth Charter (principles for building a just, sustainable, and peaceful global society):
  http://www.earthcharter.org
Earth Institute at Columbia University: http://www.earth.columbia.edu/
Earth Science World Image Bank (6,000 photos by American Geological Institute):
  http://www.earthscienceworld.org/imagebank
Earth Times: http://www.earthtimes.org

Earth Trends (environmental, social and economic trends, presented by World Resources Institute): http://www.earthtrends.wri.org
Earthwatch Institute (build sustainable future, global volunteering): http://www.earthwatch.org/
Earthweek: A Diary of the Planet: http://www.earthweek.com/
EcoNet: http://www.igc.org/
EElink-Environmental Education on the Internet: http://www.eelink.net
El Nino: http://www.pbs.org/wgbh/nova/elnino/
Entri-Environmental Treaties and Resource Indicators (by Columbia University's Center for International Earth Science): http://sedac.ciesin.columbia.edu/entri/index.isp
Envirolink: http://www.envirolink.org/
Environment Directory: http://www.webdirectory.com
Environmental Defense: http://www.edf.org/
Environmental News Network: http://www.enn.com/
Environmental Resources on the Internet: http://www.southampton.liu.edu/library/environ.htm
European Centre for Nature Conservation: http://www.ecnc.nl/
Facing the Future: People and the Planet (global issues education and action opportunities): http://www.facingthefuture.org/
Federation of American Scientists: http://www.fas.org
Food First/Institute for Food and Development Policy: http://www.foodfirst.org
Friends of the Earth: http://www.foe.org/
Gap Analysis Program (biodiversity): http://www.gap.uidaho.edu/
Global Environment Outlook (450 ecological and economic variables by UN Environment Programme): http://geodata.grid.unep.ch
Global Forest Watch (an initiative of the World Resources Institute): http://www.globalforestwatch.org
Global Recycling Network: http://grn.com/grn/
Global Volunteers: http://www.globalvolunteers.org/
Global Warming – a project of the Cooler Heads Coalition, updates by the Competitive Enterprise Institute (skeptics of global warming): http://www.globalwarming.org/
Globe Program (Global Learning and Observations to Benefit the Environment): http://www.globe.gov
Greenpeace USA: http://www.greenpeace.org/usa/
How to Compost: http://www.howtocompost.org/
The Hunger Site: http://www.thehungersite.com/
Intergovernmental Panel on Climate Change: www.ipcc.ch/
International Atomic Energy Agency: http://www.iaea.org/
International Ecotourism Society: http://www.ecotourism.org
International Food Policy Research Institute: http://www.cgiar.org/ifpri
International Institute for Sustainable Development Linkages: http://www.iisd.ca/
International Research Institute for Climate and Society at Columbia University: http://iri.columbia.edu
International Solar Energy Society: http://www.ises.org/
International Water Resources Association: http://www.iwra.siu.edu/
Izaak Walton League of America (conservation): http://www.iwla.org
Jane Goodall Institute – for Wildlife Research, Education, and Conservation: http://www.janegoodall.org
Job/Volunteer Links (jobs and volunteer opportunities): http://www.igc.org/jobs.html
Land Trusts: http://www.possibility.com/landtrust/

League of Conservation Voters: http://www.lcv.org/
Millennium Ecosystem Assessment: http://www.maweb.org
NASA World Wind (lets you zoom from satellite altitude into any place on earth): http://worldwind.arc.nasa.gov
The National Academies: Advisors to the Nation on Science, Engineering, and Medicine (includes the National Academy of Science and the National Research Council): http://www.nationalacademies.org/
National Aeronautics and Space Administration (NASA): http:www.nasa.gov/home/
National Audubon Society: http://www.audubon.org/
National Center for Atmospheric Research: http://www.ncar.ucar.edu/
National Council for Science and the Environment: http://www.ncseonline.org/
National Institutes of Health: http://www.nih.gov/
National Oceanic and Atmospheric Administration: http://www.noaa.gov/
National Renewable Energy Laboratory: http://www.nrel.gov/
National Snow and Ice Data Center: http://www.nsidc.org/
National Wildlife Federation: http://www.nwf.org/
Natural Resources Defense Council: http://www.nrdc.org/
The Nature Conservancy (land trusts): http://www.nature.org/
The Nature Conservancy's Sustainable Waters Program: http://www.nature.org/initiatives/freshwater/
Near Earth Objects Dynamic Site (asteroids etc. by University of Pisa, Italy): http://newton.dm.unipi.it/neodys
Netaid – Educating, inspiring and empowering young people to fight global poverty: http://www.netaid.org
North American Association for Environmental Education: http://www.naaee.org
Organization for Economic Cooperation and Development (OECD): http://www.oecd.org
Pew Center on Global Climate Change: http://www.pewclimate.org/
Popline (population bibliography by Johns Hopkins University with about 300,000 abstracts): http://db.jhuccp.org/popinform
Population Action International: http://www.populationaction.org/
Population Connection (formerly Zero Population Growth): http://www.populationconnection.org
Population Council: http://www.popcouncil.org
Population Index (Princeton University, Office of Population Research): http://popindex.princeton.edu
Population Reference Bureau: http://www.prb.org
Program for Monitoring Emerging Diseases, Federation of American Scientists: http://www.fas.org/promed/
Public Interest Research Groups (PIRGs): http://pirg.org/
Rainforest Action Network: http://www.ran/org/
Reef Check (coral reefs): http://www.reefcheck.org
Resources for the Future: http://www.rff.org/
Sci Dev Net (Science and Development Network) – scientific information relevant to developing nations to promote North–South and South–South collaboration: http://www.scidev.net
Scientific American.com (a popular and respected US scientific magazine): http://www.sciam.com
SciTechResources.gov – a catalog of (US) government science and technology websites (by subject): http://www.scitechresources.gov/

Scorecard – The Pollution Information Site (pollution in individual US communities by Environmental Defense): http://www.scorecard.org
ServiceLeader.org (volunteering opportunities including volunteering for online activities by University of Texas at Austin): http://www.serviceleader.org/new/
SERVIR – The Mesoamerican Regional Visualization and Monitoring System (environmental changes in Central America): http://servir.nsstc.nasa.gov/home.html
Sierra Club: http://www.sierraclub.org
Smithsonian National Museum of Natural History: http://www.nmnh.si.edu/
Smithsonian National Zoological Park – Conservation and Science: http://www.si.edu/crc
Solar News: http://www.solarnews.com/index.html
Student Environmental Action Coalition: http://www.seac.org/
Sustainable Agriculture Research and Education: http://www.sare.org
Sustainable Communities Network: http://www.sustainable.org/
SustainUS – The US Youth Network for Sustainable Development: http://www.sustainus.org/mambo
Terrorism (by the Council on Foreign Relations): http://www.terrorismanswers.com
Tree of Life Web Project (biodiversity): http://tolweb.org/tree/phylogeny.html
Union of Concerned Scientists (energy, global warming, nuclear arms, environment, food, health): http://www.ucsusa.org/
United Kingdom Department for International Development: http://www.dfid.gov.uk
United Nations Children's Fund (UNICEF): http://www.unicef.org
United Nations Development Programme (UNDP): http://www.undp.org
United Nations Educational, Scientific, and Cultural Organization (UNESCO): http://www.unesco.org/
United Nations Environment Programme (UNEP): http://www.unep.org/
United Nations Food and Agriculture Organization (FAO): http://www.fao.org/
United Nations High Commission for Refugees (UNHCR): http://www.unhcr.org
United Nations Population Fund (UNFPA): http://www.unfpa.org
United Nations Population Information Network (POPIN): http://www.un.org/popin/
United States Agency for International Development (AID): http://www.usaid.gov
United States Census Bureau (international database): http://www.census.gov/ipc/www/idbnew.html
United States Central Intelligence Agency (CIA – The World Factbook): https://www.cia.gov/cia/publications/factbook/index.html
United States Department of Agriculture, Natural Resources Conservation Service: http://www.nrcs.usda.gov/
United States Department of Energy: http://www.doe.gov
United States Department of the Interior: http://www.doi.gov/
United States Environmental Protection Agency (EPA): http://www.epa.gov/
United States Geological Survey, Water Resources of the United States: http://water.usgs.gov/
United States National Oceanic and Atmospheric Administration: http://www.noaa.gov/
United States Nuclear Regulatory Commission: http://www.nrc.gov/
Water Environment Federation: http://www.wef.org/
Wildlife Conservation Society: http://www.wcs.org
Wind Power (by the Danish Wind Industry Association): http://www.windpower.org/en/core/htm
World Affairs Council: http://www.world-affairs.org/

World Bank: http://www.worldbank.org/
World Business Council for Sustainable Development: http://www.wbcsd.org/
World Conservation Monitoring Centre/United Nations Environmental Programme: http://www.unep-wcmc.org/
World Conservation Union (IUCN): http://iucn.org/
World Energy Efficiency Association (WEEA): http://www.weea.org/
World Food Programme of the United Nations (food aid): http://www.wfp.org
World Health Organization: http://www.who.int/en
World Resources Institute: http://www.wri.org/
World Trade Organization: http://www.wto.org
World Water Assessment Programme by UNESCO: http://www.unesco.org/water/wwap
World Wildlife Fund: http://www.wwf.org/
Worldwatch Institute: http://www.worldwatch.org/
WWT – Wildfowl and Wetlands Trust: http://www.wwt.org.uk
Yale Working Papers on Solid Waste Policy: http://www.yale.edu/pswp/

Glossary

biotechnology "Use of biological processes for medical, industrial, or manufacturing purposes. Humans have long used yeast for brewing and bacteria for products such as cheese and yoghurt. Biotechnology now enjoys a wider application. By growing microorganisms in the laboratory, new drugs and chemicals are produced. Genetic engineering techniques of cloning, splicing, and mixing genes facilitate, for example, the growing of crops outside their normal environment, and vaccines that fight specific diseases. The safety of genetically modified food has been questioned by some . . ." *Source*: "biotechnology," in *World Encyclopedia* (Philip's, 2005).

carrying capacity "Refers to the maximum population that a given environment and resource base can sustain. Carrying capacity is a crucial issue in any relatively closed ecosystem; it therefore plays a vital role in economic and ecological research into the sustainability of human life at the global level. There is still little agreement about the planet's carrying capacity or what consequences approaching or exceeding it would have." *Source*: "carrying capacity," in Craig Calhoun (ed.), *Dictionary of the Social Sciences* (Oxford University Press, 2002). Human beings can increase carrying capacity, e.g. by applying fertilizer to range-lands, or can lower it, e.g. by causing erosion from deforestation. See also chapter 2, "Population," subsection "The carrying capacity of the earth."

civil society Voluntary associations of people that are between the state and the extended family. There are two branches of the use of this concept, one emphasizing economic associations (e.g. Adam Smith) and one emphasizing social and political associations (e.g. Montesquieu, Rousseau, and de Tocqueville). This book uses the latter definition with a focus on social and political organizations such as churches, clubs, interest groups, social movements, and political parties. *Source*: "civil society," in Craig Calhoun (ed.), *Dictionary of the Social Sciences* (Oxford

University Press, 2002). See also chapter 1, "Wealth and Poverty," section "The Civil Society Approach."

climate change (global warming, the greenhouse effect) See chapter 4, "Energy," section "Climate Change."

demographic transition The four basic changes the population of a country seems to go through as the country passes from being a traditional, rural and agricultural country to a modern, urban and industrial country. In the first stage, there are high birth and death rates. In the fourth stage there are low birth and death rates. Stages two and three are called the transitional period. In the early part of the transition (second stage) the death rates begin to drop quickly as modern medicine takes hold while birth rates continue to be high. The population begins to increase rapidly, a situation many developing countries recently faced or are still facing. In the third stage of the intermediate period death rates continue to fall and birth rates start to fall also. Population continues to increase but less rapidly than in the second stage. Some developing countries are at present in this state. A few demographers have recently stated that there may be a fifth demographic stage where birth rates are so low that the size of the population starts to shrink, a situation that Japan and some European countries, e.g. Austria, Germany, Sweden and Italy, appear to be in at present. See also chapter 2, "Population," subsection "Demographic transition."

developing country A relatively poor nation where agriculture or mineral resources have a large role in the economy while industry has a lesser role. The infrastructure of the country (transportation, communications, education, health, and other social services) is usually inadequate for its needs. About 80 percent of the world's people live in nations like this, also called **less developed** or **underdeveloped**. These countries are often located in Africa, Latin America, and Asia. (Some of these nations are highly developed in culture and are the homes of ancient civilizations that had great achievements in architecture, religion, and philosophy.) Since many of the less (economically) developed nations are in the Southern Hemisphere, they are at times referred to as **the South**. During the Cold War these nations were often called the **Third World**, a term still in use. (First World was capitalist, noncommunist nations, and Second World was state socialist countries). Industrialized countries are called **developed** nations. Most of them are located in the Northern Hemisphere so they are at times called **the North**. The World Bank classifies nations according to their level of income, as measured by per capita gross national income, placing low and middle income countries in the developing category and high income countries in the developed category. The poorest countries such as Bangladesh and Somalia are called the **least developed**. An intermediate category between developed and developing is the **newly industrializing countries** (such as South Korea, Taiwan, Hong Kong, Singapore, Brazil, and Mexico) which have become richer by expanding their manufacturing and exporting goods mainly to the US, Europe, and Japan. All of these terms are imprecise. There are often many differences among developing nations. Their relative poverty is often their sole similar quality.

development Economic growth plus the social changes caused by or accompanying that economic growth. When the economy expands, more goods (material objects) and services (health care, education, etc.) are produced. The economic growth causes or is accompanied by changes in the society, how people live. If pollution results from the economic growth, people may breathe harmful air and drink toxins in their water. If economic growth leads to better education in the country rather than just more consumer goods, a more highly educated society is created which is better able to understand its problems and take appropriate actions to remedy them.

foreign aid (development assistance) **Foreign aid** is "the transfer of state resources through loans, grants, or provision of goods from more to less developed countries, for development or emergency relief purposes. This can be on a bilateral basis [one nation to another nation], or through multilateral bodies [involves more than two nations] such as United Nations agencies, the European Economic Community [European Union] or World Bank. From the 1970s it has increasingly been queried as a means of promoting economic growth." *Source*: "foreign aid," in John Scott and Gordon Marshall, *A Dictionary of Sociology* (Oxford University Press, 2005). Aid given for economic development is called **development assistance**. Aid given to strengthen the recipient's military forces is called **military assistance**. "From the recipient's perspective, foreign aid adds to the resources available for investment and increases the supply of foreign exchange to finance necessary imports. From the donor's point of view, foreign aid is an instrument of foreign policy, and often comes with implicit or explicit expectations of reciprocity in areas where the recipient can be of assistance. Aid packages frequently restrict the recipient, moreover, to purchases from producers in the donor country." *Source*: "foreign aid," in Craig Calhoun (ed.), *Dictionary of the Social Sciences* (Oxford University Press, 2002).

Gaia hypothesis The theory is based on an idea put forward by the British scientist James Lovelock "that the whole earth, including both its . . . living and . . . nonliving components, functions as a single self-regulating system. Named after the Greek earth goddess, it proposes that the responses of living organisms to environment conditions ultimately bring about changes that make the earth better adapted to support life; the system would rid itself of any species that adversely affects the environment. The theory has found favor with many conservationists." *Source*: "Gaia hypothesis," in *A Dictionary of Biology* (Oxford University Press, 2004).

global issues Issues or problems that affect most nations around the world, that cannot be solved by any single nation, and that show our increasing interdependence. Often interdisciplinary knowledge is required to attack these complex problems, which at times can affect the ability of our planet to support life. See also "Introduction: The Creation of Global Issues."

globalization The increase of global economic, political, environmental, and social activities. Expanding international capitalism, mainly through the reach of multinational corporations; the activities of the more important international political organizations, such as the United Nations, World Bank, International Monetary Fund, and World Trade Organization; and growing global communications and social interactions are leading to a more interdependent world. See also chapter 1, "Wealth and Poverty," section "Globalization." The emphasis of **antiglobalization** is that the benefits of the new globalization are unevenly shared, with some parts of the world growing wealthier – such as the US, much of Western Europe, Japan, and sections of China and India – while some nations are not benefiting or being hurt – such as parts of Latin America and Africa. Public protests have been waged over working conditions in developing countries, environmental destruction, social justice, Third World development and debt, and by anti-capitalists who oppose the increasing power of large corporations.

Green Revolution The bringing of Western agricultural technology to the developing world, including highly productive hybrid seeds and the use of fertilizers, pesticides, and irrigation, which has led to vastly increased yields of rice, wheat, and corn in some countries and during some periods. See also chapter 3, "Food," section "The Green Revolution."

Millennium Development Goals Out of the Millennium Declaration adopted by the 189 nations attending the UN's Millennium Summit in 2000 came eight development goals that nations agreed they would focus on in the twenty-first century. The eight goals are (1) eradicate extreme poverty and hunger; (2) achieve universal primary education; (3) promote gender equality and empower women; (4) reduce child mortality; (5) improve maternal health; (6) combat HIV/AIDS, malaria, and other diseases; (7) ensure environmental sustainability; (8) develop a global partnership for development. See also chapter 1, "Wealth and Poverty," section "The Millennium Development Goals."

multinational corporation Business organization which has its headquarters in one country (often in a developed nation) and branches in other countries (often developing countries) where its production facilities are often located because of cheap labor, access to markets and resources, lower taxes, weaker pollution regulations, and bypassing protectionist barriers. It is the main vehicle in spreading globalization.

ozone depletion See chapter 5, "The Environment," subsection "Ozone depletion."

sustainable (sustainable development) Able to be maintained at a certain level indefinitely. A widely accepted definition of sustainable development is contained in *Our Common Future*, the report of the UN's 1987 World Commission on the Environment and Development (the so-called "Brundtland Report") as "development that meets the needs of the present without compromising the ability of future generations to meet their own needs." See also chapter 7, "Alternative Futures," section "Sustainable Development."

tragedy of the commons "This phrase arose from the ancient English custom by which villagers were allowed to graze their animals on common land; thoughtless or greedy people put too many animals on the commons, impoverishing the land and thereby the whole community." *Source*: "tragedy," in Judith Siefring (ed.), *The Oxford Dictionary of Idioms* (Oxford University Press, 2004). See also chapter 6, "Technology," section "Short-term versus Long-Term Benefits – The Tragedy of the Commons."

Bibliography

Adams, Ruth, and Susan Cullen (eds), *The Final Epidemic: Physicians and Scientists on Nuclear War* (Chicago: Educational Foundation for Nuclear Science, 1981).

Adelman, Irma, and Cynthia Taft Morris, *Economic Growth and Social Equity in Developing Countries* (Stanford: Stanford University Press, 1973).

Alley, Richard, "Abrupt Climate Change," *Scientific American*, 291 (November 2004), pp. 62–9.

Alvarez, Lizette, "Norway Leads Industrial Nations Back to Breast-Feeding," *New York Times*, national edn (October 21, 2003).

Annan, Kofi, *We the Peoples: The Role of the United Nations in the 21st Century* (New York: United Nations, 2001).

Arms, Suzanne, *Immaculate Deception: A New Look at Women and Childbirth in America* (Westport, CT: Bergin and Garvey, 1984).

Ashley, Steven, "On the Road to Fuel-Cell Cars," *Scientific American*, 292 (March 2005).

Atran, Scott, "Genesis of Suicide Terrorism," *Science*, 299 (March 7, 2003), pp. 1534–9.

Banerjee, Neela, "Pushing Energy Conservation into the Back Seat of the SUV," *New York Times*, national edn (November 22, 2003), p. B2.

Banerjee, Neela, "The SUV Is Still King, Even as Gas Prices Soar," *New York Times*, national edn (May 4, 2004), pp. C1, C6.

Barboza, David, "Development of Biotech Crops Is Booming in Asia," *New York Times*, national edn (February 21, 2003), p. A3.

Barringer, Felicity, "United States Ranks 28th on Environment, a New Study Says," *New York Times*, national edn (January 23, 2006), p. A3.

Barrionuevo, Alexei, "Mountains of Corn and a Sea of Farm Subsidies," *New York Times*, national edn (November 9, 2005), p. A1.

Baum, Warren C., and Stokes M. Tolbert, *Investing in Development: Lessons of World Bank Experience* (Oxford: Oxford University Press, 1985).

Becker, Elizabeth, "Far from Dead, Subsidies Fuel Big Farms," *New York Times*, national edn (May 14, 2001), p. A12.

Becker, Elizabeth. "Two Acres of Farm Lost to Sprawl Each Minute, New Study Says," *New York Times*, national edn (October 4, 2002), p. A19.

Becker, Elizabeth, "UN Study Finds Global Trade Benefits Are Uneven," *New York Times*, national edn (February 24, 2004), p. C5.

Becker, Elizabeth, "US Sets New Farm-Animal Pollution Curbs," *New York Times*, national edn (December 17, 2002), p. A28.

Becker, Jasper, *Hungry Ghosts: Mao's Secret Famine* (New York: Free Press, 1997).

Belson, Ken, "Why Japan Steps Gingerly in the Middle East," *New York Times*, national edn (September 17, 2002), p. W1.

Bencivenga, Jim, "Scale-Down of US Nuclear Arsenal Will Affect Widespread Industry," *Christian Science Monitor* (November 29, 1991), p. 4.

Bender, William, and Margaret Smith, "Population, Food, and Nutrition," *Population Bulletin*, 51 (Population Reference Bureau, Washington, DC) (February 1997).

Bennet, James, "Trucks' Popularity Undermining Gains in US Fuel Savings," *New York Times*, national edn (September 5, 1995), pp. A1, C3.

Bennhold, Katrin, "Love of Leisure, and Europe's Reasons," *New York Times*, national edn (July 29, 2004), p. A8.

Bernstein, Richard, "No Longer Europe's Sewer, but Not the Rhine of Yore," *New York Times*, national edn (April 21, 2006), p. A4.

Berry, Wendell, "Out of Your Car, Off Your Horse," *Atlantic Monthly*, 267 (February 1991).

Birch, Charles, *Confronting the Future* (New York: Penguin, 1976).

Blair, Bruce G., Harold A. Feiveson, and Frank N. von Hippel, "Taking Nuclear Weapons off Hair-Trigger Alert," *Scientific American*, 277 (November 1997), pp. 74–81.

Bornstein, David, "A Force Now in the World, Citizens Flex Social Muscle," *New York Times*, national edn (July 10, 1999), pp. A15, A17.

Botkin, Daniel B., "A New Balance of Nature," *Wilson Quarterly* (Spring 1991), pp. 61–72.

Bradsher, Keith, "Auto Makers Seek to Avoid Mileage Fines," *New York Times*, national edn (April 3, 1998), p. A14.

Bradsher, Keith, "China Pays a Price for Cheaper Oil," *New York Times*, national edn (June 26, 2004), pp. B1–2.

Broad, William, "Astronomers Revive Scan of the Heavens for Signs of Life," *New York Times*, national edn (September 29, 1998), pp. B11, B13.

Broad, William, "Guarding the Bomb: A Perfect Record but Can It Last?" *New York Times*, national edn (January 29, 1991), pp. B5, B8.

Broad, William, "Maybe We Are Alone in the Universe, After All," *New York Times*, national edn (February 8, 2000), p. D1.

Broad, William, "When World's Collide: A Threat to the Earth Is a Joke No Longer," *New York Times*, national edn (August 1, 1994), p. A1.

Brody, Jane, "Aggressiveness and Delinquency in Boys Is Linked to Lead in Bones," *New York Times*, national edn (February 7, 1996), p. B6.

Brody, Jane, "Another Study Finds a Link Between Excess Weight and Cancer," *New York Times*, national edn (May 6, 2003), p. D7.

Brody, Jane, "Even Low Lead Levels Pose Perils for Children," *New York Times*, national edn (August 5, 2003), p. D7.

Brody, Jane, "How to Save Your Skin in the Season of Sun," *New York Times*, national edn (May 24, 2005), p. D7.

Brody, Jane, "Studies Suggest Meats Carry Resistant Bacteria," *New York Times*, national edn (October 18, 2001), p. A12.

Brook, Edward, "Tiny Bubbles Tell All," *Science*, 310 (November 25, 2005), pp. 1285–7.

Brooke, James, "Brazil Evicting Miners in Amazon to Reclaim Land for the Indians," *New York Times*, national edn (March 8, 1993), p. A4.

Brooke, James, "Food Emergency in North Korea Worsens as Donations Dwindle," *New York Times*, national edn (December 5, 2002), p. A16.

Brooke, James, "In an Almost Untouched Jungle Gold Miners Threaten Indian Ways," *New York Times*, national edn (September 18, 1990), p. B6.

Brooke, James, "North Korea, Facing Food Shortages, Mobilizes Millions from the Cities to Help Rice Farmers," *New York*

Times, national edn (June 1, 2005), p. A6.

Brown, Elizabeth, "Bottle Bills Proliferate in States and in Congress," *Christian Science Monitor* (March 5, 1991), p. 7.

Brown, Lester R., *Building a Sustainable Society* (New York: W. W. Norton, 1981).

Brown, Lester R., *The Twenty-Ninth Day* (New York: W. W. Norton, 1978).

Brown, Lester R., Christopher Flavin, and Sandra Postel, "Picturing a Sustainable Society," in Lester R. Brown et al., *State of the World – 1990* (New York: W. W. Norton, 1990), pp. 173–90.

Buckley, Chris, "Chance to Revive Sales Draws Nuclear Industry to China." *New York Times*, national edn (March 10, 2004), pp. W1, W7.

Buckley, Chris, "China, as Summer Nears, Braces for Power Shortages," *New York Times*, national edn (April 8, 2004), pp. W1, W7.

Burns, John, "West Bengal Finds It Creates Child Beggars for Export," *New York Times*, national edn (March 13, 1997).

Burros, Marian, "High Pesticide Levels Seen in U.S. Food," *New York Times*, national edn (February 19, 1999), p. A12.

Caldwell, John C., and Pat Caldwell, "High Fertility in Sub-Saharan Africa," *Scientific American*, 262 (May 1990), pp. 118–25.

"Cancer: Causes and Prevention," *Scientific American*, 275 (September 1996), pp. 79–101.

Cantrell, Ronald, and Timothy Reeves, "The Cereal of the World's Poor Takes Center Stage," *Science*, 296 (April 5, 2002), p. 53.

"Carbon Dioxide (CO_2) Emissions for Selected Countries, and the Kyoto Pact Target," *Science*, 311 (March 24, 2006), p. 1703.

Carson, Rachel, *Silent Spring* (Greenwich, CT: Fawcett Books, 1962).

Carter, Marion, "Spotlight Brazil," *Population Today* (Population Reference Bureau, Washington, DC) (August 1996), p. 7.

Chameides, William, and Michael Bergin, "Soot Takes Center Stage," *Science*, 297 (September 27, 2002), pp. 2214–15.

Chang, Kenneth, "Scientists Say Recovery of the Ozone Layer May Take Longer than Expected," *New York Times*, national edn (December 7, 2005), p. A25.

"Chinese Official Sees Private Role on Environment," *New York Times*, national edn (June 6, 2006), p. A6.

Chowdhury, A., and A. Mushtaque, "Arsenic Crisis in Bangladesh," *Scientific American*, 291 (August 2004), pp. 86–91.

Clemens, Walter C., Jr, "Baltics Sang Their Way to Independence," *Christian Science Monitor* (September 5, 1991), p. 19.

Cohen, Joel E., *How Many People Can the Earth Support?* (New York: W. W. Norton, 1995).

Cohen, Joel, "Human Population Grows Up," *Scientific American*, 293 (September 2005), pp. 48–55.

Cohen, Marc, "Crop Circles," *Natural History*, 112 (October 2003), pp. 58–64.

Coleman, Daniel, "Amid Ethnic Wars, Psychiatrists Seek Roots of Conflict," *New York Times*, late edn (August 2, 1994), pp. C1, C13.

Commoner, Barry, *The Closing Circle* (New York: Alfred A. Knopf. 1971).

Conquest, Robert, *The Harvest of Sorrow: Soviet Collectivization and the Terror-Famine* (New York: Oxford University Press, 1986).

Council on Environmental Quality, and the Department of State, *Global Future: Time to Act* (Washington, DC: Government Printing Office, 1981).

Council on Environmental Quality, and the Department of State, *The Global 2000 Report to the President: Entering the Twenty-First Century*, vol. 1 (Washington, DC: Government Printing Office, 1980; New York: Penguin Books, 1982).

Cowell, Alan, "UN Population Meeting Adopts Program of Action," *New York Times*, national edn (September 14, 1994), p. A2.

Critchfield, Richard, *Villages* (Garden City, NY: Anchor Press/Doubleday, 1981).

"Crop Scientists Seek a New Revolution," *Science*, 283 (January 15, 1999), p. 310.

Crossette, Barbara, "UN Report Raises Questions about Small Loans to the Poor," *New York Times*, national edn (September 3, 1998), p. A8.

Cushman, John, "Michigan Balks at Tainted-Salmon Warning," *New York Times*, national edn (February 8, 1997), p. 7.

Cushman, John, "Pollution Control Plan Views Factory Farms as Factories," *New York Times*, national edn (March 6, 1998), p. Al.

Cushman, John, "Poultry Growers Unite to Address Waste Issue," *New York Times*, national edn (August 25, 1998), p. A14.

Dalai, Lama, "Foreword," in Rimpoche Gehlek, *Good Life, Good Death* (New York: Riverhead Books, 2001).

Danner, Mark, "Taking Stock of the Forever War," *New York Times Magazine* (September 11, 2005), pp. 46–87.

Dao, James, "Protesters Interrupt Powell Speech as UN Talks End," *New York Times*, national edn (September 5, 2002), A8.

Darmstadter, Joel, "Economic Growth and Energy Conservation: Historical and International Lessons," Reprint No. 154, (Resources for the Future, Washington, DC, 1978).

Davis-Floyd, Robbie E., *Birth as an American Rite of Passage* (Berkeley: University of California Press, 1992).

Dean, Cornelia, "Scientists Warn Fewer Kinds of Fish Are Swimming the Oceans," *New York Times*, national edn (July 29, 2005), p. A6.

"Defense When Money Is No Object," *Bulletin of the Atomic Scientists*, 57 (September/October 2001), pp. 36–7.

De Souza, Roger-Mark, John Williams, and Frederick Meyerson, "Critical Links: Population, Health and the Environment," *Population Bulletin*, 58 (Population Reference Bureau, Washington, DC) (September 2003).

Deutsch, Claudia, "Dupont Looking to Displace Fossil Fuels as Building Blocks of Chemicals," *New York Times*, national edn (February 28, 2006), p. C1.

Diamond, Jared, *Collapse: How Societies Choose to Fail Or Succeed* (New York: Viking, 2005).

Doyle, Rodger, "Measuring Modernity," *Scientific American*, 289 (December 2003), p. 40.

Doyle, Rodger, "Melting at the Top," *Scientific American*, 292 (February 2005), p. 31.

Dreifus, Claudia, "Tracing Evolution of Cosmos from its Simplest Elements," *New York Times*, national edn (April 28, 1998), p. B15.

Dreifus, Claudia, "With Findings on Storms, Centrist Recasts Warming Debate," *New York Times*, national edn (January 10, 2006), p. D2.

Drucker, Peter, "Trade Lessons from the World Economy," *Foreign Affairs*, 73 (January/February 1994).

Dugger, Celia, "Debate Stirs over Tiny Loans for World's Poorest," *New York Times*, national edn (April 29, 2004), p. A1.

Dugger, Celia, "Report Warns Malnutrition Begins in Cradle," *New York Times*, national edn (March 3, 2006), p. A6.

Dunlap, Thomas, *DDT: Scientists, Citizens, and Public Policy* (Princeton, NJ: Princeton University Press, 1981).

Durning, Alan B., "Mobilizing at the Grassroots," in Lester R. Brown et al., *State of the World – 1989* (New York: W. W. Norton, 1989), pp. 154–73.

Durning, Alan B., "Supporting Indigenous Peoples," in Lester Brown et al., *State of the World – 1993* (New York: W. W. Norton, 1993), pp. 80–100.

Dyson, Freeman, *Weapons and Hope* (New York: Harper & Row, 1984).

"Ecology: The New Great Chain of Being," *Natural History*, 77 (December 1968), p. 8.

Ehrlich, Paul, *The Population Bomb* (New York: Ballantine Books), 1971.

Ehrlich, Paul R., and Anne H. Ehrlich, *Extinction* (New York: Random House, 1981).

Ehrlich, Paul R., and Anne H. Ehrlich, *The Population Explosion* (New York: Simon & Schuster, 1990).

Ehrlich, Paul, Anne Ehrlich, and John Holdren, *Ecoscience: Population, Resources, Environment* (San Francisco: W. H. Freeman, 1977).

Eicher, Carl K., "Facing Up to Africa's Food Crisis," *Foreign Affairs*, 61 (Fall 1982), pp. 151–74.

Evenson, R. E., and D. Gollin, "Assessing the Impact of the Green Revolution, 1960 to 2000," *Science*, 300 (2 May 2003), pp. 758–62.

Falkenmark, Malin, and Carl Widstrand, "Population and Water Resources: A Delicate Balance," *Population Bulletin*, 47 (Population Reference Bureau, Washington, DC) (November 1992), p. 23.

Farb, Peter, and George Armelagos, *Consuming Passions: The Anthropology of Eating* (Boston: Houghton Mifflin, 1980).

Farvar, M. Taghi, and John P. Milton (eds), *The Careless Technology: Ecology and International Development* (Garden City, NY: Natural History Press, 1972).

Ferber, Dan, "Livestock Feed Ban Preserves Drugs' Power," *Science*, 295 (January 4, 2002), pp. 27–8.

Fischetti, Mark, "Turn Turn Turn," *Scientific American*, 287 (July 2002), p. 86.

Flynn, Stephen, "America the Vulnerable," *Foreign Affairs* (January/February 2002), pp. 60–74.

Foderao, Lisa, "Caution Urged in Eating of Fish from Mountains of New York," *New York Times*, national edn (April 16, 2005), p. A11.

Freedman, Ronald, "Theories of Fertility Decline: A Reappraisal," in Philip M. Hauser (ed.), *World Population and Development* (Syracuse, NY: Syracuse University Press, 1979).

French, Howard, "As China Races to Build Nuclear Power Plants, Some Experts Still See Danger," *Spartanburg Herald-Journal* (January 16, 2005), p. A16.

French, Howard, "Billions of Trees Planted, and Nary a Dent in the Desert," *New York Times*, national edn (April 11, 2004), p. 3.

French, Howard, "China's Boom Brings Fear of an Electricity Breakdown," *New York Times*, national edn (July 5, 2004), p. A4.

French, Howard, "In Search of a New Energy Source, China Rides the Wind," *New York Times*, national edn (July 26, 2005), p. A4.

French, Howard, "Safety Problems at Japanese Reactors Begin to Erode Public's Faith in Nuclear Power," *New York Times*, national edn (September 16, 2002), A8.

Gardner, Gary, "Air Pollution Still a Problem," in *Vital Sign 2005* (New York: Worldwatch Institute, 2005), p. 94.

Gargan, Edward, "On Meanest of Streets, Salvaging Useful Lives," *New York Times*, national edn (January 8, 1992), p. A2.

Gibbons, John, H., Peter D. Blair, and Holly L. Gwin, "Strategies for Energy Use," *Scientific American*, 261 (September 1989), pp. 136–43.

Gibbs, Walter, "Fish Farm Escapees Threaten Wild Salmon," *New York Times*, national edn (October 1, 1996), p. B7.

Gibbs, W. Wayt, "How Should We Set Priorities?" *Scientific American*, 293 (September 2005), pp. 108–15.

Gibbs, W. Wayt, "On the Termination of Species," *Scientific American*, 285 (November 2001), pp. 40–9.

Gibbs, W. Wayt, "Saving Dying Languages," *Scientific American*, 287 (August 2002), pp. 79–85.

Glantz, Michael, "Drought in Africa," *Scientific American*, 256 (June 1987), pp. 34–40.

Gleick, Peter, "Safeguarding Our Water," *Scientific American*, 284 (February 2001), pp. 38–45.

"Global Prevalence of Obesity," *International Obesity Task Force*, updated March 16, 2005, at http://www.iotf.org/media/globalprev.htm.

Goleman, Daniel, "Shammans and Their Longtime Lore May Vanish with the Forests," *New York Times*, national edn (June 11, 1991), B5.

Goode, Erica, "The Gorge-Yourself Environment," *New York Times*, national edn (July 22, 2003), pp. D1, D7.

Goodell, Jeff, "Cooking the Climate with Coal," *Natural History*, 115 (May 2006), pp. 36–41.

Gordon, Michael, "Maneuvers Show Russian Reliance on Nuclear Arms," *New York Times*, national edn (July 10, 1999), p. A1.

Grady, Denise, "A Move to Limit Antibiotic Use in Animal Feed," *New York Times*, national edn (March 8, 1999), p. A1.

Grady, Denise, "WHO Finds Use of Antibiotics in Animal Feed Can Be Reduced," *New York Times*, national edn (August 14, 2003), p. A5.

Greer, William, "Public Taste and U.S. Aid Spur Fish Farming," *New York Times*, national edn (October 29, 1986), p. 1.

Guerinot, Mary Lou, "The Green Revolution Strikes Gold," *Science*, 287 (January 14, 2000), pp. 241, 243.

Hakim, Danny, "EPA Holds Back Report on Car Fuel Efficiency," *New York Times*, national edn (July 28, 2005), pp. C1, C8.

Hakim, Danny, "Fuel Economy Hit 22-Year Low in 2002." *New York Times*, national edn (May 3, 2003), p. B1.

Halweil, Brian, "Grain Harvest and Hunger Both Grow," in *Vital Signs 2005* (New York: Worldwatch Institute, 2005), pp. 22–3.

Hansen, J., et al., "Climate Impact of Increasing Atmospheric Carbon Dioxide," *Science*, 213 (August 28, 1981), p. 966.

Hardin, Garrett, "Living on a Lifeboat," *Bioscience*, 24 (October 1974), pp. 561–8.

Hardin, Garrett, "The Tragedy of the Commons," *Science*, 162 (December 13, 1968), pp. 1243–8.

Hausmann, Ricardo, "Prisoners of Geography," *Foreign Affairs*, (January/February 2001), pp. 45–53.

Hegeri, Gabriele and Nathaniel Bindoff, "Warming the World's Oceans," *Science*, 309 (July 8, 2005), pp. 254–5.

Heilprin, John, "Pollution Eases, but Lead Worries EPA," *Spartanburg Herald-Journal* (July 1, 2003), p. A3.

"The Hidden Cost of Gas," *Sierra*, 87 (March/April 2002), p. 15.

Hilts, Philip, "Results of Study on Pesticide Encourage Effort to Cut Use," *New York Times*, national edn (July 5, 1993), p. 8.

Hopkins, Raymond, Robert Paarlberg, and Michael Wallerstein, *Food in the Global Arena* (New York: Holt, Rinehart, & Winston, 1982).

Houghton, J. T., G. J. Jenkins, and J. J. Ephraums (eds), *Climate Change: The IPCC Scientific Assessment* (Cambridge: Cambridge University Press, 1990).

Hubbard, Harold M., "The Real Cost of Energy," *Scientific American*, 264 (April 1991), pp. 36–42.

Hudson, Peter and Ottar Bjornstad, "Vole Stranglers and Lemming Cycles," *Science*, 302 (October 31, 2003), p. 797.

Hughes, T. P., et al., "Climate Change, Human Impacts, and the Resilience of Coral Reefs," *Science*, 301 (August 15, 2003), pp. 929–33.

Ibrahim, Youssef, "Gulf War's Cost to the Arabs Estimated at $620 Billion," *New York Times*, national edn (September 8, 1992), p. A4.

Illich, Ivan, "Outwitting the 'Developed' Countries," in Charles K. Wilbur (ed.), *The Political Economy of Development and Underdevelopment*, 2nd edn (New York: Random House, 1979), pp. 436–44.

Janofsky, Michael, "Environmental Groups Are Praising EPA for Updating Cancer-Risk Guidelines," *New York Times*, national edn (April 4, 2005), p. A18.

Janofsky, Michael, "Mercury Taints Fish across US," *New York Times*, national edn (August 25, 2004), p. A19.

Johnson, Dirk, "Growth of Factory-Like Hog Farms Divide Rural Areas of the Midwest," *New York Times*, national edn (June 24, 1998), p. A12.

Johnson, Nels, et al., "Managing Water for People and Nature," *Science*, 292 (May 11, 2001), p. 1071.

Joughin, Ian, "Greenland Rumbles Louder as Glaciers Accelerate," *Science*, 311 (March 24, 2006), pp. 1719–20.

Kaiser, Jocelyn, "Mounting Evidence Indicts Fine-Particle Pollution," *Science*, 307 (March 25, 2005), pp. 1858–61.

Kaiser, Jocelyn, "Wounding Earth's Fragile Skin," *Science*, 304 (June 11, 2004), pp. 1616–18.

Kaliish, Susan, "Culturally Sensitive Family Planning: Bangladesh Story Suggests It Can Reduce Family Size," *Population Today*, 22 (Population Reference Bureau, Washington, DC) (February 1994), p. 5.

Kaplan, Robert D., *The Ends of the Earth: A Journey at the Dawn of the 21st Century* (New York: Random House, 1996).

Karoly, David, "Ozone and Climate Change," *Science*, 302 (October 10, 2003), pp. 236–7.

Kennedy, Paul, *Preparing for the Twenty-First Century* (New York: Random House, 1993).

Kent, Mary M., "Shrinking Societies Favor Procreation," *Population Today*, 27 (December 1999), pp. 4–5.

Kerr, Richard, "A Little Respect for the Asteroid Threat," *Science*, 297 (September 13, 2002), pp. 1785–7.

Kerr, Richard, "A Worrying Trend of Less Ice, Higher Seas," *Science*, 311 (March 24, 2006), pp. 1698–701.

Kerr, Richard, "Bumpy Road Ahead for World's Oil," *Science*, 310 (November 18, 2005), p. 1106.

Kerr, Richard, "Is Katrina a Harbinger of Still More Powerful Hurricanes," *Science*, 309 (September 16, 2005), p. 1807.

Kerr, Richard, "Millennium's Hottest Decade Retains Its Title for Now," *Science*, 307 (February 11, 2005), pp. 828–9.

Kerr, Richard A., "Ozone Loss, Greenhouse Gases Linked," *Science*, 280 (April 10, 1998), p. 202.

Kerr, Richard, "Three Degrees of Consensus," *Science*, 305 (August 13, 2004), pp. 932–4.

Kerr, Richard, and Robert Service, "What Can Replace Cheap Oil – and When?" *Science*, 309 (July 1, 2005), p. 101.

Kinsella, Kevin, and David Phillips, "Global Aging: The Challenge of Success," *Population Bulletin*, 60 (Population Reference Bureau, Washington, DC) (March 2005), pp. 1–40.

Klaidman, Stephen, "Muddling Through," *Wilson Quarterly* (Spring 1991).

Kormondy Edward, *Concepts of Ecology*, 2nd edn (Englewood Cliffs, NJ: Prentice-Hall, 1976).

Krajick, Kevin, "Long-Term Data Show Lingering Effects from Acid Rain," *Science*, 292 (April 13, 2001), pp. 195–6.

LaBastille, Anne, "Heaven, Not Hell," [development in the Amazon Basin] *Audubon*, 81 (November 1979), pp. 68–103.

Lacey, Marc, "Belatedly, Africa Is Converting to Lead-Free Gasoline," *New York Times*, national edn (October 31, 2004).

Lake, James, et al., "Next-Generation Nuclear Power," *Scientific American*, 286 (January 2002), pp. 73–81.

Landsberg, Hans, et al., "Nonfuel Minerals," in Paul Portney (ed.), *Current Issues in Natural Resources Policy* (Washington, DC: Resources for the Future, 1982), p. 83.

Laurance, William F., "Gaia's Lungs: Are Rainforests Inhaling Earth's Excess Carbon Dioxide?" *Natural History*, 108 (March 1999), p. 96.

Laux, E. Vernon, "In a Vast Hungry Wave, Owls Are Moving South," *New York Times*, national edn (March 8, 2005), p. D2.

Lawton, John, "Earth System Science," *Science*, 292 (June 15, 2001), p. 1965.

"Lead Concentrations Down in Greenland Ice," *New York Times*, national edn (October 15, 1991), p. B8.

Leary, Elizabeth, "As Contraceptive Use Rises, Abortions Decline," *Popline*, 25 (Population Institute, Washington, DC) (November/December 2003), pp. 3, 8.

Lee, Jennifer, "EPA Orders Companies to Examine Effects of Chemicals," *New York Times*, national edn (April 15, 2003), p. D2.

Lee, Jennifer, "Report Says Plan to Safeguard Nuclear Material Is Lacking," *New York Times*, national edn (March 13, 2003), p. A13.

Lersner, Heinrich von, "Commentary: Outline for an Ecological Economy," *Scientific American*, 273 (September 1995), p. 188.

Linden, Eugene, "The Exploding Cities of the Developing World," *Foreign Affairs*, 75 (January/February 1996), pp. 52–65.

Livernash, Robert, and Eric Rodenburg, "Population Change, Resources, and the Environment," *Population Bulletin*, 53 (Population Reference Bureau, Washington, DC) (March 1998), pp. 1–40.

Loewenberg, Samuel, "Old Europe's New Ideas," *Sierra*, 89 (January/February 2004), pp. 40–50.

Lovins, Amory, "More Profit with Less Carbon," *Scientific American*, 293 (September 2005), pp. 74–83.

Luoma, Jon, "The $33 Billion Misunderstanding," *Audubon*, 83 (November 1981), pp. 111–27.

Luoma, Jon, "Sharp Decline Found in Arctic Air Pollution," *New York Times*, national edn (June 1, 1993), p. B7.

Luoma, Jon, "Trash Can Realities," *Audubon*, 92 (March 1990), pp. 86–97.

Lyall, Sarah, "Britons Running out of Patience as Costly Gas Runs Out," *New York Times*, national edn (September 12, 2000).

Markoff, John, "Technology's Toxic Trash Is Sent to Poor Nations," *New York Times*, national edn (February 25, 2002), pp. C1, C4.

Marsh, Bill, "The Evolving World Diet," *New York Times*, national edn (August 20, 2002), p. D4.

Marsh, Bill, "One Receipt for a (Mostly) Emissions-Free Economy," *New York Times*, national edn (November 4, 2003), p. D3.

Marybury-Lewis, David, *Millennium: Tribal Wisdom and the Modern World* (New York: Viking, 1992).

McFalls, Joseph A., Jr, "Population: A Lively Introduction," *Population Bulletin*, 53 (Population Reference Bureau, Washington, DC) (September 1998).

McIntyre, Loren, "Jari: A Billion Dollar Gamble," *National Geographic* (May 1980), p. 701.

McKibben, Bill, "A Special Moment in History," *Atlantic Monthly*, 281 (May 1998), pp. 55–78.

McManus, Reed, "Out Front in the Air Wars: California Struggles against Air Pollution, Asthma, and the White House," *Sierra*, 89 (January/February 2004), p. 12.

McManus, Tran, "Imagine a City with 30 Percent Fewer Cars," *Sierra*, 91 (January/February 2006), pp. 48–9.

McNeil, Donald, "Protests on New Genes and Seeds Grow More Passionate in Europe," *New York Times*, national edn (March 14, 2000), p. A1.

Meadows, Donella, et al., *The Limits to Growth*, 2nd edn (New York: Universe Books, 1974).

Meadows, Donella H., Dennis L. Meadows, and Jorgen Randers, *Beyond the Limits: Confronting Global Collapse, Envisioning a Sustainable Future* (Post Mills, VT: Chelsea Green, 1992).

Meller, Paul, "Europe and US Will Share Research on Hydrogen Fuel," *New York Times*, national edn (June 17, 2003), p. W1.

Mellor, John W., and Sarah Gavian, "Famine: Causes, Prevention, and Relief," *Science*, 235 (January 1987), pp. 539–45.

"Mexican Men Get the Message about Limiting Family Size," *Christian Science Monitor* (July 8, 1992), p. 11.

Miller, G. Tyler, Jr, *Living in the Environment: Principles, Connections, and Solutions*, 8th edn (Belmont, CA: Wadsworth, 1994).

Milton, Katherine, "Civilization and Its Discontents," *Natural History* (March 1992), pp. 37–43.

Moore, Barrington, Jr, *Social Origins of Dictatorship and Democracy: Lord and Peasant in the Making of the Modern World* (Boston: Beacon Press, 1966).

Moore, Charles, "Trashed: Across the Pacific Ocean, Plastics, Plastics, Everywhere," *Natural History*, 112 (November 2003), pp. 46–51.

Motavalli, Jim, "All Abroad: Good Public Transportation Need Not Be High Tech or High Cost," *Sierra*, 87 (January/February 2002), pp. 46–8.

Myers, Norman, "A Convincing Call for Conservation," *Science*, 295 (18 January 2002), pp. 447–8.

Myers, Norman, "The Exhausted Earth," *Foreign Policy*, 42 (Spring 1981).

Myers, Norman, "Room in the Ark," *Bulletin of the Atomic Scientists*, 38 (November 1982).

Nachtwey, James, "Romania's Lost Children," *New York Times Magazine* (June 24, 1990), pp. 28–33.

National Research Council, *Alternative Agriculture* (Washington: National Academy Press, 1989).

National Research Council, *Population Growth and Economic Development: Policy Questions* (Washington, DC: National Academy Press, 1986).

Neimark, Jill, "Using Flows and Fluxes to Demythologize the Unity of Life," *New York Times*, national edn (August 11, 1998), p. B8.

Noble, Kenneth, "US for Decades Let Uranium Leak at Weapons Plant," *New York Times*, national edn (October 15, 1988), p. 1.

Nordstrom, Kirk, "Worldwide Occurrences of Arsenic in Ground Water," *Science*, 296 (June 21, 2002), pp. 2143–6.

Normile, Dennis, "Variety Spices up Chinese Rice Yields," *Science*, 289 (August 18, 2000), pp. 1122–3.

Normile, Dennis, and Elizabeth Pennisi, "Rice: Boiled Down to Bare Essentials," *Science*, 296 (April 5, 2002), pp. 32–6.

"Nuclear Sites May Be Toxic in Perpetuity, Report Finds," *New York Times*, national edn (August 8, 2000), p. A12.

"Nuclear War: The Aftermath," *Ambio: A Journal of the Human Environment*, 11/2–3 (Royal Swedish Academy of Sciences and Pergamon Press) (1982), pp. 76–176.

O'Connor, Anahad, "Study Details 30-Year Increase in Calorie Consumption," *New York Times*, national edn (February 6, 2004), p. A19.

Onishi, Norimitsu, "How Do Japanese Dump Trash? Let Us Count the Myriad Ways," *New York Times*, national edn (May 12, 2005), p. A1, A4.

Ophuls, William, *Ecology and the Politics of Scarcity* (San Francisco: W. H. Freeman, 1977).

Paddock, William, and Paul Paddock, *Famine-1975!* (Boston: Little, Brown, 1967).

Parikh, Lina, "Spotlight: Bangladesh," *Population Today* (January 26, 1998), p. 7.

Pauly, Daniel, and Reg Watson, "Counting the Last Fish," *Scientific American*, 289 (July 2003), pp. 42–7.

Pearson, Frederic S., and J. Martin Rochester, *International Relations: The Global Condition in the Late Twentieth Century*, 4th edn (New York: Random House, 1998).

"Peddling Lead," *Science*, 299 (February 7, 2003), p. 795.

Perlez, Jane, "Forests in Southeast Asia Fall to Prosperity's Axe," *New York Times*, national edn (April 29, 2006), pp. A1, A7.

Pillay, Kavita, "The Politics of Famine in North Korea," *Ninth Annual Report on the State of World Hunger* (Silver Springs, MD: Bread for the World Institute, 1998).

Pimentel, David, "Changing Genes to Feed the World," *Science*, 306 (October 29, 2004), p. 815.

Pimm, Stuart and Clinton Jenkins, "Sustaining the Variety of Life," *Scientific American*, 293 (September 2005), pp. 66–73.

Pogrebin, Robin, "Now the Working Class, Too, Is Foraging for Empty Cans," *New York Times*, national edn (April 29, 1996), p. B12.

Pohl, Otto, "European Environmental Rules Propel Change in US," *New York Times*, national edn (July 6, 2004), p. D4.

Pollack, Andrew, "No Foolproof Way Is Seen to Contain Altered Genes," *New York Times*, national edn (January 21, 2004), p. A10.

Population Reference Bureau, "Transitions in World Population," *Population Bulletin*, 59 (March 2004), pp. 1–39.

Population Reference Bureau, *2005 World Population Data Sheet* (Washington, DC: Population Reference Bureau, 2005).

Potts, Malcolm, "The Unmet Need for Family Planning," *Scientific American*, 282 (January 2000), pp. 88–93.

"Poultry Growers Unite to Address Waste Issue," *New York Times*, national edn (August 25, 1998), p. A14.

Presidential Commission on World Hunger, *Overcoming World Hunger: The Challenge*

Ahead (Washington, DC: US Government Printing Office, 1980).

Prosterman, Roy, "The Decline in Hunger-Related Deaths," Hunger Project Papers No. 1, The Hunger Project, San Francisco, 1984.

Pyne, Solana, "Small Particles Add Up to Big Disease Risk," *Science*, 295 (March 15, 2002), p. 1994.

Quadfasel, Detlef, "The Atlantic Heat Conveyor Slows," *Nature*, 438 (December 1, 2005), pp. 565–6.

Rai, Saritha, "Tiny Loans Have Big Impact on the Poor," *New York Times*, national edn (April 12, 2004), p. C3.

Raven, Peter, "Tropical Rain Forests: A Global Responsibility," *Natural History*, 90 (February 1981).

Rabinovitch, Jonas, and Josef Leitman, "Urban Planning in Curitiba," *Scientific American*, 274 (March 1996), pp. 46–53.

"Reduce, Reuse, Rejoice: Heaps of Garbage Become Piles of Possibilities," *Sierra*, 90 (November/December 2005), pp. 42–3.

Reganold, John P., Robert I. Papendick, and James F. Parr, "Sustainable Agriculture," *Scientific American*, 262 (June 1990), pp. 112–20.

Renner, Michael, "Violent Conflicts Unchanged," in Worldwatch Institute, *Vital Signs 2005* (New York: W. W. Norton, 2005), pp. 74–5.

Renner, Rebecca, "Tracking the Dirty Byproducts of a World Trying to Stay Clean," *Science*, 306 (December 10, 2004), p. 1887.

Renner, Rebecca, "Scotchgard Scotched," *Scientific American*, 284 (March 2001), p. 18.

Repetto, Robert, "Accounting for Environmental Assets," *Scientific American*, 266 (June 1992), pp. 94–100.

"Reseeding the Green Revolution," *Science*, 277 (August 22, 1997), p. 1038.

"Resurgent Forests Can Be Greenhouse Gas Sponges," *Science*, 277 (July 18, 1977), p. 316.

Revkin, Andrew, "China Is Bright Spot in Dark Report on the World's Diminishing Forests," *New York Times*, national edn (November 15, 2005), p. D3.

Revkin, Andrew, "Climate Data Hint at Irreversible Rise in Seas," *New York Times*, national edn (March 24, 2006), p. A12.

Revkin, Andrew, "FDA Considers New Tests for Environmental Effects," *New York Times*, national edn (March 14, 2002), p. A20.

Revkin, Andrew, "Panel Urges US to Tighten Approval of Gene-Altered Crops," *New York Times*, national edn (February 22, 2002), p. A18.

Revkin, Andrew, "Plastics Permeate Even the Seabed," *New York Times*, national edn (May 11, 2004), p. D2.

Revkin, Andrew, "Report Forecasts Warming Effects," *New York Times*, national edn (June 12, 2000), pp. A1, A25.

Revkin, Andrew, "Scientists Say Slower Atlantic Currents Could Mean a Colder Europe," *New York Times*, national edn (December 1, 2005), p. A8.

Revkin, Andrew, "Stream Tests Show Traces of Array of Contaminants," *New York Times*, national edn (March 13, 2002), p. A14.

Riding, Alan, "A Common Culture (from the USA) Binds Europeans Ever Closer," *New York Times*, national edn (April 26, 2004), pp. B1, B5.

Rifkin, Jeremy, *Entropy: A New World View* (New York: Viking Press, 1980).

Riley, Nancy, "China's Population: New Trends and Challenges," *Population Bulletin*, 59 (Population Reference Bureau, Washington, DC) (June 2004), pp. 1–35.

Riley, Nancy E., "Gender, Power, and Population Change," *Population Bulletin*, 52 (Population Reference Bureau, Washington, DC) (May 1997).

Robey, Bryant, Shea O. Rutstein, and Leo Morris, "The Fertility Decline in Developing Countries," *Scientific American*, 269 (December 1993), pp. 60–7.

Rohde, David, "Bangladesh Wells Pumping Poison as Cleanup Lags," *New York Times*, national edn (July 17, 2005), p. 6.

Rohter, Larry, "A Mirage of Amazonian Size," *New York Times*, national edn (November 9, 1999), pp. C1, C2.

Rohter, Larry, "A New Intrusion Threatens a Tribe in Amazon: Soldiers," *New York Times*, national edn (October 1, 2002), p. A1.

Rohter, Larry, "Relentless Foe of the Amazon Jungle: Soybeans," *New York Times*, national edn (September 17, 2003), p. A3.

Rohter, Larry, "South America Seeks to Fill the World's Table," *New York Times*, national edn (December 12, 2004), p. 1.

Rohter, Larry, 'With Big Boost from Sugar Cane, Brazil Is Satisfying Its Fuel Needs," *New York Times*, national edn (April 10, 2006), pp. A1, A10.

Romero, Simon, "Demand for Natural Gas Brings Big Import Plans, and Objections," *New York Times*, national edn (June 15, 2005), p. C8.

Romero, Simon, "The $6.66-a-Gallon Solution," *New York Times*, national edn (April 30, 2005), pp. B1, B10.

Romm, Joseph, "Needed – A No-Regrets Energy Policy," *Bulletin of the Atomic Scientists*, 47 (July/August 1991), pp. 31–6.

Ross, Marc H., and Daniel Steinmeyer, "Energy for Industry," *Scientific American*, 263 (September 1990), pp. 88–98.

Rosenthal, Elisabeth, "Rural Flouting of One-Child Policy Undercuts China's Census," *New York Times*, national edn (April 14, 2000), p. A10.

Rothschild, Emma, *Paradise Lost: The Decline of the Auto-Industrial Age* (New York: Random House, 1973).

Rourke, John, *International Politics on the World Stage*, 7th edn (New York: Dushkin/McGraw-Hill, 1999).

Royston, Michael, "Making Pollution Prevention Pay," *Harvard Business Review* (November/December 1980), pp. 6–14.

Russett, Bruce, and Harvey Starr, *World Politics: The Menu of Choice*, 2nd edn (New York: W. H. Freeman, 1985), Ch. 16.

"Russian Nuclear Forces, 2004," *Bulletin of the Atomic Scientists*, 60 (July/August 2004), p. 72.

Sachs, Jeffrey, "Can Extreme Poverty Be Eliminated?" *Scientific American*, 293 (September 2005), pp. 56–65.

Sachs, Jeffrey, Andrew Mellinger, and John Gallup, "The Geography of Poverty and Wealth," *Scientific American* 284 (March 2001), pp. 70–5.

Sanchez, Pedro, and M. S. Swaminathan, "Cutting World Hunger in Half," *Science*, 307 (January 21, 2005), pp. 357–9.

Sanger, David, "Japan Joins in Embargo against Iraq," *New York Times*, national edn (August 6, 1990), p. C7.

Schell, Jonathan, *The Fate of the Earth* (New York: Avon Books, 1982).

Scheper-Hughes, Nancy, "Death without Weeping," *Natural History*, 98 (October 1989), pp. 8–16.

Scheper-Hughes, Nancy, and Daniel Hoffman, "Brazil: Moving Targets," *Current History*, 106 (July/August 1997), pp. 34–43.

Schmitt, Eric, "Bush Ordering Missile Shields at Sites in West," *New York Times*, national edn (December 18, 2002), p. A1.

Schneider, Keith, "Candor on Nuclear Peril," *New York Times*, national edn (October 14, 1988), p. 1.

Schneider, Keith, "Wide Threat Seen in Contamination at Nuclear Units," *New York Times*, national edn (December 7, 1988), p. 1.

Schumacher, E. F., *Small Is Beautiful: Economics as if People Mattered* (New York: Harper & Row, 1973).

Scott, Bruce, "The Great Divide in the Global Village," *Foreign Affairs* (January/February 2001), pp. 162–3.

Seelye, Katharine, and Jennifer Lee, "EPA Calls US Cleaner and Greener than 30 Years Ago," *New York Times*, national edn (June 24, 2003), p. A28.

Seitz, John L. "The Failure of US Technical Assistance in Public Administration: The Iranian Case," in Eric Otenyo and Nance Lind (eds), *Comparative Public Administration: The Essential Readings* (Oxford: Elsevier, 2006), pp. 321–34.

Service, Robert, "Is It Time to Shoot for the Sun?" *Science*, 309 (July 22, 2005), pp. 548–51.

"Sex Trade Is Enslaving Millions of Women, Youth," *Popline* (Population Institute,

Washington, DC) (November/December 2003), pp. 1, 6.
"Signs of a Changing Planet," *Sierra*, 90 and 91 (July/August 2005; March/April 2006), pp. 17 and 13.
Simon, Julian L., *Population Matters: People, Resources, Environment, and Immigration* (New Brunswick, NJ: Transaction, 1990).
Simon, Julian, *The Ultimate Resource* (Princeton, NJ: Princeton University Press, 1981).
Simons, Marlise, "Boats Plunder Mediterranean with Outlawed Nets," *New York Times*, national edn (June 4, 1998), p. A3.
Simons, Marlise, "Rising Iron Curtain Exposes Haunting Veil of Polluted Air," *New York Times*, national edn (April 8, 1990), p. 1.
Simons, Marlise, "Wind Turbines Are Sprouting off Europe's Shores," *New York Times*, national edn (December 8, 2002), p. 3.
Sivard, Ruth, *World Military and Social Expenditures 1993*, 15th edn (Washington, DC: World Priorities, 1993).
Sivard, Ruth, *World Military and Social Expenditures 1996*, 16th edn (Washington, DC: World Priorities, 1996).
Skidmore, Thomas, and Peter Smith, *Modern Latin America*, 2nd edn (New York: Oxford University Press, 1989).
Small, Meredith F., "Our Babies, Ourselves," *Natural History*, 106 (October 1997), pp. 42–51.
Smil, Vaclav, *Feeding the World: A Challenge for the Twenty-First Century* (Cambridge, MA: MIT Press, 2000).
Solomon, Stephen, "The Controversy over Infant Formula," *New York Times Magazine* (December 6, 1981), p. 100.
Sprout, Harold, and Margaret Sprout, *The Context of Environmental Politics* (Lexington: University Press of Kentucky, 1978).
Stevens, William, "Enormous Haze Found over Indian Ocean," *New York Times*, national edn (June 10, 1999), p. A23.
Stevens, William, "Expectations Aside, Water Use in US Is Showing Decline," *New York Times*, national edn (October 10, 1998), p. A1.
Stevens, William, "The 'Hot Spot' Approach to Saving Species," *New York Times*, national edn (March 14, 2000), p. D3.
Stevens, William, "Lost Rivets and Threads, and Ecosystems Pulled Apart," *New York Times*, national edn (July 4, 2000), p. D4.
Stevens, William, "New Eye on Nature: The Real Constant Is Eternal Turmoil," *New York Times*, national edn (July 31, 1990), pp. B5–6.
Stevens, William "New Survey Shows Growing Loss of Arctic Atmosphere's Ozone," *New York Times*, national edn (April 6, 2000), p. A19.
Stevens, William, "Ozone Loss over US Is Found To Be Twice as Bad as Predicated," *New York Times*, national edn (April 5, 1991), p. A1.
Stevens, William, "Sprawl Quickens Its Attack on Forests," *New York Times*, national edn (December 7, 1999).
Stevenson, Mark, "Sewage-laced Mexico City to Host Water Conference," *Spartanburg Herald-Journal* (March 13, 2006), p. A9.
Stobaugh, Robert, and Daniel Yergin (eds), *Energy Future: Report of the Energy Project at the Harvard Business School* (New York: Ballantine Books, 1980).
Stokstad, Eric, "Engineered Fish: Friend or Foe of the Environment?" *Science*, 297 (September 13, 2002), p. 1797.
Stokstad, Erik, "Organic Farms Reap Many Benefits," *Science*, 296 (May 31, 2002), p. 1589.
Stokstad, Erik, "Study Shows Richer Harvests Owe Much to Climate," *Science*, 299 (February 14, 2003), p. 997.
"Study Hailed as Convincing in Tying Fat to Cancers," *New York Times*, national edn (April 24, 2003), p. A23.
"Study Links DDT and Cancer," *New York Times*, national edn (April 22, 1993), p. A10.
Tarrant, John, *Food Policies* (New York: John Wiley, 1980).
Teitelbaum, Michael S., "The Population Threat," *Foreign Affairs*, 71 (Winter 1992/3), pp. 63–78.
Thurow, Lester, *The Zero-Sum Society* (New York: Basic Books, 1980).

Tien, H. Yuan, et al., "China's Demographic Dilemmas," *Population Bulletin*, 47 (Population Reference Bureau, Washington, DC) (June 1992), pp. 1–44.

Tierney, John, "Betting the Planet," *New York Times Magazine* (December 2, 1990), pp. 52–81.

Timberlake, Lloyd, "The Dangers of 'Development,'" in *Only One Earth: Living for the Future* (New York: Sterling, 1987), pp. 13–22.

Timmer, C. Peter, "Unbalanced Bounty from America's Farms," *Science*, 298 (November 15, 2002), pp. 1339–40.

"Twenty Percent of Married Women Unserved by Contraceptives," *Popline*, 25 (Population Institute, Washington, DC) (November/December 2003), p. 7.

Tyler, Patrick, "China Planning People's Car to Put Masses behind Wheel," *New York Times*, national edn (September 22, 1994), p. A1.

"A Typical Coal Plant," *Nucleus* (Spring 2000), p. 5.

Uhl, Christopher, "You Can Keep a Good Forest Down," *Natural History*, 92 (April 1983), pp. 71–9.

United Nations, *World Population Prospects: The 2002 Revision* (New York: United Nations, 2003).

United Nations Development Programme, *Human Development Report 2003* (New York: Oxford University Press, 2003).

United Nations Development Programme, *Human Development Report 2004* (New York: Oxford University Press, 2004).

United Nations Development Programme, "Overview: Globalization with a Human Face," in *Human Development Report 1999* (New York: Oxford University Press, 1999), pp. 1–13.

United Nations Development Programme, United Nations Environment Programme, World Bank, and World Resources Institute, *A Guide to World Resources 2000–2001* (Washington, DC: World Resources Institute, 2000).

United Nations Development Programme, United Nations Environment Programme, World Bank, and World Resources Institute, *A Guide to World Resources 2002–2004* (Washington, DC: World Resources Institute, 2002).

United Nations Development Programme, United Nations Environment Programme, World Bank, and World Resources Institute, *World Resources 2005* (Washington, DC: World Resources Institute, 2005).

United Nations Food and Agriculture Organization (FAO), *The State of Food and Agriculture 2005: World and Regional Review*, at www.fao.org/.

United Nations Population Division, Department of Economic and Social Affairs, "World Population Prospects: The 2004 Revision," in *Population Database 2005*, at http://esa.un.org/unpp.

United Nations Population Division, *World Urbanization Prospects: The 2001 Revision* (New York: United Nations, 2001).

"US Nuclear Forces, 2004," *Bulletin of the Atomic Scientists*, 60 (May/June 2004), p. 68.

"US Waste and Recycling," *New York Times*, national edn (August 20, 2002), p. D4.

Wade, Nicholas, "Experts Say They Have Key to Rice Genes," *New York Times*, national edn (April 5, 2002), p. A19.

Wald, Matthew, "Automakers Use New Technology to Beef Up Muscle, Not Mileage," *New York Times*, national edn (March 30, 2006), p. C4.

Wald, Matthew, "Citing Children, EPA Is Banning Common Pesticide," *New York Times*, national edn (August 3, 1999), p. A1.

Wald, Matthew, "High Radiation Doses Seen for Soviet Arms Workers," *New York Times*, national edn (August 16, 1990), p. A3.

Wald, Matthew, "Questions about a Hydrogen Economy," *Scientific American*, 290 (May 2004), pp. 68–73.

Wald, Matthew, "Report Questions Bush Plan for Hydrogen-Fueled Cars," *New York Times*, national edn (February 6, 2004), p. A19.

Wald, Matthew, "Waste Dumping that US Banned Went On at Its Own Atom Plants," *New York Times*, national edn (December 8, 1988), p. 1.

Wald, Matthew, "Wider Peril Seen in Nuclear Waste from Bomb Making," *New York Times*, national edn (March 28, 1991), p. A1.

Wald, Matthew, "Will Hydrogen Clear the Air? Maybe Not, Some Say," *New York Times*, national edn (November 12, 2003), pp. C1, C8.

Waltz, Kenneth, *Man, the State and War: A Theoretical Analysis* (New York: Columbia University Press, 1959).

Ward, Barbara, *Progress for a Small Planet* (New York: W. W. Norton, 1979).

"A Warning on Population," *Christian Science Monitor* (March 24, 1992), p. 20.

Webster, Paul, "Study Finds Heavy Contamination across Vast Russian Arctic," *Science*, 306 (December 10, 2004), p. 1875.

Weiner, Tim, "With Little Loans, Mexican Women Overcome," *New York Times*, national edn (March 19, 2003), p. A8.

Willett, Walter C., Graham A. Colditz, and Nancy E. Mueller, "Strategies for Minimizing Cancer Risk," *Scientific American*, 275 (September 1996), pp. 88–95.

Williams, Ted, "The Metamorphosis of Keep America Beautiful," *Audubon*, 92 (March 1990), pp. 124–34.

Wilson, Edward O., *The Diversity of Life* (Cambridge, MA: Harvard University Press, 1992).

Wirth, Timothy, C. Boyden Gray, and John Podesta, "The Future of Energy Policy," *Foreign Affairs*, 82 (July/August 2003), pp. 132–55.

Wolfe, Alan, "Three Paths To Development: Market, State, and Civil Society," a paper prepared for the International Meeting of Nongovernmental Organizations (NGOs) and UN System Agencies, Rio de Janeiro, 1991.

Wolfe, Alan, *Whose Keeper? Social Science and Moral Obligation* (Berkeley: University of California Press, 1989).

Wong, Genevieve, "Study: Asthma Rates Rise Dramatically," *Spartanburg Herald-Journal* (May 1, 2004), p. B5.

World Bank, World Development Indicators 2005, at www.worldbank.org/data/wdi2000/.

World Bank, *World Development Report 1990: Poverty* (New York: Oxford University Press, 1990).

World Bank, *World Development Report 2000–2001: Attacking Poverty* (New York: Oxford University Press, 2001).

World Bank, *World Development Report 2005* (New York: World Bank and Oxford University Press, 2004).

World Commission on Environment and Development, *Our Common Future* (New York: Oxford University Press, 1987).

World Health Organization (WHO), *Public Health Impact of Pesticides Used in Agriculture* (Geneva: WHO, 1990).

World Resources Institute, *The 1993 Information Please Environmental Almanac*, (Boston: Houghton Mifflin, 1993).

World Resources Institute, *World Resources 1992–1993* (New York: Oxford University Press, 1992).

World Resources Institute, UN Environment Programme, UN Development Programmme, and World Bank, *World Resources 1998–99: A Guide to the Global Environment* (New York: Oxford University Press, 1998).

Worldwatch Institute, *Vital Signs 2005* (New York: W. W. Norton, 2005).

Wright, Robin, "The Most Innovative City in the World," *Los Angeles Times* (June 3, 1996).

Yanagishita, Machiko, "Japan's Declining Fertility: '1.53 Shock,'" *Population Today*, 20 (Population Reference Bureau, Washington, DC) (April 1992), pp. 3–4.

Yardley, Jim, "Bad Air and Water, and a Bully Pulpit in China," *New York Times*, national edn (September 25, 2004), p. A4.

Yardley, Jim, "China Races to Reverse Its Falling Production of Grain," *New York Times*, national edn (May 2, 2004), p. 6.

Yeboah-Afari, Ajoa, "Male Responsibility: Still a Missing Link," *Popline*, 13 (Population Institute, Washington, DC) (May/June 1991), p. 6.

Yergin, Daniel, and Martin Hillenbrand (eds), *Global Insecurity: A Strategy for Energy and Economic Renewal* (Boston: Houghton Mifflin, 1982).

Yergin, Daniel, and Michael Stoppard, "The Next Prize," *Foreign Affairs*, 82 (November/December 2003), pp. 103–14.

Yoon, Carol, "Simple Method Found to Increase Crop Yields Vastly," *New York Times*, national edn (August 22, 2000), p. D1.

Yoon, Carol, and Melody Peterson, "Cautious Support on Biotech Foods by Science Panel," *New York Times*, national edn (April 6, 2000), p. A1.

Yunus, Muhammad, "The Grameen Bank," *Scientific American*, 281 (November 1999), pp. 114–19.

Index

NOTE: Page numbers in **bold** indicate chapters.

LIBRARY, UNIVERSITY OF CHESTER